P9-ARV-714

VALIDITY & SOCIAL EXPERIMENTATION

VALIDITY & SOCIAL EXPERIMENTATION

Donald Campbell's Legacy

Volume 1

LEONARD BICKMAN
Editor

Sage Publications, Inc.
International Educational and Professional Publisher
Thousand Oaks ■ London ■ New Delhi

For information:

Sage Publications, Inc.
2455 Teller Road
Thousand Oaks, California 91320
E-mail: order@sagepub.com

Sage Publications Ltd.
6 Bonhill Street
London EC2A 4PU
United Kingdom

Sage Publications India Pvt. Ltd.
M-32 Market
Greater Kailash I
New Delhi 110 048 India

Printed in the United States of America

Library of Congress Cataloging-in-Publication Data

Main entry under title:
Validity and social experimentation: Donald Campbell's legacy / edited by Leonard Bickman.
p. cm.
Includes bibliographical references and index.
ISBN 0-7619-1160-X (cloth: alk. paper)
ISBN 0-7619-1161-8 (pbk.: alk. paper)
1. Sociology—Methodology. 2. Social sciences—Methodology. 3. Meta-analysis. 4. Evaluation research (Social action programs) 5. Experimental design. 6. Campbell, Donald Thomas, 1916- I. Bickman, Leonard, 1941-
HM511 .V35 2000
301′.07′24—dc21 99-050757

This book is printed on acid-free paper.

00 01 02 03 04 05 06 7 6 5 4 3 2 1

Acquiring Editor: C. Deborah Laughton
Editorial Assistant: Eileen Carr
Production Editor: Wendy Westgate
Editorial Assistant: Nevair Kabakian
Typesetter: Lynn Miyata
Cover Designer: Candice Harman
Indexer: Molly Hall

Contents

PART II: Social Experiments

Introduction

I did not know Don Campbell well, but you did not have to know him well to appreciate his brilliance, his contributions to the field, and his kindness. I met Don when he was at Northwestern University and I was teaching just down the street at Loyola University of Chicago some 20 years ago. I attended seminars at Northwestern and on occasion had dinner with Don. One memorable event happened after we had dinner. The waiter approached our table and asked us to complete a form evaluating the dinner and the restaurant. Don responded that he did not believe in such things and said that it should be obvious whether we enjoyed the dinner. At first I was confused by Don's refusal; after all, here was the champion of evaluation refusing to participate in an evaluation. As we discussed his refusal, it became clear that there were several reasons for his behavior. He did not think that things that were obvious needed to be probed by written questionnaires. More important, our conversation moved into the area of data quality and accountability. He was most concerned that evaluations where accountability is involved can result in corrupted data. This is still a problem for which we do not have good solutions.

When I heard that Don had passed away, I spoke to several of his closest collaborators to find out if there was going to be some effort to memorialize Don's work. I was then president of the American Evaluation Association (AEA) and strongly believed that our organization should sponsor a project. With the AEA board's sponsorship, these two volumes were written as a memorial to Don Campbell and his work.

The effort to put together these two volumes was truly collaborative. All the authors reviewed one another's chapters. As editor, my job was very simple: remind (nag?) my colleagues about the due dates of their drafts and integrate what the reviewers noted in the feedback to authors. I encouraged the contributors to be informal in places in their chapters. In particular, I wanted them to convey to the reader a personal sense of how Don influenced their thinking and their research. I encouraged the authors to recount anecdotes that would illustrate how Don collaborated and supported colleagues and students, because that was a key but often ephemeral aspect of his contribution to the development of evaluation and research design. The introductions to each volume are meant to highlight the contributions of the chapters and provide a taste of the richness that was to come. I was greatly aided by Tom Cook and Will Shadish in the selection of the contributors and the initial review of topics.

INTRODUCTION TO VOLUMES I AND II

Don Campbell was close to 80 years old when he died on May 6, 1996. His death was premature. He had much more to contribute both in his personal life and in the social sciences. There are several excellent obituaries that document Don's life story and his contributions to the social sciences (Brewer & Cook, 1997; Stanley, 1998), as well as a part of an issue of the *American Journal of Evaluation* (vol. 19, 1998, pp. 397-426). I will not attempt to repeat or rephrase these publications. Instead, I hope that the works encompassed in these two volumes are by themselves eloquent testimonials to Don Campbell's contributions.

The two volumes, sponsored by the American Evaluation Association, are the organization's major memorial to Campbell's awesome contributions to evaluation and the social sciences. Thirty-seven accomplished researchers agreed to contribute a chapter to these volumes in tribute to Don Campbell's life. The first volume focuses on validity and the experimenting society, and the second on measurement and design. There is, however, overlap between the two volumes, because Campbell's influence is broad and difficult to compartmentalize. I have subtitled both of these volumes *Donald Campbell's Legacy.* A legacy is usually defined as gift of some property. The property here is intangible. It is the transmission of ideas, humanity, and enthusiasm for using social science research to improve the quality of people's lives.

INTRODUCTION TO VOLUME I

This first volume focuses on Campbell's contributions to the concept of validity and on the more activist side of his thinking, social experimentation. The reader should note, however, that dividing these topics into two sections is for the convenience of the editor, because in Campbell's writing validity and activism are related because social experimentation must be based on the concepts of validity and is not merely social activism. We start with validity because that concept forms the foundation for social experimentation. In Campbell's writing, four types of validity have been distinguished—internal, external, construct, and statistical conclusion. The most writing has focused on internal validity, the attribution of cause to an intervention. Because internal validity is so critical to research and evaluation design, Volume II is devoted to this issue. The other three validities are the focus of the first section of this volume.

In the first chapter of Volume I, Thomas D. Cook, a longtime friend and collaborator of Campbell, provides us with insight into one of the relatively neglected validities described by Campbell. External validity never had the prominence of internal validity in Campbell's writings. Cook points out that this was in spite of the fact that Campbell originated the term. External validity is concerned with generalizing across times, places, persons, and the specific program evaluated. High external validity contributes to scientific knowledge. For example, there is little scientific knowledge gained when learning about only one Head Start program unless we can generalize to Head Start programs in general, or even better, to other compensatory education programs. In particular, we want to generalize about the causal explanation about why the Head Start program works. Under what conditions does Head Start work?

Given that there is little scientific interest in learning about a specific local program, why was external validity not prominently featured in Campbell's writing? It was not the case that others ignored external validity—another major figure in evaluation, Lee Cronbach, featured it in his theory of evaluation. Cook believes that external validity was not prominent for Campbell because he thought external validity would be established through individual studies and replications following a natural science model. Over time, however, he came to realize that replications, as in the natural sciences, were not a significant aspect of social science research. Thus, Cook, based on Campbell's thoughts and writings, provides us with an approach to external validity that is both modest and practical. In this chapter, Cook responds to Cronbach's conceptualization of external validity using concepts derived from Campbell's perspectives. One of the major approaches to testing causal generalization is through meta-analysis. This approach is described in greater detail in the next chapter.

Elvira Elek-Fisk, Lanette A. Raymond, and Paul M. Wortman describe the use of meta-analysis within Campbell's validity framework and thus marry two important forces in modern research methods. Starting with the well-known quartet of validity constructs—internal, external, construct, and statistical conclusion—the authors show how these concepts can be used in meta-analysis much more efficiently than in traditional narrative reviews. For example, in the area of internal validity, a meta-analysis of a specific area can be used to examine if the results are biased depending on whether the design was a randomized experiment or a quasi- experiment. The authors provide a detailed example of how Campbell's validity framework can be used in meta-analysis. Moreover, they show how the validity concepts can be applied to meta-analysis to help in the selection of moderator variables and to detect bias.

The next chapter, by Norman Miller, William C. Pedersen, and Vicki E. Pollock, presents an in-depth view of discriminative validity, a form of construct validity. Here the primary concern is the fit between the operations used in a study and the conceptual definition. It is the proper labeling of either the cause or the effect that is critical in construct validity. The authors further differentiate the concept into trait, nomological, process, and theoretical discriminative validity. Moreover, they relate discriminative validity to another major Campbellian contribution, the multitrait-multimethod matrix (Campbell & Fiske, 1959), because it is critical not to draw a possible wrong conclusion based on similar methods being used. That is, the constructs may be different, but this difference can be obscured by the lack of method variance. From a practical point of view, discriminative validity is important for replication, efficiency, and what the authors term the avoidance of "irrelevant research." The authors use school desegregation as an example, pointing out that the meaning of the term *desegregation* is not clear; it is even less clear which features of desegregation would need to be replicated. Thus, it is critical to define the important components of the construct for efficient replications. The authors provide excellent examples of how to both measure and improve discriminative validity using examples from diverse literatures, but especially social psychology. The authors also provide excellent examples of the problems caused by the low discriminative validity of psychiatric diagnostic terms such as anxiety. They call for a greater sensitivity about discriminative validity so that researchers have the possibility of really knowing what they are talking about.

These first chapters primarily focus on external and construct validity. Internal validity issues are more fully discussed in the second volume, which focuses on research design. The fourth traditional form of validity is statistical conclusion validity, which concerns the probability that the results of a study could be due to chance. This validity, in some ways, is the most seductive of simplistic thinking. It is the only validity to which quantitative criteria can be applied easily. Moreover, there are conventions about what is and what is not statistically significant. As the next two chapters show, however, this is an oversimplification.

Mark W. Lipsey, one of the most prominent experts in statistical power analysis and meta-analysis, discusses the problems that surround the use of conventional statistical testing. Lipsey notes that Campbell and his collaborators in their early work did not discuss statistical validity much. Lipsey and other meta-analysts, however, have shown that there is reason to be concerned about the statistical validity of much research. Statistical conclusion validity is concerned not only with wrongly rejecting the null hypothesis when it is not true (Type I error) but also with accepting the null hypothesis when it is not true (Type II error). Lipsey, through the use of meta-analysis, has demonstrated that many intervention studies commit Type II errors because they do not have sufficient statistical power to detect an effect when it is present. In evaluations, this error results in labeling an intervention a failure when it was effective. Most of the research community is sensitive to the commission of Type I errors and guards against those by keeping alpha levels low (i.e., less than .05) but is not sensitive to the commission of Type II errors of mistakenly labeling an intervention as ineffective. Lipsey provides explanations for this imbalanced concern and, more important, describes several practical ways to deal with this problem. One possible solution to the Type II error problem is to be more explicit about effect sizes.

The next chapter, by Robert Rosenthal, provides greater details concerning the role of effect sizes in applied research. Rosenthal provides an excellent summary of many of the issues surrounding effect sizes. He describes several measures of effect sizes, provides some tools we can use to help understand how to use effect sizes in the real world, and finally suggests one effect size measure that has the most versatility. Rosenthal provides some interesting examples of very small effect sizes in health care that have important implications for the social sciences. Although Rosenthal focuses on effect sizes as a form of statistical testing, his discussion is also relevant to construct validity. He provides some rules that help interpret the meanings of effects.

The next chapter, by Melvin M. Mark, serves as an excellent transition between concerns about validity and social experimentation. Mark combines his interests in philosophy of knowledge, validity, and social experimentation in advancing Campbell's thoughts in all three areas. He carefully explains how Campbell's perspectives as a critical realist and fallibilist affect Campbell's notions of both the experimenting society and validity. Mark shows, for example, that a greater emphasis on the development of taxonomies of interventions would not only advance our understanding of construct validity but also would be of immense practical help in designing and implementing interventions in an experimenting society. Mark also places a greater emphasis on understanding underlying causal mechanisms as a way of advancing both validity and social experimentation.

Robert G. St.Pierre and Michael J. Puma titled their chapter "Toward the Dream of the Experimenting Society" thus suggesting that Campbell's utopian vision has yet to be fulfilled. They note that Campbell did not mean to suggest that society had to use randomized experiments to decide everything. What Campbell meant was to use empirical data, from randomized experiments and quasi-experiments, to help make decisions. When the data indicate that the intervention was not working, then society should try something else. It was this emphasis on "trying" that made it the experimenting society. The authors take us on a tour of efforts designed to improve outcomes of low-income families and children and show that most of these efforts have not been successful. Unfortunately, policymakers have not taken seriously Campbell's advice to move on and try something else. This reluctance to experiment, as the authors note, has several explanations, but one of them is the evaluators' reluctance to talk truth to power. St.Pierre and Puma also chide some evaluators for not being adequate methodologists and thus similar to many program and policy personnel who confuse change with evidence of program effectiveness.

The next chapter takes us into one of the practical problems of conducting social experiments. Robert Boruch and Ellen Foley take on the difficult task of providing advice on how to design randomized field trials. Boruch has been an advocate for the use of randomized experiments for more than 25 years. In this chapter, he and his coauthor focus on one of the main roadblocks to valid field experimentation. Early studies used the individual as the unit of analysis, but as the authors note, the proper unit in many circumstances is a unit larger than the individual, such as a school or community. Boruch, by developing compendiums of successful experiments, has confounded earlier critics who maintained that randomized experiments could not be conducted in the real world because of feasibility or ethical issues. In a similar fashion, Boruch and Foley take on the critics who say that larger units of analysis are not feasible by providing an extensive bibliography of such studies. The authors are quick to note, however, that the tasks of implementing and analyzing these studies are not easy. They provide some practical suggestions and also point the way to future research needed in this area.

Robert K. Yin, in his chapter, reinforces the notion that experimenting does not necessarily mean the use of randomized experiments or quasi-experiments. This chapter focuses not on the design of an evaluation to determine truth but on another key aspect of Campbell's thinking–plausible alternative explanations. Although chapters in Volume II also focus on alternative explanations, Yin is unique in applying the concept to a case study approach. To assist the reader in understanding this approach, Yin provides several good examples of this thinking. These examples also illustrate how this approach can be used when any type of experiment is not possible. Finally, Yin provides an interesting taxonomy of plausible explanations that should be considered in any study, be it a case study or an experiment.

Burt Perrin, unlike the other authors in these volumes, met Campbell while an undergraduate at Northwestern University. This chapter is a very personal account of the influence Campbell had on Perrin's career as a practicing evaluator. Perrin is quick to point out that he has never used the experimental method in his practice yet considers that Campbell had a major influence on his practice of evaluation. He cites Campbell's evolutionary epistemology, plausible rival hypotheses, and triangulation as key influences on his practice. Perrin has some very specific ideas and recommendations about a key aspect of the experimenting society. He is very concerned about the misuse of performance indicators in an accountable society.

The last chapter in this volume, by Carol Hirschon Weiss, takes a somewhat different approach than did the other chapters. While pointing out the major and significant contributions made by Campbell, Weiss also recognizes some of the limitations of his work on the experimenting society. I would say that she would consider Campbell's experimenting society as naive about the uses of evaluation and its relationship to policy. She points out that Campbell did not acknowledge the major scholars in the policy world such as March, and she attributes this to Campbell's possible disdain with "untidiness." Weiss and others familiar with how policy is made would use that and stronger words to describe the irrational aspects of policy formulation. Campbell, as Weiss notes, thought the primary route to increasing the utilization of evaluation was through improved methods. While recognizing the importance of more valid evaluations, Weiss believes that the competing goals of stakeholders, the political and democratic environment of evaluations, and the lack of direction provided by many evaluations all conspire to reduce the importance of the quality of a study. She notes that Campbell did not consider many of these forces in his thinking about the experimenting society. Weiss provides some interesting directions for future research in this exciting area.

These chapters demonstrate the critical importance of Campbell's ideas. He was truly an original thinker and a major contributor to our concepts of validity and how society can become better through applied research. His colleagues also note some of the gaps in Campbell's thinking and the major issues that have developed in the past few years. I think Campbell would be very pleased to see how those who followed him elaborated, expanded, improved, and even corrected his ideas.

REFERENCES

Brewer, M. B., & Cook, T. D. (1997). Donald T. Campbell. *American Psychologist, 52*(3), 267-268.

Campbell, D. T., & Fiske, D. W. (1959). Convergent and discriminant validation by the Multitrait-Multimethod Matrix. *Psychological Bulletin, 56*(2), 81-105.

Stanley, J. C. (1998). Biographical memoirs: Donald Thomas Campbell. *Proceedings of the American Philosophical Society, 142*(1), 115-120.

PART I

Validity Issues

CHAPTER 1

Toward a Practical Theory of External Validity

Thomas D. Cook

This chapter outlines a practical theory for exploring external validity—the generalization of causal relationships. Cronbach has framed the external validity issue in two ways, each associated with a different methodology: (1) generalizing from achieved research samples to the cause, effect, person, and setting categories they represent, for which random sampling is usually advocated as the method of preference; and (2) generalizing from the categories represented in a study to manifestly different categories, for which the usual method of preference is identifying all the effect-influencing mechanisms that mediate between when a global cause varies and the effect is observed. Formal random sampling, however, is rarely practical for generalizing to all parts of a causal connection, and social scientists rarely achieve even close approximations to full knowledge of causal mediation. Fortunately, a third scheme for generalization exists based on those theories of construct validity that are used for generalizing from specific manipulations to abstract causes and from operational measures to effect constructs. Explicating these theories results in identifying five principles that promote causal generalization in whichever way the issue is framed. Although the theory of causal generalization so achieved is not as compelling as sampling theory or causal explanation, it is more generally applicable.

PROLOGUE

It was not easy to irritate Don Campbell. His goodwill and good sense permeated all. That is why it was so strange when we once spent 3 days in his Bethlehem home working on a paper about relabeling "internal" and "external" validity (Campbell, 1986)—a paper to which I in the end did not add my name because the new labels, though more accurate, were very cumbersome. In any event, in discussing others' complaints about the secondary role he assigned to external relative to internal validity, he became quite agitated, saying something like "How can they say I think external validity is unimportant when I coined the term?" To Don, external validity was very important; nevertheless, it did play second fiddle to internal validity. One has only to consider how more numerous and better articulated are the internal validity threats, how many more pages are devoted to ruling them out than to dealing with external validity, and how salient his claim has been that internal validity is the *sine qua non* of experimentation.

Will Shadish and I are now in the last stages of rewriting Campbell and Stanley (1966) and Cook and Campbell (1979), to be published as written by Shadish, Cook, and Campbell. One innovation in that book is to take external validity as seriously as internal validity. What follows is an exposition of the theory of external validity we espouse therein. It depends heavily on systematizing insights from Don's thinking about external validity and integrating them into Cronbach's theory of generalization in evaluation studies (Cronbach, 1982; Cronbach et al., 1980). The basic outlines of our new thinking have been presented elsewhere (Cook, 1990, 1991; Shadish, 1995). We present them in somewhat revised form here as a tribute to Don, knowing we cannot match his intellectual creativity, sharpness of focus, and tenacity in pursuing the implications of insights. His shadow, though long, paradoxically continues to cast light.

EXPLICATING EXTERNAL VALIDITY

The Language of Causal Generalization

Causal generalization is concerned with drawing conclusions about four entities. One is treatments. Rarely do we want to make inferences about the specific details of a particular manipulation. Instead, we want the manipulation to represent a class or entity specified in abstract (i.e., general) language, so we refer to "external threat" as a cause of in-group cohesion and make no attempt to provide an exhaustive description of how an experimenter induced such threat. Similarly, we refer to "Head Start" as a causal agent without providing a comprehensive list of the Head Start centers in a project or of the many different activities taking place across the sampled centers. We also seek to draw general conclusions about observations, especially those representing outcomes. In studying external threat and in-group cohesion, we have little conceptual interest in the specifics of how in-group cohesion was measured. The inference we seek is about in-group cohesion per se. Similarly, we have little interest in the particular achievement test used to assess Head Start effects. We want to generalize to achievement writ large. We also need to know in which populations and settings the treatment and outcome are causally related, these being the third and fourth entities to which generalization is often sought.[1] Causal generalization is concerned with specifying the range of application of a causal connection that has been demonstrated with at least one instance of a treatment and outcome and at least one sample of persons and settings.

Causal connections are not necessarily causal explanations (Collingwood, 1940; Cook & Campbell, 1979, 1986; Gasking, 1957; Mackie, 1974). Causal connections are implicit in the statements about external threat causing in-group cohesion, Head Start increasing achievement, aspirin reducing headaches, or school desegregation causing white flight. Each example has a manipulable treatment specified in general language, a response specified in the same type of language, and a statement that the relationship between them is causal in the sense that, if the causal agent were made to vary, then the outcome would subsequently change and would not have changed the same way had the cause not been present. Causal explanations identify *how* or *why* a causal connection occurs. Why does external threat cause in-group cohesion? Why does Head Start raise achievement? Why does aspirin reduce headaches? Why do whites flee when schools desegregate? At a minimum, such explanations require specifying (a) all the mediational forces set in motion when the causally efficacious components of the treatment vary and without which the effect would not have occurred, and (b) all the moderator variables on which a cause-effect relationship is contingent, some of which influence the fidelity or strength of a treatment without necessarily instantiating novel causal mediating processes. *This chapter is about the generalization of descriptive causal connections*, though we will see that causal explanation is one way of promoting such generalization.

Cronbach's First Problem of Causal Generalization

Cronbach (1982) has claimed that causal generalization can be framed in two ways. The first involves how sample data are used to identify the cause-and-effect constructs and the populations of persons and settings involved in a causal relationship. For example, one might ask whether a particular patient education manipulation represents the general category "patient education" and whether such education then influences the general construct "physical recovery from surgery" (rather than just length of hospital stay, degree of wound healing, time to return to normal activities at home, etc.). One might also ask whether this causal connection holds with all surgical patients and not just those who have undergone a particular type of surgery, and whether it holds in all hospitals rather than just, say, teaching hospitals. To justify attaching general labels to particular samples or operations requires a valid theory of the correspondence between samples or operations and the constructs or populations they represent.

Empirical results often force researchers to limit generalization to a level lower than originally planned. Thus, a causal agent such as "patient education" might need to be respecified as *pre*surgical or *post*surgical education because data analysis strongly suggests that only one of them is causally related to recovery from surgery. Similarly, a causal connection might be supported for cholecystectomy patients but not orthopedic patients. In similar fashion, results sometimes can lead researchers to create more general constructs and categories than those originally built into a research program. Thus, patient education might come be construed as an instance of "perceived control" or recovery from surgery might be treated as a subclass of "personal coping." Whatever the level of generality, the conceptual issue is always the same: how to justify generalizing from a sample of instances and the data patterns associated with them to particular entities, populations, universes, categories, or classes (terms we use interchangeably).

Framing causal generalization as the correspondence between samples and populations suggests that such generalization is best promoted via formal sampling, selecting units with known probability from some clearly designated universe. This is the only known procedure that matches the sample and population on all measured and unmeasured attributes within known limits of sampling error. It is not a surprise, therefore, that some statisticians claim that confident causal generalization depends on such sampling. Lavori, Louis, Bailar, and Polansky (1986) opine:

> So far we have dealt with methods for ensuring that the analysis of a parallel-treatment study correctly assesses the relative effects of treatment on patients enrolled in that study. The investigator must also consider whether the results can properly be extended to patients outside the study; we refer to this issue as generalizability. . . . Unless there is random sampling from a well-defined population (not to be confused with random allocation of treatment), the question of generalizability may not be completely resolvable. Nevertheless a brief discussion of selection criteria and a concise description of the clinically relevant facts about the patients chosen for study can go a long way toward meeting this objective. (pp. 62-63)

Note this last disclaimer, which we discuss later, for it suggests that researchers need not throw in the towel if random selection is not possible. Note also that

Lavori and colleagues mention only one domain across which causal connections are to be generalized: populations of persons. What about the relevance of the formal sampling ideal for generalizing to causes, effects, settings, and times? Kish (1987) also prefers first randomly selecting units (to enhance generalization) and then assigning them to treatments at random (to promote causal inference), but Kish acknowledges that this ideal is rarely feasible and explicitly chooses not to deal with how manipulations and measures might be randomly selected so as to represent cause-and-effect constructs. If random selection is to be a general model for causal generalization, however, then it also has to apply to sampling causal manipulations and outcome measures. And what about time? It seems very unlikely that the two-step statistical theory that Kish (and I) prefer can constitute a practical framework for simultaneously generalizing a causal relationship to populations of persons, treatments, outcomes, and settings, let alone times.

Cronbach's Second Problem of Causal Generalization

Cronbach's second problem of causal generalization involves generalizing from the samples and whatever universes they represent to populations that are manifestly different. Involved here is the *transfer* of causal knowledge to *novel* cause-and-effect constructs, to *novel* classes of persons and settings, and to *future* time periods. The utility of such transfer is indisputable. Theorists assign little value to causal propositions that are hedged with many boundary conditionals (Bandura, 1986); and with valid knowledge about transfer, practitioners and policymakers can be confident that a causal connection will hold in the settings where they work, even if it has never been explicitly tested there (Cronbach et al., 1980; Cook, Leviton, & Shadish, 1985).

Causal explanation is widely invoked as the means for creating transferable knowledge (see Cronbach et al., 1980). The assumption is that understanding how or why a phenomenon comes about permits recreating the phenomenon *wherever* and *however* the causal ingredients essential for its production can be brought together. Knowing how electricity is generated allows us to provide such power even in space satellites where electricity may never have been available before, and knowing the pharmacologically active ingredients of aspirin permits any inventive person to create the drug's equivalent out of local plants that might never have been used to cure headaches before. So, knowing the complete causal system (as with electricity) or of the total set of causally efficacious components (as with aspirin) should make it easier to reproduce the causal connection in question across a wide range of previously unexamined settings and populations and even with novel ways of creating the cause and understanding the effect (Bhaskar, 1975). This potential for broad-ranged transfer is probably what makes causal explanation the "Holy Grail" of science. It is also why evaluation theorists keep reinventing the importance of testing the substantive theory undergirding a program's design (Connell, Kubisch, Schorr, & Weiss, 1995; Cronbach, 1982; Wholey, 1977), although few of them emphasize that explaining a program's effects requires first being sure they are there.

Complete explanatory knowledge is rarely available to social scientists. A question like "Why does patient education promote recovery from surgery?" supposes (a) a valid answer to the descriptive causal question, (b) well-corroborated substantive theories whose constructs are well specified and can be measured validly, and (c) the ability to reject all other theories that could fit the explanatory data as well as the preferred explanation or better. None of these is easy to find, and obtaining all three is well-nigh impossible (Glymour, Scheines, Sprytes & Kelly, 1987). It is not surprising, then, that we still do not know why patient education is effective, despite about 150 studies on the topic and many explanatory theories emphasizing quite different classes of construct—some physiological, some psychological, and others interpersonal (Devine, 1992). Complete explanation is undoubtedly useful for transfer, but because *complete* causal knowledge is not a reality, the key question is this: Under what conditions—and how often—can we gain explanatory causal knowledge that is useful despite being imperfect?

This chapter will (a) examine the roles that random selection can realistically play in generalizing to all four target entities in Cronbach's first framing of the causal generalization issue; (b) examine Cronbach's second framing of the issue to assess the roles that causal explanation can realistically play in justifying inferences to novel populations and classes; (c) explicate an alternative theory of causal generalization that is based on theories of construct validation, classification, and category membership (e.g., Lakoff, 1985; Smith & Medin, 1981) that are relevant to Cronbach's two generalization tasks but do not require either random sampling or complete causal explanation; and (d) argue that this alternative theory of causal generalization is most applicable when multiple studies of a causal proposition are available (Glass, McGaw, & Smith, 1981; Hedges & Olkin, 1985).

The principles we explicate from theoretical work on construct validity overlap with Mark's (1986) independent analysis of how similarity, heterogeneous irrelevancies, and explanation promote causal generalization. In his case and ours, there is a shared, osmotic debt to Campbell's work on external and construct validity, especially in Campbell (1969, 1978), Campbell and Fiske (1959), Campbell and Stanley (1966), Webb, Campbell, Schwartz, and Sechrest (1966), and Cook and Campbell (1979). No claim is made that all of Campbell's insights into the nonstatistical and nonexplanatory bases of causal generalization have been systematized here. Our more modest goal is to increase awareness of the need for improved theories of causal generalization and to convince theorists of method to consider nontraditional techniques for this.

CAUSAL GENERALIZATION AND SAMPLING THEORY

Random Sampling and Inferences About Cause-and-Effect Constructs

Sampling theory requires a clearly designated population that can be pointed to. It is relatively easy to develop such descriptions for many human populations and settings, such as cholecystectomy patients or American hospitals; however, the task is more difficult with the less material terms typically used to designate cause-and-effect constructs in the social sciences (e.g., patient education, recovery from surgery, anxiety, anomie). Scholars often disagree about which components to ascribe to a construct, the relative weight each component should have, and how they are interrelated.

Even if a comprehensive description were forthcoming, it would still be difficult to enumerate all the instances of a cause or effect prior to random selection. Full enumeration occasionally is possible, as when a computer generates every two-digit addition or subtraction possibility to create the universe of all two-digit arithmetic problems, but this is of interest to few. Most item enumeration involves subject-matter experts explicating a construct, identifying components, and then crafting manipulations or measures to reflect these descriptions. Judgment is of the essence, and there is no presumption of generating the universe of all possible items or manipulations.

A judgment process can result in a long list of plausible manipulations or measures, but even here random selection rarely occurs. In experiments, implementing even two planned variants of the same causal construct involves competing claims for the resources this requires. Why not, instead, implement a second type of theory-based treatment, a better control group, more measures, or more measurement waves, or add more units? Anyway, sampling theory requires many treatment variants for confident generalization to result from the selection process. If achieving two variants of the same treatment is so difficult, imagine the difficulties inherent in many more variants.

It is manifestly more feasible to implement multiple measures of a possible effect, especially when paper-and-pencil tests are used, but that has certain costs. Reflect that causal research in field settings nearly always involves outcome measures with fewer items than when the same construct is assessed in individual difference research. Field experimenters are under pressure to restrict the total testing time and to measure many different outcome, mediator, and moderator variables. Adding items to a construct uses resources to create marginal improvements in what will turn out to be a very small number of constructs. Such mundane considerations suggest that random item selection—with its requirement of a large sample of items per construct—will rarely be appropriate in social experiments. More realistic is a rational selection process based on human judgments about the domains prototypical of a construct and the items prototypical of the domains within a construct. Early in his career, Cronbach called for random item selection (Cronbach, Rajaratram, & Gleser, 1967), but he later argued that how items are selected is less important than ensuring (Cronbach, Gleser, Nanda, & Rajaratram, 1972) and justifying (Cronbach, 1985) that a construct description captures all the construct components of theoretical relevance.

Random Selection and Populations of Persons and Settings

Random samples have been used mostly to generalize to specific populations of persons and settings; however, when causal relationships are to be generalized, practical constraints limit the researcher's power to sample units at random. Not all the individuals and organizations selected for a cause-probing study will agree to participate, and the units so agreeing are likely to differ in many ways from the units refusing, leading to a mismatch between the targeted and achieved populations. A mismatch also arises when informed consent is needed, limiting generalization to the theoretically irrelevant circumstance where individuals are knowledgeable about a study and agree to be in it. Because most cause-probing field research is also longitudinal, attrition inevitably will result as individuals die, move home, or are bored by the demands of treatment or measurement. When added to initial refusal patterns, such attrition further reduces the correspondence between the intended and achieved populations. It is not that high refusal and attrition rates render random selection meaningless, or even inferior to its alternatives, but they do attenuate its advantages.

No discussion of sampling realities can be complete without mention of the financial costs of mounting an intervention at more than a handful of sites. Many interventions are expensive, and, by definition, their efficacy is not established. It rarely makes administrative sense, therefore, to fund a "representative" study with many randomly selected sites to test something that may not work. This administrative reality may explain why, except for some multi-million-dollar clinical trials in medicine and public health, few studies of ameliorative treatments have included randomly chosen samples of persons and settings that were then randomly assigned to different treatments. In the few cases of this that I know about, the random selection was either from within a circumscribed entity—households within the Palo Alto telephone book—or involved treatments with very low salience—as in testing the relative efficacy of various mass-mail marketing strategies. Financially and logistically, it is no accident that we have so few experimental studies with random selection of populations and settings, however feasible this might be in the abstract.

A further "problem" with random selection has to do with some trade-offs its implementation usually requires. Consider the trade-off between internal and external validity. Imagine drawing a random sample of respondents or sites and then randomly assigning them to different treatments. Treatment-correlated refusals would be anticipated whenever the treatments differ in intrinsic desirability, as they often do. Such differential attrition creates the very group noncomparability that random assignment is designed to avoid. To prevent this, Riecken and Boruch (1974) suggest postponing random assignment until after respondents (a) have agreed to participate in the complete measurement framework, (b) have been fully informed of the different treatments, and (c) have agreed to participate in whichever treatment group the randomization process determines for them. This sampling strategy enhances internal validity because it reduces the rates of treatment-correlated refusal and attrition, but it also compromises external validity because more respondents will refuse to be in the study after learning about the measurement burdens and the possibility of being assigned to a less desirable treatment. This trade-off is not inevitable, but it is highly likely to operate when measurement burdens are high, treatments differ in intrinsic desirability, and respondents can learn about differences between treatments.

The same trade-off also holds for quasi-experiments. Consider a study on the effects of a televised smoking cessation campaign in the Chicago SMSA (Warnecke et al., 1991; Warnecke, Langenberg, Wong, Flay, & Cook, 1992). Broadcasts were made twice each evening during the news hour on one network for 21 days, with many different kinds of follow-up provided. Two major options were considered for sampling and treatment assignment. The population could be limited to smokers who reported watching the evening news regularly—say, five or more times per week. The treatment group would then consist of regular news-watchers on the network broadcasting the antismoking materials, while the control group would be regular news-watchers at the same time on a competing network. This plan maximizes internal validity under the dual assumptions that the two groups do not differ much—each being faithful watchers of the network evening news on a single channel—and that the treatment contrast will be large because faithful viewers of the target channel would watch more of the antismoking shows. Restricting the study to such groups means, however, that the results apply only to smokers who are also faithful viewers of a particular TV channel. The second sampling option thus required identifying and sampling smokers with a lower viewing threshold—say, twice per week. A broader population now ensues, but the contrast between levels of viewing the antismoking materials is now reduced, and it is now more likely that control group members will view some of the antismoking materials on the competing channel. It is questionable whether random selection from meaningful populations can be as routinely used in cause-probing experimental studies as it is in surveys of human populations.

Random sampling is of limited utility when causal generalization is at issue because (a) the model is never relevant to making generalized inferences about manipulable causal agents and (b) it is rarely relevant to making generalized inferences about effects. As for persons and settings, (c) the model assumes that sampling occurs from some meaningful population, although ethical, political, and logistical constraints often limit random selection to less meaningful populations; (d) in many situations random selection and its goals conflict with random assignment and its goals, although generalizing a causal connection clearly supposes the primacy of identifying a causal connection over assessing its generality; and (e) realities of budget and control over the quality of treatment implementation and measurement often limit the selection of units to a small and geographically circumscribed population. It should also be remembered that random sampling is not relevant to (f) generalizing to future time periods or (g) generalizing beyond target constructs and populations. Taken together, these limitations indicate that random selection cannot be advocated as "the" model for causal generalization, despite claims on its behalf (Kish, 1987; Lavori et al., 1986). Is there a viable alternative?

FIVE CONSTRUCT VALIDITY PRINCIPLES APPLIED TO CAUSE-AND-EFFECT CONSTRUCTS

Samples of measures and manipulations are regularly used without random selection to draw conclusions about causes and effects. We explicate five principles that "somehow" make this possible and later ask whether they also apply to justifying general inferences about persons and settings.

Proximal Similarity

The precondition for construct validity is an exhaustive multivariate description of the target construct. This should be based explicitly on the prior literature, where available, and should also specify which of the construct's attributes are considered prototypical (Rosch, 1978). Operational instances are then selected to represent this theoretical description.

One criterion for selecting instances is their proximal similarity to the construct description (Campbell, 1969). This implies that, in the judgment of the focal research community, the operational instances embody most of the construct's identified prototypical components and contain as few conceptual irrelevancies as possible. This matching is a theoretical exercise and depends on capturing those attributes of a construct that are reputedly most prototypical. The matching is on manifest characteristics and not on any of those more latent causal forces that, in a mediational sense, may be more proximally responsible for effects (Bhaskar, 1975). To define proximal similarity otherwise is to restrict it to the subset of contexts where considerable explanation is already available, and in most social science fields this is quite a narrow subset. Proximal similarity thus refers to the correspondence in manifest descriptive attributes between a theoretical description and its operational representation(s). Expressed this way, proximal similarity achieves overlap only on observed variables, and it supposes a theory of the construct that is widely accepted and that specifically includes elements considered to be prototypical. Proximal similarity, however, is not enough by itself.

Heterogeneous Irrelevancies

Any single cause or effect operation underrepresents the full set of relevant components and contains its own unique irrelevancies. Thus, if attitude were assessed using a paper-and-pencil measure, the data collection method would be intrinsic to the measure but irrelevant to the concept; if Head Start were evaluated, the particular personnel in the centers that were sampled would be a conceptual irrelevancy. Knowledge of the omnipresence of irrelevancies in measurement has spurred the call for multiple operationalism—measuring a construct in many ways that are as similar as possible in their prototypical elements and as different as possible in the conceptual irrelevancies they contain (Webb et al., 1966). The aspiration is either to probe whether a cause-effect relationship is obtained *despite* variation in the pattern of irrelevancies across tests or to identify the nature and strength of whichever irrelevancies a causal relationship is found to depend on.

Brunswik (1955) is the theorist most closely associated with thinking about the heterogeneity of irrelevancies. He worked mostly in human perception and used the example of facial characteristics to illustrate the contextual dependency of human judgment. Imagine studying whether observers attribute slyness to someone who squints. Because squints exist only on faces, Brunswik proposed that they should not be abstracted from faces. But which other facial characteristics should be built into a research project, and how? Should there be details about eyes, noses, facial contours, chin forms, size of pores, length of eyelashes, and the number or depth of wrinkles? Practically speaking, a single study cannot make all these irrelevant features heterogeneous, so decisions are required about which patterns of attributes are more ecologically representative of the facial settings in which squints are naturally embedded. There is legitimate room for disagreement about this, and some features will seem indeterminate. Brunswik himself advocated sampling those combinations of facial attributes that most often co-occur in the population and are maximally different from one another. If a cause-effect relationship were to remain robust across all such combinations, he would then conclude that the relationship between squinting and slyness is independent of the most commonly occurring facial characteristics in which eyes are embedded—characteristics that are conceptually irrelevant. If the data suggest that the relationship occurs only with certain kinds of faces, then some conditions are specified under which a squint does and does not cause people to see others as sly.

Fisher (1935) also clearly noted the advantage of multiple irrelevancies when probing cause-effect relationships.

> Any given conclusion . . . has a wider inductive base when inferred from an experiment in which the quantities of other ingredients have been varied, than it would have from any amount of experimentation in which these had been kept strictly constant. Standardization weakens rather than strengthens our ground for inferring a like result, when, as is invariably the case in practice, these conditions are somewhat varied. (p. 99)

Although both Brunswik and Fisher use an argument based on ecological representativeness, a falsificationist rationale for their preference also can be adduced. Causal results that are demonstrably robust across many irrelevancies constitute many failed attempts to disconfirm the hypothesis of a *generalized* causal relationship. The implication is that only those causal connections are worth provisionally treating as general that continue to hold after researchers have deliberately built multiple sources of irrelevant heterogeneity into their research, including those sources that substantive theory suggests are most likely to make a causal connection disappear.

What are the criteria for claiming that a causal finding is "robust"? Many theorists believe that the world is very contingently ordered (Mackie, 1974) and is therefore poorly characterized in terms of causal arrows from A to B. The alternative contention is that the world is more aptly characterized, say, as a set of interconnected, multivariate pretzels with causal feedback loops within and between the pretzels. In support of part of this, Cronbach and Snow (1976) have examined many causal relationships in education and claim that complex statistical interactions are the norm, and McGuire (1984) has claimed the same for social psychology. It is entirely possible, then, that if robustness were specified as *constancy of effect sizes*, few causal relationships in the social world would be robust.

We prefer, however, to conceptualize robustness as *constancy of causal direction or sign*. Several factors argue for this. First, casual examination of many meta-analyses convinces us that causal signs tend to be similar across individual studies even when the effect sizes vary considerably. Second, it is often difficult to shape legislation or regulations to suit local contingencies. Instead, the same plan has to be promulgated across an entire nation or state lest focused inequities arise between individual places or groups. Thus, policymakers hope that a program or policy will have generally positive effects whatever the site-level variability in effect sizes with the same causal sign. Third, substantive theories require causal statements that not only are obviously novel but also are dependable. Otherwise, the theorizing is about unstable causal connections—an unfortunate commonplace in today's social science. Thus, we favor a looser criterion of robust replication based on the stability of causal signs and not effect sizes.

Discriminant Validity

Some irrelevancies are more substantive than methodological. Because nearly all theoretical constructs share components with other constructs, a decomposition of variance is required to probe whether a causal connection is truly (a) from A to B or is, instead, (b) from some of A's components that are shared with C or D, or (c) from A or some of A's components to some of B's components that are shared with E or F. Is the relationship from A to B, from A to E or F, from C or D to B, or from, say, C to F?

To resolve such questions, Campbell and Fiske (1959) proposed complementing an analysis of cross-method generalization with an analysis discriminating a target construct from its cognates. An example of their multitrait, multimethod approach to construct validation would be when nurses in patient education studies know who is receiving an intervention, making it crucial to distinguish this knowledge from the effects of education. Normally this would require a double-blind experiment or otherwise varying whether professionals know of the patient's experimental condition. Neither of these is feasible if nurses deliver the intervention face-to-face. Moreover, practicality dictates that there is usually a low ceiling on the number of cognate manipulations that can be added to a study. Adding differentiating outcomes is more feasible, though. Consider length of hospital stay as a measure of recovery from surgery. Hospital administrators are under pressure from insurers to reduce hospital stays and are now typically reimbursed for a prespecified number of days per diagnosis. If patients left the hospital prematurely rather than because of a genuine recovery, then they should be in poorer physical health and run a greater risk of being rehospitalized or delayed in their return to normal activities after convalescence. To distinguish these interpretations, valid measures are required of patients' physical condition on leaving hospital, of their rehospitalization rates, and of the time it takes them to resume normal activities at home. Discriminating target from cognate outcomes should be widespread in experimentation, although the same cannot be said for discriminating target from cognate causes—at least not in experimental work.

A component-of-variance approach does not help if the same source of bias runs through all the available operational instances. Constant bias precludes discovering whether an effect is found when the bias is absent, for it is never absent.

It also precludes testing whether the cause-effect relationship is obtained despite heterogeneity in the direction of bias, for there is only homogeneity in the direction of bias. For example, a serious threat to causal generalization would ensue if all patient education interventions had been implemented by researchers rather than by regular staff nurses or physicians. In such a case, the number of past studies and their heterogeneity on other irrelevant attributes are not very important. The most the analyst can do is invoke relevant theories, past findings, or any other form of indirect argument that might help assess the likelihood of an unexamined source of bias causing the causal connection under analysis. Constant bias is a major problem in construct validation, precluding both the variance analysis required for empirically ruling out alternative interpretations and the less direct forms of falsification based on summing data across different sources of irrelevancy.

Causal Explanation

In nomological theories of construct validation (Cronbach & Meehl, 1955), an interpretation of an entity is validated rather than the entity as such. Thus, anxiety is inferred when a theory successfully predicts (a) what the causally efficacious components of multivariate constructs are, (b) what the factors are that cause anxiety measures to vary, (c) what the *pattern* of effects are to which anxiety leads, and (d) what the processes are through which anxiety influences this pattern of effects. Theoretical predictions like these increase the number and complexity of implications associated with a particular interpretation, tending to reduce the number of viable alternative explanations. Many more predictions are involved here than just the notion that a theoretical construct description should seem similar to the manipulations or items purporting to represent it. Explanation requires researchers to specify the ways in which a target construct is related to other constructs. When theory is specific about a construct's measurable antecedents, consequences, and processes, then drawing inferences about it is less difficult than otherwise. Thus, if a theory of patient education allowed us to predict that such education will speed recovery from surgery among cancer patients but not others or that it is effective because it increases patients' feelings of self-efficacy, then we could collect the relevant data and test these hypotheses within the limits of current methods for testing mediating processes.

Achieving secure and complete explanation is not easy, however ardently social scientists seek it, so a crucial question is this: To what extent are Cronbach's two types of generalization promoted by the approximations to full explanation typically achieved? In actual research practice, theoretical concepts and their relationships are underspecified, and it is rare for R^2 values to be even close to one. It is not easy to use theory of this quality to validate constructs through specifying antecedents, moderators, mediators, and patterned outcomes, though more attempts are needed to do this. The most one can claim is that some advantages for construct validation will likely follow from the explanatory gains made in prior work and from the marginal gains in explanation that single studies typically achieve.

Interpolation and Extrapolation

Nearly all research questions are framed in terms of general cause-and-effect constructs, and no reference is made to specific levels of either. To generalize to constructs, however, requires assuming that a causal result holds across all levels of the causal construct and across the full range of the effect. Unfortunately, parametric field experiments are rare. More typical are two levels of the independent variable, with the contrast between them serving as the causal agent writ large. For example, researchers might specify a research question in terms of patient education as the causal agent but then conduct a study in which an average of only 3 hours of it is created. Because the control groups will also experience some patient education, the achieved contrast in patient education will be less than 3 hours—say, 2. How likely is it that written research conclusions will refer to the cause either as "3 hours of patient education" or as the more exact "a 2-hour differential in patient education"? Instead, the unqualified construct "patient education" will almost always be invoked as the causal agent.

The ideal solution to this problem depends on describing the function relating the full range of patient education to recovery from surgery—that is, conducting dose-response studies with independent samples that have been randomly assigned to many levels of patient education. This strategy requires an independent variable that can be expressed quantitatively and for which it is practical to vary many levels—requirements that at present are rarely met in experiments outside the laboratory. The researcher's dilemma is to decide how many levels can be varied. Two is the minimum—one as high and as powerful as practical and the other equivalent to a no-treatment control group. This limits conclusions to the effect of A as it varies between these two levels, not about the effect of A more generally. With more resources, a third, intermediate group could be added. If an effect is then observed at both treatment levels, then assuming a linear relationship between cause and effect, one might conclude that the effect would also have been observed at interpolated (and hence unsampled) levels between the two. If the data analysis reveals an effect only at the higher level, this would suggest that the treatment can be effective and that there is a causal threshold higher than the lower treatment level sampled.

Of particular interest is the level of treatment implementation characterizing an intervention in its "usual practice" form. This requires having an experimental treatment corresponding to the anticipated mode for actual practice—apparently a rarity in the many studies of the effectiveness of psychotherapy

(Shadish et al., 1997; Weisz, Weiss, & Donenberg, 1992) but a necessity in the National Institutes of Health's (NIH) model of "effectiveness research" (Flay, 1986). This sampling strategy would not inform us about the treatment's potential when implemented at its best, what NIH calls its "efficacy." Because studying interventions at their best rather than their representative has always been a concern of theorists and program developers, it is probably best to incorporate into experimental designs a no-treatment control, a level close to the modal, and one that represents high intensity and fidelity to theory. This moves closer to describing the functional relationship between independent and dependent variables, although many treatments are so multidimensional that they cannot be so neatly scaled. Judgment is usually as important as measurement.

A reality of field experimentation is that treatments are nearly always implemented more variably than anticipated. When measured, this variation results in many unplanned levels on the independent variable and permits functional analyses of how they are related to the outcomes. However, a selection problem arises here if the individuals or sites implementing at one level are systematically different from others. This natural variability can permit only gross tests of how different treatment levels are related to the outcome, yet they probably are worth doing, even if only cautiously, because without them we would know nothing about causal thresholds and other inflection points in the function relating cause and effect. Interpolation depends on such knowledge.

Extrapolation involves generalizing a causal connection beyond the treatment range actually sampled. We might, for example, ask what effects patient education would have if it were more comprehensive than it has ever been in the past, or if it were provided for 1 hour per day rather than for the 30 minutes of past studies. Similar reasoning and assumptions are involved with extrapolation as with interpolation: The more levels there are on the independent variable and the more systematic the results turn out to be across levels, the more grounded is the belief that the causal function can be extrapolated beyond either end of the sampled distribution. Confidence in extrapolation also depends on the size of the gap between the endpoint of the sampled distribution and the level of desired application. Short extrapolations are easier to justify, presumably because qualitative transmutations are less likely then; that is, when water reaches a certain temperature, it creates steam or at a different temperature it creates ice; or when a metal cracks on reaching one temperature and melts when it reaches another; or when personal stress escalates and social relations move from indignation to shouting and then to blows. Shorter extrapolations also entail a lower likelihood of exogenous variables coming into play—as when only a very high body temperature brings in the doctor or when only a bank's extreme financial woes bring in special auditors. Unless theory is good enough to specify such transmutation points—and in our opinion they are rarely so good in the social sciences—there inevitably will be uncertainty about extrapolation, especially when an impressionistically large gap exists between the sampled treatment values and any values to which generalization is sought.

Summary

The argument advanced above is that, in validating cause-and-effect constructs, a theory of generalization is involved that is subtly different from both the probability theory buttressing random sampling and the theoretical reductionism underlying most theories of explanation. This alternative uses the description of a target class to guide the selection of instances so that they all belong in the target class and *demonstrably* share most of its prototypical components. Variants among the instances are then purposively selected to rule out the interpretation that the target causal relationship is due to irrelevancies associated with (a) the methods by which measurement was made, (b) the components that a target construct shares with cognate constructs, (c) the particular levels at which a treatment was manipulated, or (d) any competing substantive interpretations that might be invoked.

The supposition undergirding this approach is that irrelevant interpretations of manipulations or measures are best ruled out using some form of Mill's Method of Difference. This implies measuring or manipulating contending interpretations. Because it is not easy to specify the full range of competing explanations, inferences about constructs are all the more solid if multiple complex implications of the construct are specified and tested. The more numerous and unique these implications are, the fewer are the plausible alternative interpretations. This modified falsificationist approach emphasizes rational (rather than random) selection as well as data analyses that probe whether a causal relationship remains robust across all the potential contingency variables examined. This is not an argument against random selection in construct validation; rather, it is an argument about such selection being neither necessary nor widely applicable when the generalization of a causal hypothesis is under test.

APPLYING THESE SAME FIVE PRINCIPLES TO PERSONS AND SETTINGS

Are the principles that facilitate generalizing to target cause-and-effect constructs also relevant for generalizing to target populations of settings and people—the domains for which random sampling was originally developed and to which the methods of construct validation have rarely, if ever, been consistently applied?

Proximal Similarity

The principle of proximal similarity is easily dealt with in individual studies. The instances of people or settings selected for study should fit into the target categories to which generalization is sought, demonstrably incorporating most of its recognized prototypical elements. Thus, if the goal were to generalize to the patients and physicians within a particular hospital and random selection is not possible, it would be desirable to select patients or physicians from within the hospital, preferably those who score at or near the hospital mode on measures of demographics, diagnoses (in the case of patients), and training and specialty (in the case of physicians). This is in distinction to selecting persons who belong in the target hospital but use it so seldom or so idiosyncratically that they are relative outliers (St.Pierre & Cook, 1984).

If the goal were the broader one of generalizing to all hospitals in the United States, then an alternative might be to select the largest number of hospitals that the research budget allows whose size, patient mix, ownership pattern, and the like most closely resemble the national mode. (In cases of multimodality, as with for-profit and not-for-profit hospitals, each mode would need to be represented.) Modes are best described from archived census sources, where available, or from formally representative samples of a not-too-distant date. If neither is available, more impressionistic sources, such as clinical wisdom, must be critically appraised and used. The purpose of accidental sampling to represent the mode is for researchers to probe the causal relationship using instances that share some of the most salient attributes of the target population. This assigns to representativeness the meaning usually found in theories of categorization rather than mathematical sampling.

An important issue is how attributes are selected to represent the target category, given that the sampling option under discussion matches the population and sample on a restricted number of measured attributes, leaving dangling the extent of similarity on unmeasured attributes. The measures used for matching are likely to be those most easily available that are also highly correlated with a study's major outcomes. Hospital size and ownership pattern probably fit this bill, as do patient age, diagnosis, and method of hospital payment. As Lavori et al. (1986) indicate, samples that have been selected purposively should be described in as much detail as possible so that readers can assess for themselves how well a sample's measured attributes impressionistically overlap with the known attributes of the population to which generalization is being sought. At a minimum, this description will let readers recognize the populations the sample does not represent, and it will give them some idea of what the sample might represent. Unfortunately, it will not indicate what the sample does represent because matching by proximal similarity involves characterizing the population with only a restricted set of variables. The population and sample might differ on other attributes that, unbeknownst to the researcher, are differently related to the size or direction of a causal connection. Careful matching nevertheless can at least ensure that the samples resemble the populations on at least a circumscribed set of characteristics that uniquely describe the populations about which inference is sought.

Heterogeneity of Irrelevancies

In sampling modal instances, priority is given to cases with attributes prototypical of a labeled universe; however, these cases will differ from one another in a host of other ways, most irrelevant to the major causal hypothesis but some of them capable of affecting the dependent variable. In large-sample research, one can probe how robust a causal relationship is across theoretical irrelevancies like the regions of the country, hospital age, average tenure of attending physicians, type of patient diagnosis, or any other patient attributes. If such stratification is not feasible, then researchers can at least document whatever heterogeneity there is and can probe whether a cause-effect relationship holds *despite* this heterogeneity.

Special problems occur, though, when the sample of cases is small, as it often is for the setting factor in particular. How many states, hospitals, schools, and the like can a research budget accommodate? With small samples the question arises of how to select cases, because heterogeneity can be achieved deliberately on only a small number of conceptually irrelevant attributes. Indeed, cases arise where no deliberate sampling for heterogeneity is possible, especially when substantively important moderator variables have been specified whose exclusion from the sampling design would undermine a study's major purpose. These must take priority in determining the sample design. For instance, imagine being able to sample only 20 hospitals—10 in each of two treatment groups. After stratifying on any potential moderator of compelling theoretical or practical interest—for example, size or patient mix—little latitude remains for further stratification. The remaining heterogeneity has to be lived with.

Formal sampling theorists will likely throw up their hands at this point, contending that small samples and a plethora of potential causal determinants make sampling theory irrelevant in cases like the above. We understand and share this exasperation, but such limitations regularly confront researchers with modest budgets who want to test a causal proposition. We cannot abandon their need for generalized causal influences solely because methodologies based on large samples do not apply in the same way to small samples and because so few variables can be made heterogeneous. We can only advise that all sampling, purposive or otherwise, is useful if it achieves variation on any factor that theoretical speculation or practical wisdom suggests will modify the sign of the relationship between a treatment and an outcome.

In this regard, it is interesting to note the suggestion of Lavori et al. (1986) that researchers should analyze why individuals and organizations have agreed to participate in a study. Were all the units volunteers? Were they selected because they seemed to be exemplary cases of good practice? Or were the units experiencing special problems and so needed special help? In trying to develop a pragmatic theory of representativeness that is applicable when the mathematical sampling theory they (and we) prefer is not, Lavori et al. assign special status to identifying the specific ways in which samples differ from their intended target populations, highlighting them as part of the sample description that partially substitutes for random selection. We prefer to express their alternative in falsificationist terms. The basic hypothesis is that a population of persons or settings is of type X. Many outcomes can disconfirm this hypothesis—for example, the sample does not have the prototypical components of population X, or it has characteristics of Y not shared with X that are related to the outcome, as when the reasons for participating in a study have nothing to do with population X but might nevertheless influence the results.

We suggested earlier that factors which universal theorists consider irrelevant targets of generalization may still be relevant to others (e.g., to universe-bound theorists or applied researchers). When resources permit extensive sampling of, say, patients or hospitals, it may be possible to generalize not only to the types of patients or hospitals specified in the guiding research question but also to some subtypes of patient or hospital that are relevant to some research consumers. In examining subtypes, it is important to differentiate the types themselves from any irrelevancies associated with the particular instances of them in the sampling frame, whether these instances have been selected randomly or not. A trade-off is involved here, though. Given constant resources, generalizing to a broad array of subtypes entails many subtypes but few instances of each, whereas generalizing to a smaller number of subtypes can entail more units per subtype and hence more stable estimation. Researchers have to decide in advance whether they want to draw strong conclusions about a smaller number of subtypes or weaker conclusions about more of them. Many basic researchers are disposed to resolve this trade-off in favor of increasing the number of subtypes, assuming that a wider range of settings and persons better approximates universality. But other theorists take a differet tack and are prepared to take the results from a single population and to assume that broader generalization is warranted until additional evidence suggests otherwise (e.g., Calder, Phillips, & Tybout, 1981, 1982). In essence, they solve the problem of causal generalization by postponing it for others. This is not an attractive solution in those areas of the social sciences that lack the natural sciences' tradition of routine, rapid, independent replication and that assume that the real world is very complexly ordered (Cronbach & Snow, 1976; McGuire, 1984). We are, therefore, apprehensive about assuming that findings obtained in a single class of settings (e.g., the laboratory) with a particular class of persons (e.g., college students) will necessarily generalize any further. We prefer to await supportive evidence from heterogeneous replications that have either submitted the causal generalization to deliberate falsification or have assessed the robustness of a causal relationship across many heterogeneous person and setting types. Unless substantive theory clearly specifies the factors most likely to condition a causal connection, in our view the best sampling strategy is to select types of persons and settings that cover as wide a range of populations as possible.

Discriminant Validity

Clinical lore suggests that cholecystectomy patients are "fair, fat, female, and over 40." If this is a true picture of relative incidence, proximal similarity suggests that most respondents in patient education studies should have these same characteristics. So it would be possible to generalize to persons who are dark, slender, male, and under 40 only if their numbers are sufficient to assess whether patient education is effective for each of these groups. If it is, then we would conclude that patient education is effective with cholecystectomy patients of all the different types examined (the target class) and not just with those who have the class's modal attributes and are fair, fat, female, and over 40. In essence, the task is to discriminate effects found with typical gall bladder patients from those found with less typical ones. Only then can we assert that patient education works "generally."

Such discrimination is especially important when claims are made that a particular human population (or setting) is causally unique. To draw conclusions about patient education's unique effects on, say, prostate cancer requires differentiating effects found with cancer patients in general from those specific to prostate patients. This requires a sampling frame with many types of cancer patient and a data analysis that probes whether the relationship between patient education and recovery from surgery is different for prostate cancer patients when compared to other cancer sufferers. To study only prostate patients might permit generalizing to such patients, but it could never provide any enlightenment about the group's unique responsiveness to patient education. The same logic holds in discriminating between cancer patients as a class and noncancer patients as a class, or in examining different types of settings. Without the appropriate contrasts, there can be no demarcation of what is causally special about a particular class or type. (Of course, if the same effect holds across all types of patients, one would then conclude that the intervention helps all patients, including those with prostate cancer, though it does not affect the latter especially.)

Causal Explanation

In the present context, causal explanation has to do with understanding why a causal relationship holds with a particular population of persons or settings or differs by population in the strength of the cause-effect link. The search is to identify the population-related factors moderating a causal relationship. Thus, if patient education were to be more effective when hospital chaplains implement it (rather than physicians or nurses), is this due to denominational similarity between the educator and patient, or to the spiritual content of the information transmitted, or to the chaplain's experience in knowing how to relate to people in need? Once the more causally important attributes of chaplains have been identified, it is then possible to ask whether patient education would be successful with other professionals with these attributes and whether these same attributes can be added to the repertoire of other hospital professionals, thus facilitating knowledge transfer.

The same reasoning applies to understanding the components of settings. If a causal connection only holds in certain kinds of American hospitals, it would then be productive to learn what it is about these hospitals that facilitates the effect. If the answer is that they leave individuals considerable time by themselves in which to ponder the advice patient educators give, then this is a feature that might be made available in other (perhaps as yet unstudied) settings, including nonhospital settings.

Interpolation and Extrapolation

Imagine a causal connection that has been replicated across a wide range on some person or setting variable—perhaps the link is found with the youngest and oldest age groups or in the smallest and largest hospitals. This gives rise to the temptation to interpolate and to conclude that the same causal connection would have been found at unsampled intermediate points on the continuum, assuming the causal relationship to be linear and that there are no other restrictions to the sampling frame. In this last connection, consider a study done in the single hospital in Greenwood, Mississippi. Interpolation would be facilitated to the extent that patients can be stratified to achieve a wide range in income, race, age, diagnosis, and comorbidity attributes.

But the respondents would still be from Greenwood and its surrounds, however they have been sampled. How to extrapolate? Extrapolation to very broad populations depends on the heterogeneity of the persons or hospital types sampled and on the consistency with which a causal connection is replicated across these types. If it can be demonstrated across hospitals and patients with many different characteristics, then by simple induction it is tempting to infer that the same connection is likely to be found with persons and settings of types not yet tested. The epistemological basis for such generalization is weak, although a falsificationist defense can be constructed on the premise that analyzing a causal relationship across different person and setting types provides multiple chances to disconfirm the proposition that a causal relationship is general. Absent disconfirmation, the causal generalization remains provisionally viable; given disconfirmation, it is rejected, and some person or setting types have now been identified on which the causal conclusion depends. As with all falsification, the hypothesis of causal generality has to be put into conflict with the strongest possible counterhypotheses derived from the theories and experiences of a heterogeneous set of commentators, including some who do not share the primary analyst's theoretical predilections. The falsification tests also have to be technically adequate, given our unfortunate predicament that substantive theories are inevitably underspecified and empirical observations inevitably theory laden (Cook & Campbell, 1986).

Summary

Useful knowledge about how well a causal conclusion generalizes to specified universes of humans and settings can be drawn even without random selection. Such generalization depends on the same five principles we used to justify general inferences about both cause and effect constructs. Random selection is better technically, but we contend that it is rarely practical in causal research because financial costs, ethics, and the desirability of quality treatment implementation incline researchers toward studies with fewer sites and respondent populations that have given their consent to be in a study.

META-ANALYSIS AND THE PROMOTION OF CAUSAL GENERALIZATION

The prior discussion has been about sampling, measurement, and data analysis in *single* studies. Few such studies have the large and heterogeneous samples of persons, settings, manipulations, and outcome measures that our discussion suggests are crucial for confident causal generalization. We must, therefore, look to systematic literature reviews for enlightenment about such generalization. We now illustrate their advantages, defending two propositions: (a) that one form of review—meta-analysis—is especially suited to testing causal generalization, and (b) that its advantages follow from the use it makes of the five principles abstracted above from theoretical work on construct validity and category formation.

Meta-analysis was originally developed to provide quantitative reviews of the very type of descriptive causal connection whose generalization is under discussion here. The technique has since been extended to cover noncausal descriptive questions (e.g., "Do boys and girls differ in science achievement or persuasibility?") and to identify factors moderating a causal connection (Cook, 1991; Cook et al., 1992). Its development was a response to three major concerns. First, most individual studies of causal phenomena are statistically underpowered (Hedges & Olkin, 1985). Second, resource constraints and control needs entail that many single studies have sampling specifics that limit the number and range of third-variable causes, especially as concerns the range of respondent types and treatment implementers, the duration of treatments, and the validity of measures. Third, traditional review methods require tabulating all the studies that do and do not support a particular causal inference and then counting whether the causal relationship is more often corroborated than not. Because statistical significance tests usually determine the verdict assigned to each study, underpowered tests come to be classified as valid failures to reject the null hypothesis. With greater statistical power and the same effect size, however, the substantive conclusion might have been exactly the opposite.

To conduct a meta-analysis requires at least one effect size per study—usually a standardized difference between, say, a treatment and a control group. Such standardization puts study-specific outcome measures onto the same scale, allowing effect sizes to be averaged across studies. Tests of this average usually have considerable power because the units are group means, which are more reliable than scores from individual respondents. This may be why meta-analyses regularly result in average effect sizes reliably different from zero, in some research areas creating a more positive impression of global effectiveness than "box count" reviews of the same studies.

In good meta-analyses, effect sizes are manipulated like any other observations. Tests can assess whether the distribution of effect sizes is normal, individual studies can—and should—be weighted to reflect sample size differences, multivariate tests can hold constant all study-level irrelevancies that have been measured, and stratification can assess how a potential modifier has influenced the size or direction of a causal relationship. Meta-analysis substitutes group-level observations for individual-level ones; otherwise, little about it is statistically unique.

Effect sizes are the product of myriad judgments by both primary analysts and meta-analysts—judgments about whether to do a study at all, how to design it, whether to write it up, what to include in the write-up, whether to archive the results, and whether to include the study in a review given its topic and methodological quality. Meta-analyses assume that the effect sizes under analysis are drawn at random from the population of all possible effect sizes for a given research question. This is an assumption that cannot be tested sensitively and is, anyway, likely to be false. Meta-analysts therefore emphasize finding "all" the studies ever conducted on a topic, hunting out fugitive studies buried deep in files so as to create a census of past studies. Even this is not the most relevant population, which is of all possible studies.

Reviews are of little use if every published and unpublished study is found but all share the same constant bias. For example, in meta-analyses of how school desegregation affects the academic achievement of black children, children potentially can stay in school for up to 12 years, yet the available studies cover only the first 2 years after desegregation (Cook, 1985), almost certainly leading to underestimates of desegregation's total impact. Meta-analysts work with purposive but heterogeneous samples of studies where the population is rarely clear and not necessarily devoid of constant bias. So how can they be useful for causal generalization, as we have asserted they are?

Proximal Similarity

Meta-analysts regularly use proximal similarity, albeit without naming it as such. In the absence of a massive consensus in favor of measuring a construct one single way, different researchers tend to measure the same construct differently. This is not much of a problem if understanding of the construct is widely shared within a language community, for then its prototypical components will be represented in most operational instances and its less central components in fewer of them. This enhances construct validity by assigning more weight to the prototypical components that define proximal similarity. In this connection, consider the approximately 150 studies of patient education (Devine, 1992) where the intervention is sometimes provided prior to surgery, sometimes after it, and sometimes at both periods. In some instances, the intervention involves components from only one of the principal domains theorists discuss—providing information, skills training, and social support. In other studies, elements from two or three domains are combined. Even so, many more studies involve presurgical than postsurgical interventions, and many more studies combine intervention components than treat them singly, thereby re-creating the very emphases represented in most theoretical explications (descriptions/understandings) of the generic patient education construct. There is no exact pattern match, of course. No Platonic construct description exists against which the composite of many single operations can be compared to see if its distribution of component weights matches what has been magically specified somewhere. At issue is only fidelity in the gross correspondences between a construct's description and the frequency with which its more and less prototypic components are captured in a literature review. We assume that this correspondence is higher in reviews because they contain many more operational representations of a construct than a single study.

Similar reasoning holds for generalizations about target populations of persons or settings. Proximal similarity is achieved because inclusion in the literature review depends on membership in the target class (i.e., being a gall bladder patient). Within limits set by any errors in recording or diagnosis, all the respondents in each study are likely to belong in this class. Moreover, the attributes most highly correlated with needing gall bladder surgery (i.e., being fair, fat, female, and over 40) should be distributed more densely across the sample of studies when compared to less prototypical attributes. But how can we know this? In the cholecystectomy example, knowledge of population attributes comes from clinical lore, although a census or random sample of hospital records would be better. Whatever the source of this knowledge, some evidence is needed of the extent to which the pattern of patient characteristics in a meta-analysis deviates from the expected profile. A discrepancy suggests the possibility of bias in the sample of studies under review, whereas a lack of discrepancies reduces concerns about representativeness. To show that the achieved samples fit into the target category and densely represent its known prototypical attributes, it is sometimes possible to make explicit comparisons between the population and sample attributes, as Furstenberg, Brooks-Gunn, and Morgan (1987) did to support the claim that their sample of Baltimore teenage mothers resembled the national profile of such mothers on certain measured attributes.

The review format does not guarantee generalization to target cause, effect, person, and setting universes. If all the researchers on a topic share the same understanding of a cause or effect construct, a bias may result and future generations may come to see all their understandings as flawed. Recent claims about the widely shared sexist, racist, ageist, or first-worldist biases of past social scientists illustrate this possibility. On a more mundane level but illustrating the same basic point, if all the research on patient education has been performed in Veterans Administration hospitals, this would constitute a constant bias, however the V.A. hospitals had been sampled. Likewise, if nearly all the studies of psychotherapy effectiveness research are done by graduate students testing some pure version of a therapy in a university student health center, what results is a literature on the efficacy of psychotherapy and not on its effectiveness in ecologies of routine application (Shadish et al., 1997; Weisz et al., 1992). Reviews are superior to individual studies in their potential to represent prototypical attributes better, but this potential is not a necessary achievement and researchers must continually search to identify hidden sources of constant bias.

Heterogeneity of Irrelevancies

As far as settings are concerned, a review of cause-probing studies would rarely, if ever, be based on a random sample of hospitals representing, say, the entire United States. Instead, reviewers would have to make do with whatever sample of hospitals turned up after an exhaustive search of published and unpublished sources. Any one hospital—even all the hospitals in a particular state or region—constitutes an irrelevancy for most research consumers who prefer that the accidental sample of hospitals contains instances coming from all regions and "representing" many different patterns of hospital ownership, financial health, number of beds, and so on. Because heterogeneity is much greater in the typical review, the potential exists to deal with many more sources of conceptual irrelevancy and hence promote more confident causal generalizations.

When the number and range of setting variables are impressionistically large, it is feasible to conduct probes of whether a causal relationship is found within each type of hospital, region, patient, and way of conceptualizing the cause or effect (see Devine, 1992). If a causal connection is suggested within one particular type of, say, hospital but not another, this suggests a causal contingency that deserves further elaboration and explanation. The particular hospital type so identified is no longer a causal irrelevancy; it has become a moderator variable. The perfect state of affairs is when the sample of units in each stratum/type is not only large but also selected at random. The latter rarely holds, particularly with large units like schools or hospitals where research budgets rarely permit introducing experimental interventions into many units. Still, analyses of a carefully scrutinized opportunistic sample of a particular type of setting or person can help probe whether a causal connection holds across all the types examined, provided there is sensitivity to statistical power concerns. Devine (1992) did such a probe, demonstrating that the positive benefits of patient education held across various types of patient and hospital and across various time periods and ways of conceptualizing patient education and recovery from surgery. With fewer studies than were at Devine's disposal, it is often impossible to do separate breakdowns by types, but it may still be possible to use the review to test whether an average effect size computed across whatever small sample of hospital types is available is reliably different from zero. This at least permits concluding that the intervention makes a difference despite the existing variability in settings (or persons or times or ways of operationalizing constructs).

When considering possible temporal contingencies limiting the application of a causal relationship, the short history of the empirical social sciences curtails a review to the last half century. This range is still far greater than is found in most individual studies. For instance, Devine (1992) was able to show that the effect size from patient education to recovery from surgery was broadly stable between 1964 and 1978 but decreased thereafter without ever shrinking to zero. Computing an average effect across the entire period hides this change in effect sizes (but not causal signs!) and obscures the fact that the later years are particularly important for current deliberations about future health care policy. (Fortunately, Devine was also able to go some way toward explaining the temporal change in effect sizes, demonstrating that the smaller ones in more recent years probably were due to later studies incorporating fewer components of the global intervention, presumably because researchers felt little need to demonstrate once again the success of global patient education. Instead, they wanted to identify its most efficacious components.)

The enhanced number and range of irrelevancies in reviews helps promote inferences, not only about target populations of units, settings, and times but also about target cause-and-effect constructs. This is because, in a review, any one cause or effect construct is operationalized many ways, each depending on researcher-specific understandings of both the construct and the irrelevancies inevitably associated with the researcher's preferred way of operationalizing the construct. Also evident in the typical review are multiple modes of measurement, different time lags between the manipulation and outcome measurement, different kinds of data collectors, and so forth. Any causal finding achieved despite such heterogeneity of irrelevancies is all the more firmly anchored conceptually, especially if individual irrelevancies have been used as stratification variables and the causal connection fails to disappear when the irrelevancy is or is not present.

Even this level of confidence can be illusory unless accompanied by vigorous attempts to identify sources of constant bias. For instance, Devine and Cook (1986) showed that nearly all studies had used researchers rather than regular staff nurses to provide the patient education treatment. Staff nurses may not be as intensively trained in the intervention as researchers, and they have many more competing responsibilities to contend with in the hospital. Can they ever implement the treatment enough for it to be useful? Regular staff nurses had been used in 4 of the 102 studies reviewed at the time, and effect sizes were smaller in these studies than in those where researchers provided the intervention. Fortunately, independent evidence showed that the interventions in these four studies were less comprehensive than optimal patient education protocols, and a later study conducted with regular nurses and a presumptively powerful treatment achieved the same effect as found in the review (Devine, O'Connor, Cook, & Wenk, 1988). Even so, the fact that the type of person providing patient education was systematically correlated with treatment intensity across 102 studies suggests that the number of studies in a review is less important than their relevance to conceptual issues. Director (1979) has made the same point about job training, contending that it had been systematically provided to applicants with worse education and work histories, resulting in a constant source of bias that could not have been controlled for no matter how many similar studies had been conducted.

Some types of meta-analysis welcome more heterogeneity of irrelevancies than others. Chalmers and his colleagues (e.g., Chalmers et al., 1988) routinely conduct meta-analyses of the available randomized clinical trials of particular drugs or surgical procedures. Their meta-analyses are fairly standardized with respect to independent variables, outcome measures, experimental designs, diagnoses, physician specialties, and criteria of methodological adequacy for entering the review. Given this strategy, small numbers of studies are likely to result. The small numbers can be tolerated, however, because the researchers' major interest is in main effects of the drug or surgical procedure. This research model is close to an exact replication ideal, with the intent being to achieve the clearest possible picture of the causal efficacy of the treatment under conditions designed to maximize detecting an effect should there be one. Little need seems to be felt to explore factors moderating the effects of a drug or surgical procedure.

Contrast this priority with meta-analyses on more social topics in medicine, where the desire is often to assess effectiveness under conditions of real-world application, as with such issues as patient education, compliance with drug prescriptions, or psychotherapy. A different attitude toward sources of heterogeneity is then warranted. In patient education, the diagnoses studied meta-analytically are more variable (from transurethral resections of the prostate to limb amputations); the outcome measures legitimately vary from study to study (length of hospital stay, the amount of pain medication taken, satisfaction with the hospital stay, etc.); and the interventions vary from study to study depending on local understandings of patient education and the time and resources available for more comprehensive versions. In meta-analyses of psychotherapy, the types of therapy and outcome are even more numerous, as are the patient and therapist types and the institutions in which therapy takes place. Where heterogeneity is the rule, meta-analysts aspire either to generalize across the irrelevancies or to use multivariate procedures to examine how particular sources of heterogeneity influence average effect sizes (see Lipsey, 1992; Shadish, 1992). There is no pretense here of exact replication or of "pure" tests. Heterogeneity is welcomed to the extent it maps the heterogeneity in complex social settings of application. Moreover, the greater the heterogeneity in persons, settings, measures, and times, the greater is the confidence that a causal relationship will also be obtained in still-unexamined contexts. Reviews of more standardized studies cannot have this result.

Discriminant Validity

In their meta-analysis, Devine and Cook (1986) found a subset of studies that permitted discrimination of the target outcome—recovery from surgery—from some of its cognates. This was because some studies measured the time it took patients after discharge to return to work or resume normal activities in the home. These measures showed that patient education decreased the period required to resume such normal activities, thus reducing the plausibility of the argument that the recovery measures tapped into the hospitals' need to release patients early to reduce costs rather than into a genuine physical recovery from surgery. Thus, these discriminant data indicated that individuals exposed to patient education were not released prematurely. The larger the sample size of studies, the greater is the likelihood of finding some that help unconfound interpretations of an effect construct. Even if a single study was explicitly designed to achieve such unconfounding, it would inevitably have many other invariant features that would more likely vary in a literature review. The same stratification process also helps interpret causes. For example, a small number of patient education studies included discharge physicians who were blinded to the experimental status of respondents. Analysis showed that this blinding made no difference in the size of the link between patient education and recovery from surgery, indicating that physicians' expectations about length of stay did not constitute the causal construct and that patient education might have. To conduct such analyses requires variation in physician expectations and, all things being equal, this is more likely in a review than in any single primary study.

Although multiple studies provide many of the sources of variance needed to discriminate cognate constructs, the number of studies with a particular attribute of interest may be small, as happened with studies of regular staff nurses delivering the patient education treatment. Such small samples make it difficult to estimate whether the available studies differ from others in systematic but irrelevant ways related to a causal connection. This is why the analysis of studies with and without an important conceptual attribute should include study-level correlates of the attribute's availability. Otherwise, a selection problem can result because the studies with the attribute needed for discrimination might be systematically different from other studies. Multiple regression analyses are often desirable in meta-analyses if they seek to remove the effects of study-level methodological and theoretical irrelevancies prior to examining how a causal relationship might be conditioned by discriminating factors that are more substantively relevant, such as hospital size or the nature of the persons implementing a treatment (see Lipsey, 1992, for an example of this). The selection models will never be "perfect"—at best they will be reasonable approximations.

Causal Explanation

Researchers explain causal connections in part by identifying the causally effective components of global interventions and the causally affected components of global effects. Meta-analysis allows the researcher to take advantage of the many planned and unplanned sources of variability to explore some explanatory possibilities of this sort. For instance, Devine and Cook (1986) were able to decompose the patient education treatment and to isolate the studies that varied in all possible combinations of its three components: providing information, skills training, and social support. The effects of these different combinations were then assessed, and analyses showed that each component was minimally effective by itself but that combining the components increased the size of effect in seemingly additive fashion.

The major limitations to using meta-analysis for identifying causally relevant treatment components are practical rather than theoretical. They include (a) the paucity of detail about treatment components in many journals and books (dissertations prove to be a godsend here!), (b) published treatment descriptions that are based on what researchers intended to do as an intervention rather than what they actually accomplished, and (c) the need for large samples of studies if sensitive analysis of individual cause-and-effect components is to take place. Nevertheless, there are many published meta-analyses with reasonable probes to identify crucial components of the more global cause-and-effect constructs (e.g., Lipsey, 1992).

Full explanation goes far beyond identifying causally efficacious or impacted components. It also requires analysis of the micro-mediating causal processes that take place after a cause has varied and before an effect has occurred (e.g., Connell et al., 1995). Although there is nothing in theory to prevent meta-analyses resulting in strong inferences about micro-mediation, in practice exemplary studies are few and far between. (For an exception, see Harris and Rosenthal, 1984; and for a broader discussion, see Cook et al., 1992.) The reasons for the dearth of positive examples are that (a) one researcher rarely measures the same explanatory processes as another, given how underspecified many theories are and given how strong the social pressures are on researchers to be original; (b) some researchers prefer a black-box theory of experimentation and do not want to specify or measure intervening processes; (c) substantive theories are dynamic and change with time, often going through cycles in which old explanatory theories (and their constructs) are disparaged and hence not measured in the next wave of studies; and (d) when data on explanatory variables are collected, they are not always reported in journals in a correlational form that lends itself to synthesis. Such data usually are presented as regression coefficients whose value depends on other variables in the causal model, and models are rarely identical across studies. Hence, meta-analysts cannot use the data to assign *comparable* descriptive relationships to an explanatory construct from one study to the next. These limitations are severe and not likely to go away soon.

Coarse probes of micro-mediations nevertheless are possible and promise to provide some clues about causal mediation. Many more, and better, clues are possible, though, about person, setting, time, and method factors that moderate a causal relationship. Today's explanatory payoff from meta-analysis should be expected in identifying causal moderator variables rather than causal mediators.

Empirical Interpolation and Extrapolation

With regard to interpolation and extrapolation, meta-analysis once again holds more promise than any individual study. This is partly because of the wider range typically achieved on many of the person, setting, and time variables worth study. Thus, if patient education were effective with medical diagnoses that lead to both very short and very long-term hospital stays, we would be confident in interpolating to types of surgery between these extremes. This would also be the case if basic causal finding held with older and younger patients and with the most financially sound and imperiled hospitals. Because the extremes are likely to be farther apart in a review, the levels represented in any one study are more likely to fall within the extremes achieved across all the studies. This increases the role that interpolation plays in generalizing about causal connections and decreases the role of the more problematic extrapolation.

Individual studies differ from one another on such internal criteria as the levels of an independent variable, patient characteristics such as socioeconomic status or age, and organizational features such as hospital size or maturity. Such differences can be used to approximate dose-response relationships to probe how effect sizes are related, not only to levels of a treatment but also to levels of variables assessing patient and organizational characteristics. The aim is to describe the overall function and any thresholds where the size of the causal relationship suddenly changes, under the assumption that all study-level methodological irrelevancies have been statistically removed from the effect sizes. Because we can never know when this has been accomplished successfully, special interpretive weight needs to be given to the subset of individual studies where multiple levels of the treatment were experimentally varied or where multiple levels of a person or setting contingency variable were deliberately measured. Even when such within-study comparisons cannot be made, it is still desirable to take advantage of the greater variability reviews offer to probe the functional form of a cause-and-effect relationship across person and setting attributes as well as across treatment levels.

There is a sense in which all causal generalization is about interpolation and extrapolation. Rubin (1992) has suggested that causal generalization is about estimating response surfaces. Coarsely stated, it is about plotting the function relating a cause to an effect across a broad range on some third variable, repeating this for all possible third variables that theory or practice indicates are worth analysis. This procedure has a lot to recommend it conceptually. Practically, it is difficult to attain, particularly across a broad range of third variables and particularly if these variables are presumed to influence effect sizes in some interactive fashion. Then, one has to conceive of a multidimensional space in which the effects of a given cause are plotted. This is certainly an ideal, but it is hardly a reality at this time in the history of the social sciences, when interpretable parametric studies are rare but valued where they exist (e.g., in meta-analyses of the effects of variation in class size; see Glass and Smith, 1979). The work on class size, however, points to obvious restrictions in applicability, for only two variables are involved—class size and achievement. No single third variable is introduced, nor is any set of third variables. Even so, estimating response surfaces is a useful way to think of causal generalization, and it illustrates why interpolation is a crucial component of any theory of causal generalization, as is extrapolation, though in that case the inferential dangers are greater.

Summary

A literature review technique like meta-analysis enhances causal generalization because it has built into it the five principles that we contend promote such generalization. It deals with proximal similarity better than any single study could because—all other things being equal—it distributes the prototypical elements of the cause, effect, person, and setting constructs more densely throughout the sample of studies. Reviews also contain many more sources of methodological and substantive heterogeneity that permit analysts to assess how robust a causal relationship when these sources of heterogeneity are combined or are separately examined across a wide range of person, setting, and time factors as well as across many different ways of measuring a cause or effect. Also, depending on the range sampled, they can permit more interpolation of a causal relationship between extremes, and their wider base of explanatory measures permits some explanation, especially of the causally efficacious components of a treatment and of the causally impacted components of an effect. When meta-analytic findings are robust, they do more than merely demonstrate the populations to which generalization is warranted; they also increase the chances of being reproducible in as-yet-unstudied circumstances, thereby facilitating Cronbach's second type of causal generalization as well as his first. The *empirical* cross-population generalization that meta-analysis emphasizes, however, is quite different from the more theory-based cross-population generalization that Cronbach's theory treats as the most important method for generalizing to unsampled populations and constructs.

CONCLUSIONS

Cronbach's Framing of the First Causal Generalization Issue

One purpose of this chapter was to examine the role of random selection in facilitating generalized inferences about the connection between a treatment and an outcome. Such inferences require us to make valid statements about treatments, outcomes, persons, settings, and times as well as the nature of the link between a treatment and an outcome. Our analysis led to a somewhat pessimistic conclusion. When probing causal connections in individual studies, random sampling is sometimes practical with persons, but it is rarely, if ever, possible with treatments, settings, outcomes, or times. Even with persons, random selection is usually from within a small and circumscribed population that has itself been selected more for logistical and financial convenience than for formal representativeness of a large and theoretically meaningful population.

The second purpose of this chapter was to describe an alternative theory of causal generalization built around five principles abstracted from research on construct validation, where the task is to generalize from measures and empirical relationships to theoretical constructs, particularly cause-and-effect constructs. Much already can be done within individual studies to promote causal generalization through (a) purposively sampling proximally similar instances; (b) making irrelevancies more heterogeneous and stratifying to see if they make a difference, thereby assessing some of the person, setting, time, and method factors across which a causal connection can and cannot be obtained; (c) stratifying on cognate variables to differentiate them from the target populations of research interest; (d) identifying the ranges within which a causal relationship does and does not hold; and (e) describing the components of the cause and effect and any other variables that might moderate or mediate a causal connection, thereby helping explain why a cause and an effect are linked.

These five principles do not enjoy as strong a logical warrant as random sampling. To produce interpretable results, they require that (a) target categories are clearly described, (b) prototypical attributes can be delineated clearly, and (c) alternatives to any preferred interpretation of the populations and constructs can be ruled out. This last is the most difficult to achieve; however, it is approximated when a causal conclusion is robust across all known irrelevancies, when the influence of target constructs can be differentiated from those of cognate constructs with overlapping attributes, and when target constructs or populations are demonstrably not confounded with particular values on these entities. This is a daunting set of conditions that are better met in literature reviews than in individual studies. Literature reviews should be the primary source for confident conclusions about causal generalization.

This theory emphasizes purposive sampling for theoretical ends rather than random sampling to represent a population. The sampling is designed to capture various features that promote generalization—prototypical attributes, heterogeneity on theoretically peripheral attributes, discriminating target from cognate constructs, identifying important components and moderator variables, and extending the range on many variables where interpolation is called for. Random selection is much less important in this conception, though welcome when it can be achieved. Because random selection is the linchpin of sampling theory, what we have advocated lies outside the bounds of formal sampling theory. Just as Campbell had to explicate what it was that random assignment to treatments controlled for as a step to developing his theory of quasi-experimentation, so we have had to explicate what random selection achieves to explore alternative (and messier) ways of bringing about the same ends.

In our analysis, what random selection does is match a sample and a population on prototypical attributes, rule out alternative interpretations because irrelevancies are distributed in the sample as in the population, discriminate between target and cognate populations because only the target enters the sampling frame, and allow researchers to probe whether various subpopulations relate to research outcomes in the same way, thus promoting causal explanation. Whereas random selection accomplishes these goals principally via a procedure, the sampling designs advocated here depend on theoretical stratification and on selecting a purposive sample of cases within each stratum. No pretense can be made, therefore, that causal generalization is "automatic." It is a difficult inference, related more to attempts to disconfirm the hypothesis of causal generalization than to claims of generalization solely by virtue of the sampling procedure used. In this sense, ours is a quasi-sampling theory of causal generalization.

We do not want to claim too much for the five principles outlined. No theory of causal generalization is currently available that is so logically warranted, so empirically probed, so comprehensive in coverage, and so practically superior to the currently available alternatives that it merits hegemony for research practice. Instead, multiple methods for causal generalization will be needed to meet the criteria listed above, and much theoretical work will need to be undertaken to modify and extend these criteria. Given that no simple but practical theory of causal generalization currently exists, critical multiplism (Cook, 1985; Shadish, 1994) should be the motto and methodological dogmatism the fear.

Cronbach's Second Framing of the Causal Generalization Issue

Many scholars and practitioners are interested in Cronbach's second framing of the causal generalization issue: how to justify generalizations from entities with certain characteristics to entities with quite different characteristics. This is not the same as the traditional meta-analysis question: "Across which examined population is a particular causal conclusion warranted?" Although both questions involve cross-population generalization, in only one case does the intended inference go beyond characteristics of the available samples to include conclusions about novel causal connections and novel contexts where a causal connection may never have been studied before. Transfer of knowledge to new circumstances lies at the heart of the second framing of causal generalization.

Explanation is usually invoked as the preferred method for learning about causal transfer. The presumption is that once we know why or how a cause-effect relationship comes about, we can then identify the conditions that seem necessary and sufficient for the effect, permitting us to bring these conditions together in novel treatment configurations that are tailored to the specifics of particular local populations and local settings. For those scholars who reject the language of necessary and sufficient conditions (Cook & Campbell, 1979), causal transfer is enhanced whenever we learn about the moderator and mediator variables that condition a causal connection. Such knowledge should help practitioners decide whether the major causal facilitants are present in the settings for which they are responsible.

We do not want to question the premise that causal explanation promotes the transfer of research findings. Indeed, we claimed as much, even suggesting that it can promote causal generalization in both Cronbach's first and second framings. We did contend, however, that some uncertainty is warranted about whether, in actual social science practice, causal explanation can reliably play the transfer role ascribed to it. It seems to us that few social science theories are precise enough in the moderator and mediator constructs they invoke, and the gains they typically achieve in individual studies identify only one or two causal contingencies rather than all or most of the necessary and sufficient conditions for an effect. Cronbach is so sensitive to this shortfall between the promise and achievement of causal explanation in the social sciences that he advocates pursuing causal explanation through a variety of methods, many of which are more akin to common sense than to the quantitative causal modeling techniques whose results he finds so model-dependent.

We suggest that meta-analysis offers a different route to causal generalization in both of Cronbach's senses. Given a heterogeneous array of studies, meta-analysis permits empirical probes of the robustness of a causal connection across subtypes of the cause and the effect and across different types of persons, settings, and times. The greater the heterogeneity in such variables and the greater the consistency of results achieved, the stronger is the presumption that the same relationship will be found with quite different populations of persons, settings, and times and with variants of the cause and the effect that are not identical to those studied to date. Although many scientists assume broad causal generalization after only a few demonstrations, our preference is to undertake heterogeneous replications of the causal proposition under test. If the hypothesis of causal generalization remains robust despite all these falsification attempts, and if no plausible alternative interpretations remain to the generalization hypothesis, then this is a stronger base for inferring that results can be transferred elsewhere.

The starting point for this chapter was Don Campbell's fairly scattered writings on external validity, with the dual goals being to systematize them a little more and to relate them to Cronbach's framing of external validity, where more emphasis is placed on causal explanation than on proximal similarity. Cronbach therefore came to be as important in writing this chapter as Campbell, resulting in an imagined dialogue between the two that was predicated on each agreeing in advance to integrate ideas (where possible) rather than to stake out unique positions. I hope it is an imagined dialogue that takes each of these grand theorists seriously and that, however modest its results, reveals grounds for a synthesis that will one day result in a theory of external validity that is systematic and practically implementable. These were certainly the goals Don set himself for external validity, but being of the generation he was, he tended to think of it more in terms of what could be achieved in individual studies or the sequential replications that dominate in the natural sciences. He therefore assigned external validity less weight than internal validity, although in his later years he welcomed the many and heterogeneous replications that characterize most meta-analyses and saw that, through methods based on such replication, external validity could be placed on a firmer footing. His suspicion of causal modeling was so strong, however, that he could never bring himself to advocate it as a means of learning about causal explanation. He was apprehensive about specification errors (including measurement error) and about the virulence of the temptation to explain effects that were themselves not well documented—a problem being re-created today in the advocacy of using program theories as the central focus in program evaluation (Connell et al., 1995). Like Cronbach, Campbell preferred qualitative techniques for learning about the causal mediation of well-demonstrated program effects, and we have tried to portray here the tension between the clear utility of complete explanatory knowledge and the practical difficulties of achieving such. Herein lies the challenge for any future thinking about causal generalization that privileges Cronbach's second meaning and substantive theory. It is not a challenge, though, for those who are willing to live with stubborn replication across multiple sources of heterogeneity pertinent to the cause, to the effect, and to populations of persons, settings, and times.

NOTE

1. Campbell has added a fifth—time—because researchers sometimes want to generalize to a particular historical period (say the 1970s) or want to conclude that a causal relationship is universal and so holds for all times. Given the space constraints of this chapter and the somewhat different set of circumstances surrounding generalization over time, I stay with four entities in this chapter.

REFERENCES

Bandura, A. (1986). *Social foundations of thought and action: A social cognitive theory.* Englewood Cliffs, NJ: Prentice Hall.

Bhaskar, R. (1975). *A realist theory of science.* Leeds, UK: Leeds.

Brunswik, E. (1955). *Perception and the representative design of psychological experiments* (2nd ed.). Berkeley: University of California Press.

Calder, B. J., Phillips, L. W., & Tybout, A. M. (1981). Designing research for application. *Journal of Consumer Research, 8,* 197-207.

Calder, B. J., Phillips, L. W., & Tybout, A. M. (1982). The concept of external validity. *Journal of Consumer Research, 9,* 240-244.

Campbell, D. T. (1969). Artifact and control. In R. Rosenthal & R. L. Rosnow (Eds.), *Artifact in behavioral research.* New York: Academic Press.

Campbell, D. T. (1978). Qualitative knowing in action research. In M. Brenner & P. Marsh (Eds.), *The social contexts of method.* London: Croom Helm.

Campbell, D. T. (1986). Relabeling internal and external validity for applied social scientists. In W. M. K. Trochim (Ed.), *Advances in quasi-experimental design and analysis.* San Francisco: Jossey-Bass.

Campbell, D. T., & Fiske, D. W. (1959). Convergent and discriminant validation by the multitrait-multimethod matrix. *Psychological Bulletin, 56,* 81-105.

Campbell, D. T., & Stanley, J. C. (1966). *Experimental and quasi-experimental designs for research.* Chicago: Rand McNally.

Chalmers, T. C., Berrier, J., Hewitt, P., Berlin, J., Reitman, D., Nagalingam, R., & Sacks, H. (1988). Meta-analysis of randomized controlled trials as a method of estimating rare complications of nonsteroidal anti- inflammatory drug therapy. *Alimentary and Pharmacological Therapy, 25,* 9-26.

Collingwood, R. G. (1940). *An essay on metaphysics.* Oxford, UK: Clarendon.

Connell, J. P., Kubisch, A. C., Schorr, L. B., & Weiss, C. H. (Eds.). (1995). *New approaches to evaluating community initiatives: Concepts, methods and contexts.* Washington, DC: Aspen Institute.

Cook, T. D. (1985). Postpositivist critical multiplism. In L. Shotland & M. M. Mark (Eds.), *Social science and social policy.* Beverly Hills, CA: Sage.

Cook, T. D. (1990). The generalization of causal connections: Multiple theories in search of clear practice. In L. Sechrest, E. Perrin, & J. Bunker (Eds.), *Research methodology: Strengthening causal interpretations of nonexperimental data* (DHHS Publication No. PHS 90-3454). Rockville, MD: Department of Health and Human Services.

Cook, T. D. (1991). Clarifying the warrant for generalized causal inferences in quasi-experimentation. In M. W. McLaughlin & D. C. Phillips (Eds.), *Evaluation and education: At quarter century.* Chicago: National Society for the Study of Education.

Cook, T. D., & Campbell, D. T. (1979). *Quasi-experimentation: Design and analysis issues for field settings.* Boston: Houghton Mifflin.

Cook, T. D., & Campbell, D. T. (1986). The causal assumptions of quasi-experimental practice. *Synthese, 68,* 141-180.

Cook, T. D., Cooper, H., Cordray, D., Hartmann, H., Hedges, L., Light, R., Louis, T., & Mosteller, F. (Eds.). (1992). *Meta-analysis for explanation: A casebook.* New York: Russell Sage Foundation.

Cook, T. D., Leviton, L., & Shadish, W. R., Jr. (1985). Program evaluation. In G. Lindzey & E. Aronson (Eds.), *Handbook of social psychology* (3rd ed.). New York: Random House.

Cronbach, L. J. (1982). *Designing evaluations of educational and social programs.* San Francisco: Jossey-Bass.

Cronbach, L. J. (1985, May). *Construct validation after thirty years.* Paper presented at the symposium "Intelligence: Measurement, Theory, and Public Policy," Urbana, IL.

Cronbach, L. J., Ambron, S. R., Dornbusch, S. M., Hess, R. D., Hornick, R. C., Phillips, D. C., Walker, D. F., & Weiner, S. S. (1980). *Toward reform of program evaluation.* San Francisco: Jossey-Bass.

Cronbach, L. J., Gleser, G. C., Nanda, H., & Rajaratram, N. (1972). *The dependability of behavioral measurements: Theory of generalizability for scores and profiles.* New York: John Wiley.

Cronbach, L. J., & Meehl, P. E. (1955). Construct validity in psychological tests. *Psychological Bulletin, 52*, 281-302.

Cronbach, L. J., Rajaratram, N., & Gleser, G. C. (1967). *The dependability of behavioral measurements: Multifacet studies of generalizability.* Stanford, CA: Stanford University Press.

Cronbach, L. J., & Snow, R. E. (1976). *Aptitudes and instructional methods.* New York: Irvington.

Devine, E. C. (1992). Effects of psychoeducational care with adult surgical patients: A theory-probing meta-analysis of intervention studies. In T. Cook, H. Cooper, D. Cordray, H. Hartmann, L. Hedges, R. Light, T. Louis, & F. Mosteller (Eds.), *Meta-analysis for explanation: A casebook.* New York: Russell Sage Foundation.

Devine, E. C., & Cook, T. D. (1986). Clinical and cost relevant effects of psychoeducational interventions: A meta-analysis. *Research in Nursing and Health, 9*, 89-105.

Devine, E. C., O'Connor, F. W., Cook, T. D., Wenk, V. A., & Curtin, T. R. (1988). Clinical and financial effects of psychoeducational care provided by staff nurses to adult surgical patients in the post-DRG environment. *American Journal of Public Health, 78*, 1293-1297.

Director, S. M. (1979). Underadjustment bias in the evaluation of manpower training. *Evaluation Quarterly, 3*, 190-218.

Fisher, R. A. (1935). *The design of experiments.* London: Oliver and Boyd.

Flay, B. (1986). Efficacy and effectiveness trials (and other phases of research) in the development of health promotion programs. *Preventive Medicine, 15*, 451-474.

Furstenberg, F. F., Jr., Brooks-Gunn, J., & Morgan, S. P. (1987). *Adolescent mothers in later life.* Cambridge, UK: Cambridge University Press.

Gasking, D. (1957). Causation and recipes. *Mind, 64*, 479-487.

Glass, G. V., & Smith, M. L. (1979). Meta-analysis of research of the relationship of class size and achievement. *Educational Evaluation and Policy Analysis*, *1*, 2-16.

Glass, G. V., McGaw, B., & Smith, M. L. (1981). *Meta-analysis in social research.* Beverly Hills, CA: Sage.

Glymour, C., Scheines, R., Spirtes, P., & Kelly, K. (1987). *Discovering causal structure: Artificial intelligence, philosophy of science and statistical modeling.* San Diego, CA: Academic Press.

Harris, M. J., & Rosenthal, R. (1984). Mediation of interpersonal expectancy effects: 31 meta-analyses. *Psychological Bulletin*, *97*, 363-386.

Hedges, L. V., & Olkin, I. (1985). *Statistical methods for meta-analysis.* Orlando, FL: Academic Press.

Kish, L. (1987). *Statistical design for research.* New York: John Wiley.

Lakoff, G. (1985). *Women, fire, and dangerous things.* Chicago: University of Chicago Press.

Lavori, P. W., Louis, T. A., Bailar, J. C., & Polansky, H. (1986). Designs for experiments: Parallel comparisons of treatment. In J. C. Bailar & F. Mosteller (Eds.), *Medical uses of statistics.* Waltham, MA: New England Journal of Medicine.

Lipsey, M. W. (1992). Juvenile delinquency treatment: A meta-analytic inquiry into the variability of effects. In T. Cook, H. Cooper, D. Cordray, H. Hartmann, L. Hedges, R. Light, T. Louis, & F. Mosteller (Eds.), *Meta-analysis for explanation: A casebook.* New York: Russell Sage Foundation.

Mackie, J. L. (1974). *The cement of the universe: A study of causation.* Oxford, UK: Oxford University Press.

Mark, M. M. (1986). Validity typologies and the logic and practice of quasi-experimentation. In W.M.K. Trochim (Ed.), *Advances in quasi-experimental design and analysis* (New Directions for Program Evaluation No. 31). San Francisco: Jossey-Bass.

McGuire, W. J. (1984). Contextualism. In L. Berkowitz (Ed.), *Advances in experimental social psychology.* New York: Academic Press.

Riecken, H. W., & Boruch, R. F. (Eds.). (1974). *Social experimentation.* New York: Academic Press.

Rosch, E. H. (1978). Principles of categorization. In E. H. Rosch & B. B. Lloyd (Eds.), *Cognition and categorization.* Hillsdale, NJ: Lawrence Erlbaum.

Rubin, D. B. (1992). Meta-analysis: Literature synthesis or effect-size surface estimation? *Journal of Educational Statistics*, *17*, 363-374.

Shadish, W. R., Jr. (1992). Do family and marital psychotherapies change what people do? A meta-analysis of behavioral outcomes. In T. Cook, H. Cooper, D. Cordray, H. Hartmann, L. Hedges, R. Light, T. Louis, & F. Mosteller (Eds.), *Meta-analysis for explanation: A casebook.* New York: Russell Sage Foundation.

Shadish, W. R., Jr. (1994). Critical multiplism: A research strategy and its attendant tactics. In L. B. Sechrest & A. J. Figueredo (Eds.), *New directions for program evaluation*, San Francisco: Jossey-Bass.

Shadish, W. R., Jr. (1995). The logic of generalization: Five principles common to experiments and ethnographies. *American Journal of Community Psychology*, *23*, 419-428.

Shadish, W. R., Matt, G., Navarro, A., Siegle, G., Crits-Christoph, P., Hazelrigg, M., Jorm, A., Lyons, L. S., Nietzel, M. T., Prout, H. T., Robinson, L., Smith, M. L., Svartberg, M., & Weiss, B. (1997). Evidence that therapy works in clinically representative conditions. *Journal of Consulting and Clinical Psychology, 65*, 355-365.

Smith, E. E., & Medin, D. E. (1981). *Categories and concepts.* Cambridge, MA: Harvard University Press.

St.Pierre, R. G., & Cook, T. D. (1984). Sampling strategy in the design of program evaluations. In R. F. Connor, D. G. Attman, & C. Jackson (Eds.), *Evaluation studies review annual* (Vol. 9). Beverly Hills, CA: Sage.

Warnecke, J. R., Flay, B. D., Kviz, F. J., Gruder, C. L., Langenberg, P., Crittenden, K. S., Mermelstein, R. J., Aitken, M., Wong, S. C., & Cook, T. D. (1991). Characteristics of participants in a televised smoking cessation intervention. *Journal of Preventive Medicine, 20*, 389-403.

Warnecke, R. B., Langenberg, P., Wong, S. C., Flay, B. R., & Cook, T. D. (1992). The second Chicago televised smoking cessation program: A 24-month follow-up. *American Journal of Public Health, 82*, 835-840.

Webb, E. J., Campbell, D. T., Schwartz, R. D., & Sechrest, L. (1966). *Unobtrusive measures.* Skokie, IL: Rand McNally.

Weisz, J. R., Weiss, B., & Donenberg, G. R. (1992). The lab versus the clinic: Effects of child and adolescent psychotherapy. *American Psychologist, 47*, 1578-1585.

Wholey, J. S. (1977). Evaluability assessment. In L. Rutman (Ed.), *Evaluation research methods: A basic guide.* Beverly Hills, CA: Sage.

CHAPTER 2

Validity Applied to Meta-Analysis and Research Synthesis

Elvira Elek-Fisk

Lanette A. Raymond

Paul M. Wortman

Yi [Confucian validity/righteousness]

- *To be unbiased and impartial*
- *To conduct matters in an appropriate manner*
- *To respect people of virtue and to be tolerant toward others*
- *To be upright and selfless*

— [From a marker in a Japanese garden]

This chapter is dedicated to the legacy of Donald T. Campbell, to both the man and his work. In particular, it addresses the relevance of Campbell's concept of "validity" to the conduct of meta-analyses or research syntheses (Cooper & Hedges, 1994). As such, it is not a primer or tutorial on meta-analysis (cf. Durlak & Lipsey, 1991), but instead a simple testimonial to the importance of Campbell's work for the proper conduct and understanding of such syntheses. This chapter has three objectives, each addressed separately in the following three major sections: (a) to present a brief overview of Campbell's validity taxonomy or framework (see Matt & Cook, 1994, and Wortman, 1994, for more detail) and its inclusion in an integrated model of meta-analysis, (b) to provide

AUTHORS' NOTE: The authors thank Professor William R. Shadish, Jr., and four anonymous reviewers for their many helpful suggestions.

an example of its applicability to the synthesis of correlational research in a substantive area where Don also made a major contribution, and (c) to discuss specific controversies related to its use in syntheses of both randomized and quasi-experiments.

Since the first modern, substantial "meta-analysis" (Smith & Glass, 1977), researchers have tried to control for the quality of studies included in such syntheses (Landman & Dawes, 1982). Although Don Campbell never conducted a meta-analysis, many of his students and colleagues have applied his ideas concerning validity, or research design quality, to this area (e.g., Devine & Cook, 1983; Orwin & Cordray, 1985; Shadish, 1996; Wortman, 1992; Wortman & Yeaton, 1983). The validity framework that Campbell and his associates formulated (Campbell & Stanley, 1963, 1966; Cook & Campbell, 1979) raises important methodological issues regarding the inclusion or exclusion of studies in meta-analyses; it also helps organize any subsequent decisions, analyses, and conclusions.

THE VALIDITY FRAMEWORK

The publication in 1963 of Campbell and Stanley's original chapter both revolutionized and simplified establishing the validity of a scientific study (Campbell & Stanley, 1963). Their "validity framework" provided a methodological set of research design characteristics. These characteristics aid the understanding, identification, and elimination of alternative or "plausible rival" hypotheses that compromise the interpretability of causal relationships in scientific research. The original 1963 framework divided validity into "internal" and "external" validity. Internal validity refers to the confidence with which a causal relationship between two variables can be inferred. External validity refers to the confidence with which that presumed causal relationship can be generalized beyond the specific sample population, setting, and time studied.

Cook and Campbell (1979) modified and expanded the validity framework to include "statistical conclusion" validity and "construct" validity (see Table 2.1). Statistical conclusion validity refers to the confidence with which covariation can be inferred. It addresses issues such as whether a study has sufficient sensitivity, or statistical power, to reveal a relationship between two variables. According to Cook and Campbell, construct validity "refers to the possibility that the operations which are meant to represent a particular cause or effect construct can be construed in terms of more than one construct" (p. 59). In other words, construct validity becomes compromised when the variables of interest are confounded with some other variables in the study (e.g., when a researcher cannot ascertain whether placebo effects or the treatment of interest is the true underlying cause of the apparent effectiveness of a treatment). The resulting "validity framework" identified and defined a set of threats specific to each of the four

TABLE 2.1 Integration of Campbell's Validity Framework With Cooper's Five-Stage Integrative Review Model[a]

Cooper's Five Stage Integrative Review Model	*Campbell's Validity Framework*			
	Relevance		*Acceptability*	
Stage of Research (primary function)	*Construct Validity*	*External Validity*	*Internal Validity*	*Statistical Conclusion Validity*
Problem formation: "Constructing definitions that distinguish relevant from irrelevant studies"	Poorly defined cause and effect constructs obscure interactions	Narrow definition of constructs limits generalizability of conclusions.		
Data collection: "Determining which sources of potentially relevant studies to examine"	Constructs may be defined at different levels across accessible studies, resulting in erroneous negative conclusions	Populations sampled in studies obtained may differ from the target population, restricting possible generalization		
Data evaluation: "Applying criteria to separate valid from invalid studies."			Studies that have failed to rule out alternate rival hypotheses may be biased and may compromise the validity of conclusions	Inappropriate use of statistical tests or interpretation of statistical results may make conclusions unreliable
Analysis and interpretation: "Synthesizing valid retrieved studies"	Identify confounds and mediating variables	Cross-validation; cross-design synthesis	Comparison of randomized and nonrandomized studies	Path and causal models
Public presentation: "Applying editorial criteria to separate important from unimportant information"	Failure to provide thorough discussion of procedures used in determining studies relevant and acceptable to the meta-analysis may make results irreproducible; this, along with selective omission of review findings, makes conclusions obsolete			

a. The five-stage integrative review model appears in Cooper (1984). All quotations in the first column are taken from that source (p. 13).

categories and provided general descriptions of study designs that avoid some of the threats (cf. Wortman, 1994, for an extensive tabular description of all these 33 threats to validity as well as Matt & Cook, 1994, for a more in-depth discussion of some of these threats).

The Research-Replication-Review Cycle

Scientific knowledge ideally arises from a cumulative cyclical process that begins with methodologically sound individual research studies replicated and extended in subsequent well-designed studies. This process should conclude with a subjective, narrative review, inspiring new research studies. Social scientists, however, rarely achieve this ideal of methodologically sound research, replication, and reviews culminating in consistent conclusions. Instead, individual studies and syntheses vary both in quality and in their findings.

Don Campbell's validity taxonomy had its greatest initial impact in resolving some of the significant discrepancies within the research-replication-review cycle, especially at the level of individual studies. It only required a slight extension, in the "external validity" sense, to apply the Campbellian framework to meta-analysis. From the validity perspective, the necessity for the responsible methodological integration of a research literature seems obvious. Meta-analysis (or research synthesis) itself utilizes the scientific process (Cooper & Hedges, 1994) in the quantitative aggregation of a set of similar studies. Researchers should therefore approach it with concerns analogous to those confronted when conducting an individual research study. Consequently, the techniques for assessing methodological soundness that constitute the Campbellian validity framework should be applicable to meta-analysis.

The explosion in social science research and publishing that occurred after World War II and the advances in computer technology that dramatically increased access to scientific information together made the traditional research-replication-(subjective) review process unwieldy and unworkable. As Glass (1976) noted, the size and complexity of research domains caused the traditional subjective literature review to become obsolete and required a new method for summarizing the scientific literature. Although the validity framework might provide some guidance in this area, this method required an even more systematic approach to reviewing and integrating the large number of individual studies in any one specific domain. Cooper (1982, 1984) explained that such "integrative reviews" could be viewed through a five-stage research model providing systematic guidelines for evaluating the validity, or "trustworthiness," of review outcomes (see the leftmost column of Table 2.1).

Cooper's Five-Stage Integrative Review Model

Following the basic hypothetico-deductive scientific method, Cooper (1984) conceptualized the process of conducting an integrative research review as consisting of (a) problem formation, (b) data collection, (c) evaluation of data points, (d) analysis and interpretation, and (e) presentation of results. Each stage of the review serves a parallel function to one in basic or primary research. Similarly, methodological choices made at each stage can undermine the validity of conclusions drawn by the review, much as they can in primary research. In framing each of the five stages, Cooper's model addressed some of the validity issues elucidated by Campbell and others (Bracht & Glass, 1968; Campbell, 1975; Campbell & Stanley, 1963; Cook & Campbell, 1979) in relation to basic research.

Application of the organizational principles presented by Cooper both circumvent some of the controversy surrounding meta-analysis itself and maximize its qualitative strengths. For example, Cooper discusses how two major problems faced by meta-analysts—"garbage in, garbage out" (i.e., including poorly designed or weak studies) and "apples and oranges" (i.e., mixing different constructs)—can be avoided. He indicates points in his stages where, to address the first problem, reviewers can assess the validity of studies ("data evaluation" stage), and where, to address the second problem, they can take steps to ensure the careful specification of research questions and interventions and the uniformity of measures and designs across studies ("problem formation" stage).

Other researchers also considered the importance of the data evaluation stage of a meta-analysis. Although considerable debate revolved around the pros and cons of excluding studies from a meta-analysis, agreement now generally exists that all meta-analysts should carefully examine the quality of a study before including it in a synthesis (Glass & Kliegl, 1983; Hasselblad et al., 1995; Lau, Ioannidis, & Schmid, 1997; Mansfield & Busse, 1977; Wortman, 1994). In particular, poorly designed studies should either be excluded or analyzed separately.

Applying the Validity Framework to Meta-Analysis

While Cooper (1984) was delineating the "integrative review" process, in the early 1980s two of Don Campbell's former colleagues, Tom Cook and Paul Wortman, first proposed and demonstrated the specific application of the Campbellian validity framework to meta-analysis (Devine & Cook, 1983; Wortman, 1983; Wortman & Yeaton, 1983). This work compared randomized and quasi-experimental studies in meta-analyses of treatment effectiveness on health outcomes. In particular, the Wortman and Yeaton (1983) synthesis provided empirical support for the hypothesis that a single threat to (internal) validity can systematically bias an entire body of literature.

In this synthesis of the benefit of coronary artery bypass graft surgery, all the randomized studies (or clinical trials) suffered from a significant problem involving medically treated (control) patients crossing over into the surgical condition. On average, 21% of these control group patients ended up receiving the surgical treatment. This combination of attrition or "experimental mortality" of patients in the control (nonsurgery) group and "diffusion," where they then received the (bypass surgery) treatment, significantly biased the synthesis both medically and statistically by underestimating the effectiveness of surgery (Wortman & Yeaton, 1987).

The validity framework can further assist in locating other threats to the conclusions of meta-analyses just as it does for the conclusions of primary research studies; however, the framework applies prior to the "analysis and interpretation" stage of the meta-analytic process as well. For example, Thomas Chalmers and his associates (Chalmers et al., 1981) focused on the meta-analyses of randomized studies in medicine that routinely employ entry criteria for patient inclusion and exclusion in a study. They recommended that meta-analysts report both the list of studies excluded as well as those included (Sacks, Berrier, Reitman, Ancona-Berk, & Chalmers, 1987). The validity framework is ideally suited to describing/determining this entry criterion process. The Campbellian validity framework thus becomes relevant for the proper conduct of a meta-analysis (by providing inclusion/exclusion criteria) in a way somewhat different from its original focus on the assessment of experimental designs (Bryant & Wortman, 1984).

Decisions of Relevance and Acceptability

Wortman and Bryant (1985) proceeded to illustrate this new application of the validity framework empirically in their meta-analysis of school desegregation. The four validity categories described by Cook and Campbell (1979) were combined into two groups, the first consisting of construct and external validity and the second of internal and statistical conclusion validity. Wortman and Bryant (1985) used the two groups to devise entry criteria for Cooper's (1984) "problem formation" (hypothesis) and "data evaluation" (reporting quality) stages, respectively (see Table 2.1). These two groups indicate that decisions facing meta-analysts when they evaluate evidence for inclusion or exclusion fall into two types, those of "relevance" and those of "acceptability" (Bryant & Wortman, 1984).

As exhibited in Table 2.1, meta-analysts can assess both types of decisions according to criteria derived from the Campbellian validity framework. Criteria derived from the definitions of construct and external validity guide decisions of relevance (or inclusion). Thus, for the Wortman and Bryant (1985) school desegregation example, studies of busing and those using IQ scores were deemed nonrelevant for construct validity reasons, whereas studies from other countries and those focusing on minority populations other than African Americans were excluded based on external validity reasons. Internal and statistical conclusion validity assist the assessment of the acceptability (or exclusion) of a study. A study already determined relevant must be evaluated for acceptability before it is integrated into the meta-analysis. Wortman and Bryant considered relevant studies unacceptable in terms of statistical conclusion validity if they utilized small sample sizes, insufficient/inappropriate statistical information, or heterogeneous samples (i.e., mixed grades or ethnic groups). The employment of regional surveys or inappropriate controls (e.g., national or statewide norms) resulted in the classification of studies as unacceptable based on internal validity concerns.

This combination of Campbell's validity framework with Cooper's five-stage model of (integrative) review or research synthesis provides a comprehensive and elegantly simple perspective on the decisions facing the meta-analyst (see Table 2.1). Appropriate application of the validity framework and the concepts of relevance and acceptability give definition and direction to the process of quantitative review, much as they do to primary research. Meta-analysts can employ different components of this new model when integrating studies utilizing various methodologies, including experimental, quasi-experimental, and correlational research designs.

APPLYING THE COMBINED MODEL TO THE META-ANALYSIS OF CORRELATIONAL RESEARCH

Although more of the components of the validity framework as combined with Cooper's model apply to the meta-analysis of experimental and quasi-experimental data, many of them also provide guidance for the meta-analyses of correlational data (Hunter, Schmidt, & Jackson, 1982). In addition to entry (or inclusion/exclusion) criteria, elements of the combined model (see Table 2.1) assist the synthesizer with selecting important moderator variables. A recent meta-analysis (Elek-Fisk, 1997) examining the relationship between ethnic identifications and out-group acceptance demonstrates how the combined model can direct the process of a synthesis project. The correlational design of the studies in the synthesis almost by definition limited the application of internal validity characteristics because these types of studies lack an explicit claim of causal inference. Construct, external, and statistical conclusion validity issues, however, definitely concerned the meta-analyst.

Problem Formation

In addition to the validity framework, Don Campbell also contributed in a very important theoretical way (LeVine & Campbell, 1972) to the Elek-Fisk (1997) meta-analysis. Although he maintained an interest in research on intergroup relations throughout his career (Campbell, 1947, 1967; Campbell & McCandless, 1951), Don's collaboration with Robert LeVine on ethnocentrism specifically influenced many of the primary studies included in the meta-analysis (LeVine & Campbell, 1972). Elek-Fisk (1997) relied heavily on this work in the "problem formation" stage of the synthesis.

In the Elek-Fisk meta-analysis, the "problem formation" stage began with the understanding that most social psychological theories concerning intergroup relations developed from perspectives that attempt to explain why those relations often result in conflicts. Social identity theory (SIT; Brewer & Miller, 1984; Tajfel & Turner, 1986) and realistic group conflict theories (RGCT; LeVine & Campbell, 1972) address intergroup relations in such a way. These two sets of theories are consistent with the ethnocentrism hypothesis (LeVine & Campbell, 1972; Sumner, 1906), which states that strong ethnic or racial self-identifications lead to increased negativity toward ethnic or racial groups other than one's own (out-groups). Whereas SIT concentrates on the group identifications themselves, RGCT focuses on the perceptions of threat from out-groups toward a group with which one identifies (the in-group). RGCT expects those perceived threats to result in increased negativity and hostility toward the threatening out-groups.

In contrast to SIT and RGCT, multiculturalism-based policies, such as the affirmative action and diversity programs utilized by universities and governments, do not generally address the negative consequences of ethnic or racial group identifications. Instead, they emphasize the positive side of interethnic relations and identifications. Most of these policies either purposely or inadvertently encourage ethnic or racial identifications with the expectation of increasing the acceptance of members of other groups. In fact, Ponterotto and Pederson (1993) describe the task of gaining a "healthy and positive racial/ethnic identity" (p. 62) as a necessary component of prejudice reduction programs.

Berry (1984a, 1984b, 1991) and Lambert, Mermigis, and Taylor (1986) characterized such policies as consistent with or following from a multiculturalism "assumption" or "hypothesis." This hypothesis requires identification with one's own ethnic group along with the idea that "only when members of one group feel confident and secure in their own identity will they be open to, and accepting of members of out-groups" (Taylor & Moghaddam, 1994, p. 190). Clearly, intergroup relations policies based on the multiculturalism hypothesis encourage positive and secure ethnic or racial identifications, as might RGCT if group identifications could not be eliminated. On the other hand, SIT discourages any emphasis on ethnic or racial group identifications if interethnic harmony is the desired outcome.

Applying Construct and External Validity: Relevance Decisions

For the Elek-Fisk (1997) meta-analysis, the two opposing sets of theories/hypotheses presented an interesting dilemma. How should ethnic identification and out-group acceptance be measured? Where would measures of ethnic security fit into this picture? A focus on construct validity (Cook & Campbell, 1979, p. 59ff) forced the meta-analyst to ask which operational or research definitions most appropriately generalize to the theoretical constructs of interest (i.e., the variables at the heart of the multiculturalism- and ethnocentrism-based hypotheses). Answers to these questions guided the process of choosing relevant studies for the meta-analysis.

Ethnocentrism-based theories emphasize the importance of actual identifications with one's group, whereas multiculturalism and RGCT point to the relative importance of feeling secure in that identification. To avoid limiting the external validity of the synthesis conclusions, Elek-Fisk accepted studies assessing either ethnic security or ethnic identification (as defined through a sense of belonging/attachment, levels of ethnicity-related behaviors, or positive attributions for one's group). Subsequent moderator-variable analyses, in the "analysis and interpretation" stage, separated the two types of studies to address the potential threat to construct validity (i.e., "confounding constructs and levels of constructs"; Cook & Campbell, 1979, p. 67). Similarly, the construct of out-group acceptance was operationalized in many ways, with the target consisting of either one or multiple ethnic/racial out-groups, and acceptance defined in terms of desirability of contact, ratings on traits or stereotypes, or a more general acceptance of diversity. As with the own-ethnic-group-related variable, Elek-Fisk used a broad definition of the out-group acceptance construct to include studies and enhance the generalizability of conclusions, and in addition addressed related construct validity issues through moderator analyses of this variable in the "analysis and interpretation" stage.

Data Evaluation

After formulating the problem and deciding which of the collected studies to include based on their "relevance," Elek-Fisk (1997) needed to determine the "acceptability" of the studies in the "data evaluation" stage. Mullen, Brown, and Smith (1992) found support for in-group bias (a more general term for ethnocentrism) in a meta-analysis of experimental studies that included few ethnic/racial-group-focused studies, however. The nature of ethnic/racial identifications makes them difficult to manipulate experimentally, necessitating the study of their relationship with other variables in a correlational context. Elek-Fisk (1997) found that almost all studies examining the specific relationship between ethnic/racial identifications and out-group acceptance use nonexperimental methods. This presented a problem because correlational designs are inherently low in internal validity and thus do not usually claim to assess causality (see Cook & Campbell, 1979, p. 296). Considering the difficulty of directly manipulating ethnic/racial identifications, however, a meta-analysis of such correlational studies could add an important element to the overall understanding of intergroup relations. Elek-Fisk (1997) therefore concentrated on the available correlational studies, and internal validity appropriately was not used to exclude studies in the "data evaluation" stage.

Statistical conclusion validity criteria proved more helpful with decisions of acceptability of the correlational studies at this stage. For example, studies utilizing measures with low reliability were excluded. Elek-Fisk (1997) discovered that many studies in the ethnic identification/out-group acceptance field completely failed to report the reliability of their measures. At the "analysis and interpretation" stage, an interest in statistical conclusion validity encouraged corrections of effect sizes based on biases in the research. The existence of a sufficient subset of studies reporting these values allowed for the estimation of the reliability-related bias and subsequent correction for it (Hunter et al., 1982).

Analysis and Interpretation

In the "analysis and interpretation" stage, Elek-Fisk (1997) also calculated an overall effect size based on the included, acceptable studies and conducted moderator- (i.e., third-) variable analyses. For all types of research, including correlational designs, the validity framework informs the selection of potential moderator variables, or confounds of the studied relationship. Moderator variables can and should be examined through meta-analyses because of their potential influence on the relationship of interest (Shadish, 1996; Shadish & Sweeney, 1991). Elek-Fisk (1997) addressed the construct validity-related moderators already described. In addition, the external validity portion of the framework specifically pointed in the direction of moderators potentially threatening the generalizability of the relationship between ethnic/racial identifications and out-group acceptance.

These moderators included age or the majority/minority group status of participants (which fall under the "interaction of selection—of participants—and treatment" threat; Cook & Campbell, 1979, p. 73), the country of the study (i.e., "interaction of setting and treatment"; Cook & Campbell, 1979, p. 74), and publication date (which touches on problems of the "interaction of history and treatment"; Cook & Campbell, 1979, p. 74). Meta-analysis optimally evaluates the impact of those moderator variables suggested by the validity framework (Shadish & Sweeney, 1991). Specifically, Elek-Fisk used the moderator variables to separate studies into subset meta-analyses and determined if effect sizes differed among the subsets.

Examination of moderator variables through meta-analysis enabled the more accurate assessment of the relationship between ethnic identification and out-group acceptance, and it informs future studies of programs attempting to build or induce ethnic identity. For example, the minority or majority status of a participant's ethnic group may affect that person's level of ethnic identification (Brewer & Miller, 1984) and cause other complications for multiculturalism (Berry, 1984b). The Elek-Fisk (1997) meta-analysis found small but significant differences between the effect sizes of minority ($r = +.22$) and majority ($r = +.10$) group members, with ethnically identified minority group individuals expressing more acceptance of out-group members. Age, especially in terms of the categorizations of child, adolescent, and adult, plays a large factor in the likelihood and stability of ethnic identifications (Black-Gutman & Hickson, 1996; Phinney, 1992; Phinney & Alipuria, 1990). Elek-Fisk (1997) found a positive relation between ethnic identification and out-group acceptance for adults ($r = +.18$) but little relationship for adolescents ($r = -.01$), possibly because of the instability of their identifications.

The validity framework also suggested location as a possible moderator. The historical differences between and within various nations/regions should result in the specific comparison of studies occurring in those locations; this might clarify the question of which conditions foster a positive or negative relationship between ethnic identification and out-group acceptance. Although Elek-Fisk (1997) found the greatest positive relationship between the two variables for Canadian studies ($r = +.30$) and a negative relationship for studies conducted in the Netherlands ($r = -.17$), the limited number of studies from each country prevented drawing definitive conclusions. Along with the moderating effects of age, majority/minority status, and location, the ethnic identification/prejudice relationship has evolved over time with changes in the societal acceptance of prejudice and the encouragement of ethnic identification (Sears, 1996). Unfortunately, the studies included in the Elek-Fisk (1997) meta-analysis were within a fairly restricted time range, so no conclusions could be drawn based on this moderator.

The meta-analyst derived the above moderators based on construct and external validity characteristics, in the accepted studies that could be separated easily using those variables. Other researchers may consider different potential moderators more important, such as self-esteem (Abrams & Hogg, 1988) or mobility of status (Brewer & Miller, 1984). The accepted studies could not be subdivided based on these potential moderators, but these and other important moderators can and should face examination when a sufficient number of relevant primary studies have been conducted.

The Elek-Fisk example demonstrates the importance of the validity framework in meta-analyses of correlational studies and the helpfulness of Cooper's (1984) five-stage integrative review model. In particular, this synthesis raised the (often neglected) significance of theoretical issues (construct validity) along with the importance of assessing the generalizability (external validity) of meta-analytic findings. The final section focuses on the validity framework's more frequent application in meta-analyses of experimental and quasi-experimental studies and the importance of internal validity in assessing the acceptability of those studies (see Table 2.1, "Data Evaluation"). This has been the center of more heated discussion and debate.

UTILITY OF THE VALIDITY FRAMEWORK FOR META-ANALYSES OF EXPERIMENTS AND QUASI-EXPERIMENTS

In the late 1970s, Carol Weiss (1977) questioned whether all the methodological squabbling over the validity of the research designs used to evaluate social programs was critical. Her argument rested on the claim that policymakers did not make direct use of evaluations anyway. Those following in the Campbellian tradition, however, held as an article of faith that bad evaluation design was bad science and, as in the debate over Head Start (Campbell & Erlebacher, 1970; Cicirelli, 1971), potentially bad policy as well. As previously noted, some researchers transferred this trust in the pure science approach of the Campbellian validity framework to research on the meta-analytic method despite a troubling early warning that the results were independent of study quality or validity (cf. Landman & Dawes, 1982).

Not until Lipsey and Wilson (1993) conducted a large meta-evaluative survey of the meta-analytic literature did researchers unequivocally raise the question of the utility of the validity framework. Out of 302 meta-analyses, Lipsey and Wilson reported only a trivial .03 difference in effect size for the 27 meta-analyses comparing studies of "high" ($d = 0.40$) and "low" ($d = 0.37$) quality ratings. This reported difference, however, may be artificially low, because, as Wortman (1994) pointed out, the intervals on such quality ratings scales are by no means equivalent. Such ratings can miss or underestimate systematic bias caused by a single threat to validity that is *both* serious (i.e., large) and pervasive (i.e., present in most of the studies). Wortman and Bryant's meta-analysis (1985) examining the impact of U.S. school desegregation on the achievement levels of African American children exemplifies this single-threat problem. In this case, pretest differences in effect size resulting from systematic selection bias in the nonequivalent (i.e., nonrandom, quasi-experimental) control-group studies accounted for just more than half the posttreatment effect size.

Much more provocative was Lipsey and Wilson's (1993, p. 1193) examination of "methodological quality," which found only a .05 effect-size difference for 74 meta-analyses reporting results from both "random" ($d = 0.46$) and "nonrandom" ($d = 0.41$) designs. They state, "What these comparisons do indicate is that there is no strong pattern or bias in the direction of the difference made by lower quality methods" (pp. 1192-1193). In a series of related studies, Shadish and his associates (Heinsman & Shadish, 1996; Shadish & Ragsdale, 1996) probed this implicit contention that the validity framework did not matter (i.e., produce systematic bias) in meta-analysis.

Adjusting for Bias in Nonrandomized Studies

Specifically, Shadish and his colleagues sought to determine whether or not variables confounded with the nominal specification of "random" and "nonrandom" mediated the no-difference meta-analytic results. Unlike Lipsey and Wilson (1993), who did not code any individual studies to check the accuracy of the random-nonrandom designation, they used similar study-level coding schemes in two separate syntheses that covered a total of five different substantive areas. Shadish and his associates found that statistically significant differences in effect sizes between the two design categories largely developed from confounds rather than from the act of randomization. For example, Heinsman and Shadish (1996) reported an initial statistically significant difference in effect sizes between randomized and nonrandomized experiments ($d = 0.28$ and 0.03, respectively). Application of a regression model to control for confounds reduced this difference to 0.08. In particular, "the pretest effect size" and "the activity in the control group" defined as "passive" (i.e., no treatment or wait list) and "active" (i.e., placebo or treatment as usual) "most reduced assignment method differences." Heinsman and Shadish (1996) concluded that "nonrandomized experiments are more like randomized experiments if one takes confounds into account" (p. 162).

Similarly, Shadish and Ragsdale (1996) reduced the initial large significant difference between randomized ($d = 0.60$) and nonrandomized ($d = 0.08$) psychotherapy studies to .27. Their more circumscribed conclusion was that much but not all of this discrepancy was the result of confounds in assignment with other variables (e.g., "passive versus active control group," "pretest effect size"). In keeping with the research approach implicit in Campbell's validity framework, they find these results "important" and consistent with its agenda "to explore the possibility that an adjusted estimate from nonequivalent control group designs may adequately approximate the results from randomized experiments under some circumstances" (p. 1299).

Adjusting for Bias in Randomized Studies

The Campbellian research agenda considered randomized or "true" experiments a perfect Socratic-like "form of the good" (all the "+" signs indicating no internal "sources of invalidity" for all three randomized designs in Table 1 of Campbell & Stanley, 1966, p. 8); thus, the focus shifted to imperfections (or the "–" signs indicating "a definite weakness") in the nonrandomized or "quasi-experiments." In contrast to this view, the late Thomas Chalmers, a relentless advocate of randomized experiments, was also a stern methodological scold of researchers who imperfectly conducted studies utilizing these true experimental designs (Chalmers, 1981, 1982). For example, he and his associates (Chalmers, Celano, Sacks, & Smith, 1983) found a threefold increase in the number of studies with significant between-condition differences in mortality rates from "blinded" (8.8%) to "unblinded" (24.4%) randomized studies.

Other researchers noted that selection bias or lack of "allocation concealment" can "subvert" randomization and produce "larger treatment effects" (Schulz, 1995). Schulz reports anecdotal evidence for "deciphering" and changing participant allocation (e.g., opening "unsealed assignment envelopes") and thus in effect turning randomized into nonrandomized studies. Greenberg, Bornstein, Greenberg, and Fisher (1992), in a meta-analysis of 22 double-blind studies of antidepressants, reported strong empirical evidence of such experimenter bias by clinicians because of their awareness of the form and significance of side effects. Overall, clinicians rated both new and standard antidepressant drugs as effective ($d = 0.31$ and 0.25 respectively, with $p < .0001$ for both), while patients rated them (statistically) significantly less effective ($d = 0.12$ and 0.06, respectively). Clearly, clinicians could tell to which condition—treatment or placebo—researchers assigned patients in the double-blind studies. This form of experimenter bias fits the characteristics of "hypothesis guessing" and "experimenter expectancies," both threats to construct validity (Cook & Campbell, 1979, pp. 66-67). These threats invalidate the randomization construct even though the researchers initially conducted the randomization properly. To guard against this bias in primary research, Greenberg et al. (1992) recommend the use of "an active-placebo group" that mimics the drug's side effects. Otherwise, given the well-established bias associated with such experimenter expectations, a meta-analysis would result in an inflated estimate of effect size.

Meta-analysts of randomized experiments also should not underestimate the impact of experimenter expectancies on subjects in unblinded studies. In a synthesis of five different, discredited medical treatments, 65% to 89% (with a mean of 70%) of both physicians and patients involved in such studies rated (what was later definitely shown in double-blind studies to be) an ineffective treatment as effective (Roberts, Kewman, Mercier, & Hovell, 1993). The expectations of participants about the benefit of the experimental treatment thus considerably affected not only the experimenter (as noted above in Greenberg et al., 1992) but also the study subjects. This awareness of treatment assignment subsequently can be compounded by a threat to internal validity caused by the "resentful demoralization of respondents receiving less desirable treatments" (Cook & Campbell, 1979, p. 55) when blinding fails. Camille Wortman, a colleague of Campbell's at Northwestern, and her students examined the problem of control subjects becoming unblinded in a series of experiments (C. B. Wortman, Hendricks, & Hillis, 1976). Such control (or "becoming-aware") subjects felt worse about the study ($p < .01$), were angrier ($p < .01$), and were less motivated to make the project successful ($p < .01$) than those receiving the treatment. An example of an actual statistical adjustment for bias resulting from problems with the internal validity of randomized studies is presented in the following section.

The Two Prongs of Bias Adjustment

The approach advocated by Chalmers and his associates mirrors the Campbellian agenda. Each uses a different terminology for essentially the same validity framework (see Wortman, 1994, for a comparison), both advocate adjustments to improve research design, and both concern themselves with the differences between randomized and nonrandomized designs. The difference between the two approaches lies in their relative emphasis on the two prongs of randomized and nonrandomized studies. Campbell sought to improve the validity of nonrandomized quasi-experiments that are often an inevitable consequence of applied field research. Chalmers addressed this problem from the other prong, focusing on the randomized (clinical) trial common in medical research. Following Shadish and Heinsman (1997), we might label the former (emphasizing nonrandomized studies) the Type I, and the latter (emphasizing randomized studies) the Type II, "empirical program of methodology."

Researchers can combine these two programs of research in true Hegelian fashion via the modern approach to synthesis (Cooper & Hedges, 1994). Adjustments to both prongs of the design dichotomy—randomized and nonrandomized studies—are critical. In either case, meta-analysts or synthesizers must pay careful attention to the validity framework. Within the constraints of reporting quality, the meta-analyst/research synthesizer may attempt to adjust for the bias in both types of studies. Yeaton, Wortman, and Langberg (1983) developed a statistical adjustment for the systematic bias in the synthesis of randomized studies of bypass surgery (discussed earlier). With the "experimental mortality" and "diffusion" (internal validity) threats stemming from 21% of the medical control group patients crossing over to the treatment condition, a quantitative bias adjustment resulted in a statistically and medically significant increase in the mortality benefit for bypass surgery from 6.8% to 11.1% (Wortman & Yeaton, 1987). The "cross-design" synthesis pioneered by Droitcour and her colleagues at the U.S. General Accounting Office (Droitcour & Larson, 1994; Droitcour, Silberman, & Chelimsky, 1993) also illustrates this new, more comprehensive form of synthesis, which combines both research programs.

In conclusion, it appears likely that Lipsey and Wilson's (1993) findings suggesting that the validity framework may not be useful in meta-analysis cannot be supported. In fact, careful validity analyses at the level of the individual study reveal that biases influence both true and quasi-experimental designs. These biases may be randomly distributed among the design types (see Figure 2 in Lipsey & Wilson, 1993, p. 1193), thus producing the illusion of no difference. The meta-analyst/synthesizer's glass is simultaneously both half full because of bias in randomized experiments and half empty because of bias in quasi-experiments. Thus an additional (i.e., the Type II) task of the Campbellian research program begun by Cook and Campbell (1979) should focus on the biases and adjustments of randomized studies.

EPILOGUE

This chapter clearly demonstrates the multiple uses of the validity framework for the proper conduct of meta-analysis. First, the validity framework greatly assists with inclusion/exclusion criteria of relevance and acceptability decisions in the "problem formation," "data collection," and "data evaluation" stages of a meta-analysis and influences the choice of examined moderator variables. Second, the framework can pinpoint specific, systematic threats that permeate a synthesis and suggest ways of correcting for such bias. Clearly, Don Campbell's concept of "validity" is essential to the impartial conduct and comprehensive understanding of meta-analyses.

> There you were gently designing the future—
> one true experiment at a time.
> And it was the present then and now—
> from you to me to all of us.
> It was and is a truly timeless time—
> forever endowed with the validity
> of a spirit freed from all false threats
> to craft a method to end our social madness.
>
> —[From "To Don: A Design for Living,"
> a poem written by Paul Marshall Wortman
> in honor of Don Campbell's 75th birthday]

REFERENCES

Abrams, D., & Hogg, M. A. (1988). Comments on the motivational status of self-esteem in social identity and intergroup discrimination. *European Journal of Social Psychology, 18*, 317-334.

Berry, J. W. (1984a). Cultural relations in plural societies: Alternatives to segregation and their sociopsychological implications. In N. Miller & M. B. Brewer (Eds.), *Groups in contact: The psychology of desegregation* (pp. 11-27). Orlando, FL: Academic Press.

Berry, J. W. (1984b). Multicultural policy in Canada: A social psychological analysis. *Canadian Journal of Behavioural Science, 16*, 353-370.

Berry, J. W. (1991). Understanding and managing multiculturalism: Some possible implications of research in Canada. *Psychology and Developing Societies, 3*, 17-49.

Black-Gutman, D., & Hickson, F. (1996). The relationship between racial attitudes and social-cognitive development in children: An Australian study. *Developmental Psychology, 32*, 448-456.

Bracht, G., & Glass, G. (1968). The external validity of experiments. *American Educational Research Journal, 5*, 437-474.

Brewer, M. B., & Miller, N. (1984). Beyond the contact hypothesis: Theoretical perspectives on desegregation. In N. Miller & M. B. Brewer (Eds.), *Groups in contact: The psychology of desegregation* (pp. 281-302). Orlando, FL: Academic Press.

Bryant, F. B., & Wortman, P. M. (1984). Methodological issues in the meta-analysis of quasi-experiments. *New Directions for Program Evaluation, 24*, 5-24.

Campbell, D. T. (1947). The differential ordering of minority groups on five aspects of prejudice. *American Psychologist, 2*, 413.

Campbell, D. T. (1967). Stereotypes and the perception of group differences. *American Psychologist, 22*, 817-829.

Campbell, D. T. (1975). "Degrees of freedom" and the case study. *Comparative Political Studies, 8*, 178-193.

Campbell, D. T., & Erlebacher, A. (1970). How regression artifacts in quasi-experimental evaluations can mistakenly make compensatory education look harmful. In J. Hellmuth (Ed.), *Compensatory education: A national debate* (Vol. 3, pp. 185-210). New York: Brunner/Mazel.

Campbell, D. T., & McCandless, B. R. (1951). Ethnocentrism, xenophobia and personality. *Human Relations, 4*, 185-192.

Campbell, D. T., & Stanley, J. (1963). Experimental and quasi-experimental designs for research on teaching. In N. L. Gage (Ed.), *Handbook of research on teaching* (pp. 171-246). Chicago: Rand McNally.

Campbell, D. T., & Stanley, J. C. (1966). *Experimental and quasi-experimental designs for research.* Chicago: Rand McNally.

Chalmers, T. C. (1981). The clinical trial. *Milbank Memorial Fund Quarterly, 59*, 324-339.

Chalmers, T. C. (1982). A potpourri of RCT topics. *Controlled Clinical Trials, 3*, 285-298.

Chalmers, T. C., Celano, P., Sacks, H. S., & Smith, H., Jr. (1983). Bias in treatment assignment in controlled clinical trials. *New England Journal of Medicine, 309*, 1358-1361.

Chalmers, T. C., Smith, H., Jr., Blackburn, B., Silverman, B., Schroeder, B., Reitman, D., & Ambroz, A. (1981). A method for assessing the quality of a randomized control trial. *Controlled Clinical Trials, 2*, 31-49.

Cicirelli, V. (1971). The impact of Head Start: Executive summary. In F. G. Caro (Ed.), *Readings in evaluation research* (pp. 307-401). New York: Russell Sage.

Cook, T. D., & Campbell, D. T. (1979). *Quasi-experimentation: Design and analysis issues for field settings.* Chicago: Rand McNally.

Cooper, H. M. (1982). Scientific guidelines for conducting integrative research reviews. *Review of Educational Research, 52*, 291-302.

Cooper, H. M. (1984). *The integrative research review: A systematic approach.* Beverly Hills, CA: Sage.

Cooper, H., & Hedges, L. V. (Eds.). (1994). *The handbook of research synthesis.* New York: Russell Sage.

Devine, E., & Cook, T. D. (1983). A meta-analytic analysis of effects of psychoeducational interventions on length of postsurgical hospital stay. *Nursing Research, 32*, 267-274.

Droitcour, J. A., & Larson, E. (1994, November). *Breast conservation versus mastectomy: Patient survival in day-to-day medical practice and in randomized studies* (GAO/PEMD-95-9). Washington, DC: United States General Accounting Office.

Droitcour, J. A., Silberman, G., & Chelimsky, E. (1993). Cross design synthesis: A new form of meta-analysis for combining results from randomized clinical trials and medical-practice databases. *International Journal of Technology Assessment in Health Care, 9*, 440-449.

Durlak, J. A., & Lipsey, M. W. (1991). A practitioner's guide to meta-analysis. *American Journal of Community Psychology, 19*, 291-332.

Elek-Fisk, E. (1997). *A meta-analysis of the relationship between ethnic identification and outgroup acceptance.* Unpublished manuscript, State University of New York at Stony Brook.

Glass, G. V. (1976). Primary, secondary, and meta-analysis of research. *Education Researcher, 5*, 3-8.

Glass, G. V., & Kliegl, R. M. (1983). An apology for research integration in the study of psychotherapy. *Journal of Consulting and Clinical Psychology, 51*, 28-41.

Greenberg, R. P., Bornstein, R. F., Greenberg, M. D., & Fisher, S. (1992). A meta-analysis of antidepressant outcome under "blinder" conditions. *Journal of Consulting and Clinical Psychology, 60*, 664-669.

Hasselblad, V., Mosteller, F., Littenberg, B., Chalmers, T. C., Hunink, M. G., Turner, J. A., Morton, S. C., Diehr, P., Wong, J. B., & Powe, N. R. (1995). A survey of current problems in meta-analysis: Discussion from the Agency for Health Care Policy and Research Inter-PORT Work Group on Literature Review/Meta-Analysis. *Medical Care, 33*, 202-220.

Heinsman, D. T., & Shadish, W. R. (1996). Assignment methods in experimentation: When do nonrandomized experiments approximate answers from randomized experiments? *Psychological Methods, 1*, 154-169.

Hunter, J. E., Schmidt, F. L., & Jackson, G. B. (1982). Cumulating correlations across studies. In *Meta-analysis: Cumulating research findings across studies* (pp. 35-94). Beverly Hills, CA: Sage.

Lambert, W. E., Mermigis, L., & Taylor, D. M. (1986). Greek Canadians' attitudes toward own group and other ethnic groups: A test of the multiculturalism hypothesis. *Canadian Journal of Behavioural Science, 18*, 35-51.

Landman, J. T., & Dawes, R. M. (1982). Psychotherapy outcome: Smith and Glass' conclusions stand up under scrutiny. *American Psychologist, 37*, 504-516.

Lau, J., Ioannidis, J. P. A., & Schmid, C. H. (1997). Quantitative synthesis in systematic reviews. *Annals of Internal Medicine, 127*, 820-826.

LeVine, R. A., & Campbell, D. T. (1972). *Ethnocentrism: Theories of conflict, ethnic attitudes, and group behavior.* New York: John Wiley.

Lipsey, M. W., & Wilson, D. B. (1993). The efficacy of psychological, educational, and behavioral treatment: Confirmation from meta-analysis. *American Psychologist*, *48*, 1181-1209.

Mansfield, R. S., & Busse, T. V. (1977). Meta-analysis of research: A rejoinder to Glass. *Education Research*, *6*, 3.

Matt, G. E., & Cook, T. D. (1994). Threats to the validity of research studies. In H. Cooper & L. V. Hedges (Eds.), *The handbook of research synthesis* (pp. 503-520). New York: Russell Sage.

Mullen, B., Brown, R., & Smith, C. (1992). Ingroup bias as a function of salience, relevance, and status: An integration. *European Journal of Social Psychology*, *22*, 103-122.

Orwin, R. G., & Cordray, D. S. (1985). Effects of deficient reporting on meta-analysis: A conceptual framework for reanalysis. *Psychological Bulletin*, *97*, 134-147.

Phinney, J. S. (1992). The multigroup ethnic identity measure: A new scale for use with diverse groups. *Journal of Adolescent Research*, *7*, 156-176.

Phinney, J. S., & Alipuria, L. L. (1990). Ethnic identity in older adolescents from four ethnic groups. *Journal of Adolescence*, *13*, 171-183.

Ponterotto, J. G., & Pederson, P. B. (1993). *Preventing prejudice: A guide for counselors and educators*. Newbury Park, CA: Sage.

Roberts, A. H., Kewman, D. G., Mercier, L., & Hovell, M. (1993). The power of nonspecific effects in healing: Implications for psychosocial and biological treatments. *Clinical Psychology Review*, *13*, 375-391.

Sacks, H. S., Berrier, J., Reitman, D., Ancona-Berk, V. S., & Chalmers, T. C. (1987). Meta-analyses of randomized controlled trials. *New England Journal of Medicine*, *316*, 450-455.

Schulz, K. F. (1995). Subverting randomization in controlled trials. *Journal of the American Medical Association*, *274*, 1456-1458.

Sears, D. O. (1996). Presidential address: Reflections on the politics of multiculturalism in American society. *Political Psychology*, *17*, 409-420.

Shadish, W. R. (1996). Meta-analysis and the exploration of causal mediating processes: A primer of examples, methods, and issues. *Psychological Methods*, *1*, 47-65.

Shadish, W. R., & Heinsman, D. T. (1997). Experiments versus quasi-experiments: Do you get the same answer? In W. J. Bukoski (Ed.), *Meta-analysis of drug abuse prevention programs* (NIH Publication No. 97-4146, pp. 147-164). Washington, DC: Superintendent of Documents.

Shadish, W. R., & Ragsdale, K. (1996). Random versus nonrandom assignment in controlled experiments: Do you get the same answer? *Journal of Consulting and Clinical Psychology*, *64*, 1290-1305.

Shadish, W. R., & Sweeney, R. B. (1991). Mediators and moderators in meta-analysis: There's a reason we don't let dodo birds tell us which psychotherapies should have prizes. *Journal of Consulting and Clinical Psychology*, *59*, 883-893.

Smith, M. L., & Glass, G. V. (1977). Meta-analysis of psychotherapy outcome studies. *American Psychologist*, *32*, 752-760.

Sumner, W. G. (1906). *Folkways*. Boston: Ginn.

Tajfel, H., & Turner, J. C. (1986). The social identity theory of intergroup behavior. In S. Worchel & W. G. Austin (Eds.), *Psychology of intergroup relations* (2nd ed., pp. 7-24). Chicago: Nelson-Hall.

Taylor, D. M., & Moghaddam, F. M. (1994). *Theories of intergroup relations: International social psychological perspectives*. Westport, CT: Praeger.

Weiss, C. H. (1977). *Using social research in public policy making*. Lexington, MA: Heath.

Wortman, C. B., Hendricks, M., & Hillis, J. W. (1976). Factors affecting participant reactions to random assignment in ameliorative social programs. *Journal of Personality and Social Psychology, 33*, 256-266.

Wortman, P. M. (1983). Evaluation research: A methodological perspective. *Annual Review of Psychology, 34*, 223-260.

Wortman, P. M. (1992). Lessons from the meta-analysis of quasi-experiments. In F. B. Bryant, J. Edwards, R. S. Tinsdale, E. J. Posavac, L. Heath, E. Henderson, & Y. Suarez-Balcazar (Eds.), *Methodological issues in applied social psychology* (pp. 65-81). New York: Plenum.

Wortman, P. M. (1994). Judging research quality. In H. Cooper & L. V. Hedges (Eds.), *The handbook of research synthesis* (pp. 97-109). New York: Russell Sage.

Wortman, P. M., & Bryant, F. B. (1985). School desegregation and black achievement: An integrative review. *Sociological Methods and Research, 13*, 289-324.

Wortman, P. M., & Yeaton, W. H. (1983). Synthesis of results in controlled trials of coronary artery bypass graft surgery. In R. J. Light (Ed.), *Evaluation studies review annual* (Vol. 8, pp. 536-551). Beverly Hills, CA: Sage.

Wortman, P. M., & Yeaton, W. H. (1987). Using research synthesis in medical technology assessment. *International Journal of Technology Assessment in Health Care, 3*, 509-522.

Yeaton, W. H., Wortman, P. M., & Langberg, N. (1983). Differential attrition: Estimating the effect of crossovers on the evaluation of a medical technology. *Evaluation Review, 7*, 831-840.

CHAPTER 3

Discriminative Validity

Norman Miller
William C. Pedersen
Vicki E. Pollock

Donald Campbell has made many contributions of broad methodological, epistemological, and practical importance to psychology as a science. Within his discussions of methodological issues, validity was a prominent theme. This chapter discusses the more specific notion of discriminative validity.

Among hindrances to the progress of scientific psychology, *inventing new names for old concepts* is one of the more troublesome (Miller & Pollock, 1994). In many areas of psychology, researchers have attached their own idiosyncratic term to a concept, thereby making it more distinctive (and self-referring). Giving a previous well-studied concept, process, or idea, a new label is akin to "reinventing the wheel." Rosenthal (1994), with a deft turn of phrase, introduced the term *concept capture* to describe this not uncommon tendency. There may well be positive features of this practice, such as calling attention to a concept or process, which in turn stimulates research on it and thereby increases scientific understanding. At the same time, however, it clutters the field with unnecessary terms and is antithetical to the scientific ideals of parsimony and conceptual integration. *Put bluntly, instances of concept capture amount to a false assertion*

AUTHORS' NOTE: Preparation of this chapter was facilitated by National Science Foundation Grant SBR 931-9752 and a University of Southern California Pre-Doctoral Fellowship to William C. Pedersen. We thank Brad Bushman and two reviewers for their helpful commentary.

of discriminative construct validity, or one made with an absence of confirming evidence.

Our general discussion of the notion of discriminative validity will start with Campbell's fundamental notion of trait validation (Campbell & Fiske, 1959). Trait validation is then distinguished from nomological or construct validity, as Cronbach and Meehl (1955) used the latter term. We will use the terms *trait validity* and *nomological validity* to distinguish the concerns of Campbell and Fiske from those of Cronbach and Meehl. The latter used the terms *nomological validity* and *construct validity* interchangeably. Here, we consistently use *nomological validity* to refer to Cronbach and Meehl's approach, reserving the latter's term *construct validity* as a more overarching term that refers to the issue of concern in a more abstract or general manner.

We then consider the relation between discriminative trait validity and discriminative nomological validity. Next, we extend these notions to discriminative process validity and discriminative theoretical validity. Throughout, we give examples of contemporary instances of questionable attainment of each. Finally, in a brief concluding section, we make some constructive recommendations.

How would Donald Campbell have viewed this chapter? In buttressing our argument, we give examples that seemingly chide researchers for exhibiting questionable parsimony by failing to provide adequate evidence for the discriminative construct validity of the terms they introduce. We suspect that this aspect of the chapter might have rubbed Don Campbell the wrong way. His approach was always constructive, never critical. As graduate students, the first author of this chapter and fellow graduate student in sociology Kiyoshi Ikeda once approached Don for advice about an idea we had for writing a critical article on the literature concerned with worker morale and productivity. We had believed that much of this literature inadvertently (or uncritically) confounded morale with status. Don, never directly telling us to avoid an article that in its critical analysis would disparage the work of others, urged us, instead, to do some research on the issue.

Separate from the degree to which this anecdote illustrates features of Don's characterological makeup, his broader epistemological view also emphasizes a perspective that downplays the importance of the instances of questionable discriminative construct validity that we cite throughout the chapter. That is, in accord with his constructive, noncritical interpersonal approach, an upbeat optimism also characterized his epistemological view. It emphasized the self-corrective features of science—aspects of the peer review process and the critical interchange among scholars that might function to expose instances of questionable discriminative construct validity (Campbell, 1986). In citing instances of such scholarly exchange, in which authors directly attempt to deal with questions raised about the discriminative construct validity of concepts they have introduced (e.g., Eagly & Chaiken, 1993; Neuberg, Judice, & West, 1997; Petty, 1994; Webster & Kruglanski, 1994), we provide evidence that Campbell would have taken as supportive of his epistemological view. Without belaboring our own perspective, however, we believe that the downside of inadequate demonstration of discriminative construct validity is substantial—that it outweighs such ancillary benefits as drawing renewed attention to an abandoned area of research and/or stimulating new research. Instead, it promotes an overdifferentiated depiction of knowledge that obfuscates the bearing of previously established empirical relationships on the new work that is being presented. It slows development of integrated, broader theoretical organization. Additionally, however, we think the lack of adequate evidence of discriminative construct validity is more efficiently dealt with prior to publication rather than after published instances of its occurrence.

Finally, despite his fondness for neologisms, in his own research Don seemed to emphasize integration, synthesis, and overarching inclusiveness in preference to differentiation, distinctiveness, and analysis. In his empirical work on social projection (Campbell, Miller, Lubetsky, & O'Connell, 1964), conceptually related to our discussion below concerning the discriminative construct validity of *assumed similarity*, *false consensus*, and other related traits, he had hoped to find an empirical basis for integrating behavioristic, Gestalt, and clinical psychology traditions regarding the understanding of human behavior. Thus, to the degree that we are correct in seeing slippage or shortfall regarding Don's optimistic belief that the self-corrective features of science will overcome the consequences of the routine failure by scientists to provide evidence of the discriminative construct validity of new terms that they introduce, Don's penchant for and admiration of efforts to integrate knowledge is undermined.

DEVELOPING VALID SCIENTIFIC CONCEPTS

Trait Validation

Campbell and his associates (e.g., Campbell & Fiske, 1959; Cook & Campbell, 1979) were interested in the fit between operations and conceptual definitions. In their general discussion of trait (or concept) validity, Campbell and Fiske (1959) advised that it is necessary to achieve agreement among multiple measures that represent diverse approaches to assessment. They saw heterogeneity among the methods used to assess a trait as critical. They argued that to ensure that common method variance does not mistakenly contribute to the obtained convergence between measures of what was thought to be a single concept or trait, one must assess a trait with methods that are relatively distinct from one another (Campbell & Fiske, 1959). If similar methods of measurement are used to assess a trait, positive correlations will arise, at least in part, as a consequence of their shared method variance. In such a case, distinct concepts may mistakenly be viewed as reflecting the same latent variable. Thus, contrary to Duncan (1984), who in his erudite essays on measurement in the social sciences came to question the wisdom of combining *any* measures, Campbell and his colleagues advocated multiple, maximally heterogeneous assessments to measure an underlying, conceptually single entity or trait. They labeled their approach *trait validation* (Campbell, 1960).

Trait validation, as developed by Campbell and Fiske, shares conceptual correspondence with internal consistency reliability, as reflected in Cronbach's alpha (Cronbach, 1951) or the Kuder-Richardson reliability formula #20 (Kuder & Richardson, 1937; Richardson & Kuder, 1939), which is a special case of alpha. Internal consistency reliability (or single-factoredness) reflects the interrelation among items on a single test. By contrast, trait validation considers diverse tests (items) designed to measure a single concept. The explicit identification of method variance is the key conceptual insight of Campbell and Fiske that separates internal consistency among diverse trait measures (or latent trait measurement) from internal consistency as assessed by application of Cronbach's alpha to a set of items composing a single measure designed to assess X. As typically applied, the latter will almost invariably produce internal consistency estimates that are inflated by a shared format among the items (including their shared response scale), the spatial/temporal adjacency of subsets of items, and other meta-cues for a consistency in responding. For instance, important among such cues in the typical implementation of experimental social psychology is the presence of another set of differently formatted questions that are presented separately to the participant on a previous page—thereby implicitly indicating something distinctively shared by those among the second (and first) set. As an example, a set of mood manipulation check items might appear on one page and a set of attitude measurement items on a second page.

At the same time, Campbell and Fiske (1959) explicitly recognized the fact that *discriminative* construct validity (discriminative trait validation) goes hand in hand with construct validity. "One cannot define without implying distinctions, and the verification of these distinctions is an important part of the validation process" (Campbell, 1988, p. 40). Thus, measures of construct A must be examined in the context of measures of other distinct concepts. That is, it makes sense to argue that within the context of any single type of measurement situation, A should correlate more highly with other measures of A than with measures of B, C, D, and E.[1]

Nomological Validity

Separate from the manner in which Campbell and Fiske deal with adequate identification of a trait or scientific concept in their 1959 paper is the confirmation of theoretically important and hypothesized relationships between the construct (the trait, or the scientific concept) of interest and other constructs. Cronbach and Meehl (1955) viewed such nomological relationships—the frequency with which a construct (concept) exhibited lawful relationships with other constructs—as evidence that a construct had scientific validity. Hence, to them, construct validity meant evidence of predictable, meaningful, and replicable relationships with other concepts.

Decades ago, the cephalic index was used widely by physical anthropologists to assess the head size of racial/ethnic and national groups. It was thought that the measure reflected brain size. Measurements of dimensions of head size by means of physical measurement with a caliper can be highly accurate and reliable. Using such measures, Scots, for instance, were shown to have, on average, a small head size. Were head size also assessed by means of amount of water displaced by immersing the head in a bucket of water, and were the two indices shown to correlate highly (as they undoubtedly would), there would be evidence of trait validity in the sense prescribed by Campbell and Fiske. Furthermore, were such measures shown to be relatively less well related to similar measures of hand or foot size than they were to each other, they would simultaneously manifest evidence of discriminative trait validity. Despite the high likelihood of obtaining such confirmations, as best as is known, measures of the cephalic index have never been shown to be related to anything else. Thus, valid measurement of a trait or construct in the sense of Campbell and Fiske is not sufficient.[2] To be an important contribution to science, a concept must also be shown to have nomological validity—that is, construct validity in the sense of Cronbach and Meehl. In their discussion of trait validation, Campbell and Fiske (1959) gave little consideration to nomological validity as discussed by Cronbach and Meehl (1955).

Broader Treatments of Concept Validity

Twenty years later, Cook and Campbell (1979) discussed the link between construct validity and the idea of confounding. Here, their presentation incorporates aspects of both trait validation (Campbell & Fiske, 1959) and nomological validity (Cronbach & Meehl, 1955) under the single term *construct validity*.[3] Specifically,

> Construct validity is what experimental psychologists are concerned with when they worry about "confounding." This refers to the possibility that the operations which are meant to represent a particular cause or effect construct can be construed in terms of more than one construct. . . . Confounding means that what one investigator interprets as a causal relationship between theoretical constructs labeled A

> and B, another investigator might interpret as a causal relationship between constructs A and Y or between X and B or even between X and Y. (p. 59)

These "reinterpretations" of the causal mechanisms at work are not mere translations of the same concept into slightly different terminology but instead represent conceptually different and rival explanations of the "facts."

Why is construct validity important to the scientific researcher? Cook and Campbell (1979) provide several reasons. First, on a very general level, "researchers would like to be able to give their presumed cause and effect operations names which refer to theoretical constructs" (p. 38). Although they concede that this desire might be stronger for those working on basic theoretical issues, they note that researchers working on applied issues also want to define their variables in a more abstract manner, even while wishing to avoid the relatively burdensome task of giving precise and technical operational definitions.

Beyond this basic concern, there are practical reasons why construct validity is crucial. Specifically, they concern issues of replication, efficiency, and the avoidance of "irrelevant" research. With respect to replication, Cook and Campbell (1979) point out that experimental treatments in applied settings often involve a multitude of variables that do not necessarily reflect one single construct. Consider, for instance, the meaning of *school desegregation* as an experimentally introduced social reform. By comparison with control conditions, the situations in which it was introduced varied in size of unit (class, school, cluster of grade levels, or district), voluntary versus court-ordered implementation, one-way versus two-way busing of students, level of racial balance within classrooms, two-group versus multiple-group racial/ethnic mixing, level of parental involvement in schools, the amount of community conflict that preceded it, the degree to which increased monetary resources were coordinated with the implementation of the desegregation plan, teacher/student ratios, and other factors. Replication of research will be difficult if the relevant individual components in a multicomponent treatment are not clearly delineated and specified ahead of time. Second, efficiency dictates that if a subset of all the components is actually responsible for the overall effect, one would want to concentrate on that subset—in both future research and remedial intervention—rather than attempting to reproduce the larger overall treatment. Finally, having a tight fit between operationalizations and the constructs they reflect allows one to avoid irrelevant research.

Although it is clear that construct validity is important, it is less clear how it is achieved. Cook and Campbell (1979) first advocate rigorous attention to defining the constructs concisely and doing so in a manner that is clearly understood by the relevant scientific audience. Second, in terms of data analysis, the evidence must clearly show that the independent variable(s) affected the dependent measure(s) in the predicted manner (nomological validity). In addition, it is important to assess whether the independent variable covaries with related but conceptually distinct constructs (trait validation). Specifically, one wants to ascertain that the experimental impact on the dependent measure is due to the construct under manipulation and not some other unforeseen variable. Obviously, the trait validity of the dependent measure needs to be considered as well.

Claims of Discriminative Trait Validity

In the next sections, we present some illustrative instances in which researchers implicitly seem to claim discriminative trait validity in the absence of adequate empirical assessment.

Dissonance and Related Constructs

Festinger (1957) introduced the term *cognitive dissonance* to refer to the state produced by the holding of two salient cognitions, one of which is the obverse of the other. For example, if I am aware that the surgeon general's report says smoking causes cancer and I am a smoker, I should experience cognitive dissonance. With the possible exception of attribution theory, no other theoretical development within modern social psychology has stimulated as much research as Festinger's theory of cognitive dissonance (Bagby, Parker, & Bury, 1990).

Aronson (1969) noted the importance of the self-concept with respect to the obverse cognitions involved in a state of dissonance, thereby arguing that the dissonant state required three cognitions: "I believe the task is dull"; "I told someone that the task was interesting"; and "I am a decent, truthful human being" (Festinger & Carlsmith, 1959). Scher and Cooper (1989) argue that the arousal involved in cognitive dissonance is not simply due to the presence of obverse cognitions but instead arises because one's actions have aversive outcomes for others. They describe the internal state produced by these circumstances as consisting of a sense of *impaired self-efficacy.* This impaired self-efficacy reflects the psychological disparity implicit in the second and third cognitions, along with the additional cognitive ramifications of these thoughts. Specifically, the disparity produces a new cognition of self-dissatisfaction (or the momentary cognitive representation of self as having impaired self-efficacy). *Self affirmation,* as described by Steele in his self-affirmation theory (Steele, Spencer, & Lynch, 1993), appears to be the converse of impaired self-efficacy. It is the restoration of that which was damaged. Is *self-affirmation* (Steele et al., 1993) distinct from dissonance reduction? Is there discriminative validity among *impaired self-efficacy* and the states described by other common terms, such as *responsibility, guilt,* or *feeling bad*? These questions remain unexamined.

Did the initial introduction of the term *cognitive dissonance*, conceptualized as a state that emerged as a consequence of obverse cognitions, indeed represent a new theoretical concept that had not previously been studied? Balance theory (Cartwright & Harary, 1956) is concerned with the affective relations among elements of a triad: self, object, and other. In the notation of Jordan (1953), it examines the affective relations between p-o (I to other), p-x (I to object), and o-x (my perception of other to object). This notation can be applied to the situations studied to test fundamental hypotheses within dissonance theory. Consider again the classic dissonance experiment in which participants were induced to lie about a boring task to a fellow student (Festinger & Carlsmith, 1959). Incorporating Aronson's notion about the self, its dissonance-producing ingredients are these: (a) I like my fellow students (+ relation of I to other), (b) I say I like the boring task (+ relation of I to object), and (c) I believe that my fellow students will dislike this boring task (– relation of other to object) (Sears, 1983). Among the eight permutations of signs linking these three components, this ++– triad is one of the four viewed as unbalanced (unpleasant) by balance theory researchers. Does cognitive dissonance (an unpleasant state) differ from the arrangements of the triadic components in balance theory that are experienced as unpleasant? To our knowledge, dissonance theorists have never addressed this issue.[4]

Assumed Similarity and the False Consensus Effect

Assumed similarity refers to a judgmental bias in which the similarity between self and others is exaggerated. As shown by meta-analytic evidence, it is a consistent and reliable finding (e.g., Gross & Miller, 1997; Mullen & Hu, 1988). It applies to attitudes, personality traits, interests, and values. It was noted by Francis Bacon (1620/1853), who, in discussing its various manifestations as a conspicuous bias in human social perception, mentioned in particular the tendency to project one's own worldview onto others. Freud's (1937) discussion of paranoid projection provides an instance from the domain of personality trait attribution. Within scientific psychology, the history of research on assumed similarity is so extensive that incisive methodological suggestions concerning its quantitative analysis were raised more than three decades ago (e.g., Cronbach, 1955).

The *false consensus effect* (FCE) is defined as the difference in consensus estimates by those agreeing with and opposing an opinion position (Ross, Greene, & House, 1977). The data that constitute the FCE and the data concerning the magnitude by which consensus estimates exhibit bias from reality (viz. data on assumed similarity) are inextricably linked, being facets of the same data set (Gross & Miller, 1997). More than 150 references to work on assumed similarity had been published prior to the introduction of the new term, FCE (Miller &

Pollock, 1994). Some of these (e.g., Travers, 1941; Wallen, 1943) had used a paradigm identical to that employed specifically for assessing the magnitude of the FCE. To describe the entirety of antecedent work related to the FCE, however, its originators mention only five references.

Should the term *FCE* be singled out as a unique instance of concept capture within this specific research domain? It hardly appears to be. Instead, more than 15 other distinct labels previously had been used by various researchers to describe what is seemingly a single underlying concept (Miller & Pollock, 1994). More important, there has been little attempt to present evidence on whether any of these 15-plus labels for this judgmental bias obey laws that differentiate one member of this family of terms from another. In other words, evidence for discriminative construct validity appears to be lacking.

Need for Closure

The *need for closure* is an important theoretical construct in Kruglanski's lay epistemic theory (1989, 1990a).[5] Moreover, it provides an instance from contemporary social psychology in which both the major proponent of the importance and distinctiveness of a concept, as well as others, have examined its discriminative trait validity.

In constructing an individual difference measure to assess need for closure, Webster and Kruglanski (1994) described five separate aspects or constructs that captured the broad scope of the underlying theory: order, predictability, decisiveness, ambiguity, and closed-mindedness (Kruglanski, 1989, 1990a, 1990b). The five items that constitute the first of these constructs, preference for order and structure, were taken from a previously published scale (Thompson, Naccarato, Parker, & Moskowitz, 1993) called the Personal Need for Structure (PNS) Scale. In addition, three of the items that composed the third construct, urgency of striving for closure in judgment and decision making, were taken from the Personal Fear of Invalidity (PFI) Scale—also previously published by Thompson, Naccarato, Parker, and Moskowitz (1993). Finally, an additional three items from the fourth construct of the Need for Closure Scale (NFCS), dealing with predictability, were also taken from Thompson et al.'s (1993) Personal Need for Structure Scale.

To provide evidence of the discriminative trait validity of the NFCS, Webster and Kruglanski (1994) examined the relation between participants' scores on the NFCS with other previously developed measures that fell into three distinct conceptual categories. The first category contained scales hypothesized a priori to assess overinclusiveness: the F Scale (Sanford, Adorno, Frenkel-Brunswik, & Levinson, 1950), measuring authoritarianism; the Dogmatism Scale (Rokeach, 1960); the Intolerance of Ambiguity Scale (Eysenck, 1954); the Bieri REP Test

(Bieri, 1966), designed to measure cognitive complexity; and the MPQ Control Subscale (Tellegen, 1982), which assesses impulsivity. The second category of scales were viewed as assessing exclusiveness: the Personal Need for Structure (PNS) Scale (Neuberg & Newsom, 1993; Thompson et al., 1993) and the Personal Fear of Invalidity (PFI) Scale (Thompson et al., 1993).

Although both these sets of measures were viewed as tapping conceptual aspects of the NFCS, they also were seen as measuring other conceptual ingredients thought to be unrelated to the key conceptual aspects of lay epistemic theory. Finally, in the spirit of trait validation as discussed by Campbell and Fiske, the third category contained scales assumed to measure constructs unrelated to the NFCS: the Social Desirability Scale (Crowne & Marlowe, 1964), the Need for Cognition Scale (Cacioppo & Petty, 1982), and the Quick Test of Intelligence (Ammons & Ammons, 1962).

When Webster and Kruglanski (1994, Table 4, p. 1054) correlated each of these scales with the NFCS (both as a unitary scale and separately for its five subscales), they obtained low correlations. They report no comparisons showing differential average intra- and intercategory correlations among measures composing their three categories; nevertheless, they argue that the NFCS is conceptually distinct from the concepts measured by this host of other scales and that it "appears to possess acceptable discriminative and convergent validity with respect to other relevant psychological measures" (p. 1056).[6]

Neuberg and his colleagues provided independent evidence on these issues (Neuberg, Judice, & West, 1997; Neuberg, West, Judice, & Thompson, 1997). In all six of their samples, they obtained substantial overlap between the NFCS (when used as a unitary scale) and the PNS Scale (median $r = .79$). Neuberg, Judice, and West (1997) concluded that "the evidence is overwhelming that when used in a unidimensional manner, the NFCS is operationally redundant with the PNS Scale" (pp. 1403-1404). Moreover,

> when the NFCS is used more appropriately as a multifactorial scale, three of the five NFCS subfacets also fail to demonstrate discriminant construct validity from preexisting measures. The two strongest NFCS subfactors are highly redundant with the two PNS Scale subfactors (Neuberg & Newsom, 1993), with correlations in the .80 range. And, again despite Webster and Kruglanski's position to the contrary, a third NFCS subfactor is highly redundant with the Personal Fear of Invalidity (PFI) Scale (Thompson, Naccarato, & Parker, 1989), with correlations between the two in the .75 range. Our analyses reveal, then, that the NFCS possesses little, if any, discriminant validity. (p. 1397)

Moreover, in more recent work, Kruglanski et al. (1994) found a similar relation between the NFCS and the Personal Need for Structure Scale (mean $r = .72$).

Bogen (1975) lists more than 35 sets of dichotomous terms that theorists have used to differentiate individual differences in cognitive styles reflecting a propensity for being focused, narrow-minded, and desirous of cognitive structure, as opposed to being general and open-minded. Do the differences described by these 35 pairs of terms also correspond to individual differences in need for closure?

Having discussed issues concerning the development of valid scientific concepts and, conversely, having presented examples of the conceptual repackaging of "old wine in new bottles" from the areas of social and personality psychology, we now broaden our scope. Specifically, we next consider the extension of our concerns to distinctive lawfulness.

DISCRIMINATIVE NOMOLOGICAL VALIDITY

When does one have evidence of a new law? Cronbach and Meehl skirted the issue of *discriminative* nomological validity, as did Cook and Campbell. The Campbellian notions of discriminative trait validity, however, can be applied to the problem of distinct lawful relationships—lawful relationships whose uniqueness amounts to more than instances of concept capture. If a researcher claims to have discovered a new law, the claim may rest on (a) a new independent variable that is linked to a well-established dependent variable, (b) a new dependent variable that has never before been associated with a well-studied independent variable, or (c) a relationship that involves both new independent and new dependent variables. Consequently, distinctive nomological lawfulness can be compromised when either the A variable, the B variable, or both variables lack discriminative trait validity in the sense of Campbell and Fiske (1959).

Within the context of experimental social psychology, the steps to be taken to address questionable discriminative construct validity are less problematic when a dependent variable, as opposed to an independent variable, is the source of concern. The reason for this is that dependent variables manifestly consist of a measure of a trait, state, or entity. Specifically, dependent measures consist of questionnaire items, codings of observed behavior, or other measures that operationalize a trait or concept. Of course, in line with the Campbell and Fiske program, for the comparative examination of intra- and intertrait relationships it is necessary that there be more than a single item for assessing both the key dependent concept of interest (B) and other potentially relevant but conceptually distinct dependent variables.

When independent variables ("A variables") are manipulated variables, as in experimental social psychology, establishing their discriminative validity in a lawful relationship (if A, then B) does not on first thought fall as readily into the Campbell and Fiske mold. How might one establish discriminative trait (or state) validity for manipulated variables? One approach is to develop trait (or state) measures that correspond to the manipulated variables. Such measures, termed manipulation checks, in fact are commonly incorporated into good experimentation. Here, however, it may make sense to draw a distinction between manipulation checks that merely assess knowledge of a manifest event from those that assess successful (temporary) induction of an internal state. The former might include (a) knowledge that the experimenter said "Your task performance will be *evaluated*," (b) awareness that one was asked to read a passage on "X," or (c) acknowledgment that one listened to a tape recording of "steady rainfall." The latter might include measures, respectively, of (a) skin conductance, (b) memory of a specific substantive content, or (c) a subjective state of boredom or relaxation.

Clearly, measures of the second type correspond more closely to the conceptual variable that a researcher might hope to have manipulated. Here, there are two important points. First, to allow assessment of discriminative trait validity, and hence assessment of discriminative nomological validity as well, one needs not only two or more items that assess a (or b, or c) but also two or more items that assess allegedly distinct but relevant rival concepts, so as to justify the scientist's claim of discriminative nomological validity for a new law relating A to B. In addition, however, in accord with the distinction drawn above, the manipulation check measures must assess whether the underlying state (or process) and other related states (or processes) have been induced, rather than there merely being knowledge of the experimenter's behavior or instructions.

Claims of Discriminative Nomological Validity

The literature on *terror management* (Greenberg, Pyszczynski, & Solomon, 1986; Solomon, Greenberg, & Pyszczynski, 1991) provides a contemporary instance in which discriminative nomological validity has been claimed. Its key idea is that reminding people of death (viz., mortality salience) increases the strength of their faith in their own particular cultural worldview and that, as a consequence, it not only increases their rejection or negative evaluation of those who violate their cultural norms or beliefs but also augments positive evaluation of those who support them (e.g., Greenberg et al., 1990; Rosenblatt, Greenberg, Solomon, Pyszczynski, & Lyon, 1989). Thus, the research examines the relationship between mortality salience as an independent variable and evaluative bias as a dependent variable.

To what degree does this posited relationship exhibit discriminative nomological validity? The first thing to note is that Greenberg and his colleagues do not claim any uniqueness with respect to their conceptual dependent variable—typically, out-group derogation, but sometimes the positivity of in-group evaluation. They assess out-group negativity by one of several specific evaluative measures of target persons who threaten (or support) the actor's worldview. Their three principal operationalizations are (a) setting a monetary value of a bail bond to be required from a deviant (viz., a prostitute) to gain release from jail, pending trial (e.g., Rosenblatt et al., 1989, Study 1); (b) trait evaluations of a target's personality (e.g., Greenberg et al., 1990, Study 1); and (c) completion of the Interpersonal Judgment Scale (Byrne, 1971), which asks participants to rate a target person's intelligence, knowledge of current events, desirability as a work partner, and other characteristics.

The independent variable, mortality salience, typically has been manipulated by asking participants to write a response to the following prompts: "Please briefly describe the emotions that the thought of your own death arouses in you" and "Jot down, as specifically as you can, what you think will happen to you as you physically die and once you are physically dead." The control condition typically required participants to respond to similar prompts concerning emotional reactions to watching television. Other less frequent manipulations of mortality salience included requiring participants to respond to a Death Anxiety Questionnaire (Conte, Weiner, & Plutchik, 1982), to view a film of a gory accident, and to complete key dependent measures while adjacent to a funeral home.

It seems clear that in much of their initial work, their independent variable—mortality salience—was conceptualized as a specific form of anxiety that is distinct from other forms of anxiety, or even more generally, distinct from other forms of negative affect. Consistent with this interpretation, Greenberg, Solomon, and Pyszczynski (1997) state that "mortality salience effects are engendered specifically by concerns about one's own mortality rather than in response to *any* anxiety-provoking or self-threatening event" (p. 98, emphasis added). By contrast, however, a substantial amount of research within the large literature on intergroup relations links the rejection of out-groups to antecedent inductions of either integral anxiety—anxiety that is elicited directly by the salient out-group (e.g., Stephan & Stephan, 1996)—or incidental anxiety or general negative affect—anxiety or negative affect that is produced by an extraneous source (e.g., Urban & Miller, 1998; Wilder, 1993). This larger literature challenges the alleged uniqueness or specificity of mortality salience (death anxiety) in producing out-group rejection. That is, it questions the discriminative nomological validity of the relation between mortality salience, as a specific form of anxiety, and inductions of other types of anxiety that are empirically linked to the rejection of out-groups or those who challenge one's worldview.

To their credit, Greenberg and colleagues have attempted to rule out such alternative interpretations and show that mortality salience effects are indeed distinct from those produced by general anxiety. First, each of their studies has included an affect measure that typically contains a specific subscale for measuring anxiety or fear. Results generally have shown no difference in negative affect or anxiety between the mortality salience and control conditions, and, in instances in which such differences arose, their use of affect or anxiety level as a covariate has not altered their overall finding on their main dependent variable. Second, worldview defense effects have not emerged in additional control conditions in which participants were asked to think about aversive events other than their own death (e.g., giving a public speech, experiencing intense physical pain, taking an exam). For instance, Greenberg, Pyszczynski, Solomon, Simon, and Breus (1994) showed that whereas mortality salience elicited derogation of a target who wrote an anti-U.S. speech compared to a pro-U.S. speech, inductions of other types of anxiety such as thinking about the possibility of experiencing intense physical pain or giving a talk in front of a large audience did not. Similarly, they have shown that remembering past failures, or currently experiencing failure, did not produce outgroup bias (Greenberg, Simon, Pyszczynski, & Solomon, 1996).[7] Finally, the fact that mortality salience increased negative evaluations of a deviant but did not manifest itself on a measure of negative affect, whereas another anxiety-provoking manipulation unrelated to death had just the opposite effect (i.e., no increase of derogation but an increase in negative affect), was seen as further evidence of discriminative trait validity of mortality salience and, hence, nomological validity for its relation to outgroup bias (Greenberg et al., 1995).

Although we applaud these efforts, we wonder where they take us with respect to our concerns about discriminative nomological validity. What does it mean if mortality salience conditions fail to evidence effects that differ from those of control conditions on manipulation check measures of anxiety or negative affect? Not only are these specific comparisons irrelevant to our concern with discriminative nomological validity, but also, because the measures that they have used as manipulation checks (e.g., PANAS, MAACL) have been well validated in the past, the outcomes of such comparisons lead one instead to question whether *any* form of anxiety has been manipulated. One response to such absence of differential effects on manipulation check measures is to postulate that experimental subjects defended against expressing the death anxiety elicited by mortality salience inductions and thereby masked differential effects on manipulation checks. More recently, terror management theorists have in fact postulated that such defense does indeed occur (e.g., Greenberg et al., 1995). If so, assessment of discriminative nomological validity requires comparative examination of the effects of other types of anxiety (e.g., sexual arousal) that also elicit defense on manipulation checks. Additionally, the induction of such other defense-arousing anxieties needs to be matched in strength with the induction of mortality salience (e.g., as judged by external judges).

Their second point, that manipulations of anxiety other than the induction of mortality salience have failed to elicit rejection of out-groups (whereas mortality salience does), seems on first thought to support the contention of discriminative nomological validity of mortality salience. With these comparisons, it is incumbent on these researchers to explain why their own outcomes for incidental negative affect do not yield out-group rejection, whereas other researchers typically do find evidence that out-group rejection is increased by the arousal of negative incidental affect (e.g., Forgas, 1995; Forgas & Fiedler, 1996; Wilder & Shapiro, 1989, 1991; Wilder & Simon, 1996). Moreover, to the degree that these comparisons show greater out-group rejection for mortality salience than do other anxiety inductions, there are two other considerations that are important and need to be controlled. First, the differential importance of death by comparison with other potential worries, such as failing an exam, needs to be considered.

Second, other research suggests that suppression can produce intensified rebound effects (Wegner, 1994). As suggested in the previous paragraph, if the anxiety elicited by mortality salience is suppressed and controlled, its effects on out-group rejection are likely to be intensified. Consequently, the appropriate comparison condition for demonstrating discriminative nomological validity is not the induction of an anxiety (other than mortality salience) that is not suppressed but, rather, the induction of one that is.

Finally, another issue concerning discriminative nomological validity of terror management theory is the specificity of the findings to thinking about one's own death and not about death in general. Greenberg and colleagues state that "participants writing about the death of a loved one exhibited worldview defense only to the extent that this reminded them of their own mortality" (Greenberg et al., 1997, p. 98). This was interpreted as showing that not only is mortality salience the key component but also that it is specifically one's own death (and not death in general) that underlies their findings. The absence of relevant manipulation checks touching on this point remains a problem. Specifically, with respect to their comparison of the effects of writing about one's own death versus that of a loved one, they provide no evidence suggesting that the two manipulations differentially reminded participants of their personal mortality. Moreover, although own mortality salience induced stronger worldview defense than did writing about the death of a loved one, the latter also differed from the control. This specific finding differs from that seen in other research by Greenberg and colleagues wherein the anxiety-provoking conditions that were not related to the participant's own death never produced outcomes that differed from those in the control condition. Why should thinking about the death of a loved one be different from contemplating an impending exam or the receipt of mild electric shock? Although Greenberg et al. (1994) suggest that thinking about the death of a loved one might remind participants of their own mortality, there are no direct assessments of this possibility.

Our point in this section is not to diminish the accumulated work on mortality salience and worldview defense. Greenberg, Pyszczynski, Solomon, and colleagues are to be commended for their systematic work on the effects of mortality salience on worldview defense and their good efforts toward a process understanding of this relationship. Instead, our purpose was to use this body of research to illustrate issues with respect to discriminative nomological validity.

ASSESSMENT OF DISCRIMINATIVE PROCESS VALIDITY

Separate from trait and nomological validity is the issue of process distinctiveness. Researchers have sometimes claimed that distinct processes underlie the events that occur in response to one or another type of stimulus or setting or in the responses of one or another type of person, or group. For instance, Moscovici (1980) has argued that distinct processes account for influence by numerical minorities and majorities. Taking another example, some research suggests a discontinuity between effects at the individual level, by comparison with the group level (Brown & Turner, 1981; Schopler & Insko, 1992), suggesting separate processes. The notion that there are separate routes of persuasion (e.g., Chaiken, 1980, 1987; Petty & Cacioppo, 1986) also suggests distinct processes. Parsimony, however, requires that postulation of multiple distinct explanatory processes be eschewed when a single process model can be shown to provide adequate explanation (e.g., Kruglanski & Mackie, 1990; Kruglanski & Thompson, 1999). In this vein, researchers have frequently urged against invocation of motivational explanations when it appears that cognitive processes are sufficient for explanation (e.g., Kreuger & Clement, 1997).

The issue of discriminative process validity is more subtle and complex than is discriminative construct validity. On one hand, process assessment requires that one establish a relationship not only between (a) the situation, as operationalized, and the outcome but also between (b) the situation and the process and between (c) the process and the outcome. At the same time, it requires that one rule out to a reasonable degree (a) the influence of prior or concurrent conditions that correlate with the situation as operationally defined, (b) concurrent conditions induced inadvertently by the manipulations, and (c) alternative processes. Seven procedures have been viewed as relevant to, or useful for, assessment of process uniformity or distinctiveness (Harrington & Miller, 1993). They differ in their diagnostic strength. Two, ecological validity and experimentation, are too weak to be considered useful for the task. Both, in their bald form—that is, when implemented without the addition of process measures or manipulations—solely examine differences in outcomes. The usefulness of a third, examination of statistical interactions, requires a modification of prior interpretation (Kruglanski & Mackie, 1990) of its usefulness.

Rejected Procedures for Establishing Process Distinctiveness

Ecological Validity

Ecological validity, or natural covariation, is the most primitive approach for inferring process distinctiveness. As indicated, some have proposed that different principles apply to intergroup and interpersonal behavior (e.g., Brown & Turner, 1981). In daily life, people meet as individuals (perhaps to exchange personal information), or as members of two or more groups (perhaps to resolve a dispute). One could observe the array of behaviors emitted by the actors in the two settings and ask whether their relative frequencies differ. Observed differences in competitiveness, for instance, might seemingly support the idea of process distinctiveness between interpersonal and intergroup settings. Alternatively, however, they may reflect selection effects among those who enter each setting, as well as different motives in the same individual when entering each setting. Thus, although different frequencies of competitive behavior within each setting *may* reflect distinct underlying processes, nevertheless, empirically observing the differential occurrence of such covariation cannot provide confirming evidence of process distinctiveness.

Experimentation

Experiments provide circumstances for stronger inference, but they too ordinarily do not speak strongly on process distinctiveness. Returning to our previous example, individuals experimentally assigned to interact as a member of a dyad (interpersonal behavior) exhibit lower rates of competitiveness than those assigned to one of two groups (Schopler & Insko, 1992), an effect consistent with Brown and Turner's (1981) contention of different underlying processes in the two conditions. The experience of differential threat within the two settings, however, may provide a single-process explanation that, if controlled, will eliminate the effect.

Interactions

The previous approach can be extended by experimentally examining interactions to assess whether the effect of relevant independent variables on behavior differs across settings. One may be inclined to infer process uniformity if a variable has similar effects in different contexts. By contrast, when independent variables interact with context features, one may be inclined to infer process distinctiveness. Consider, for instance, settings that vary in cognitive overload, operationalized perhaps by a secondary task such as digit counting or memorization (e.g., Gilbert & Hixon, 1991; Gilbert & Osborn, 1989). Cognitive overload typically is viewed as interfering with encoding and retrieval of information relevant to a primary task because it reduces capacity within working memory. On first thought, its manipulation appears to have little connection with experimental inductions of negative affect (e.g., instructing participants to recall an extremely sad personal experience; Baker & Guttfreund, 1993). Further consideration, however, suggests overlap. Specifically, the heightened effort required by cognitive overload may induce negative affect (Marco & Suls, 1993; Repetti, 1993). If so, the two variables, ordinarily believed to be conceptually distinct, may yield parallel effects because they induce the shared underlying state of negative mood.

On the other hand, at least some forms of negative mood appear to induce more careful and accurate processing (Pacini, Muir, & Epstein, 1998). By contrast, cognitive overload, by taxing cognitive resources, induces a reliance on stereotypes and categorical judgments as compensatory strategies for the insufficient capacity available for careful processing—and thereby is likely to reduce accurate depiction of exemplars that deviate substantially from their category prototype. Seemingly, experiments could be designed to examine the interaction between the two variables in a context in which their potentially confounded effects were separated. Perhaps this could be achieved, for instance, by using an overload task that required processing of words sufficiently associated with positive affect to counter the negative affect induced by the increased task difficulty of the high-load condition. If, under these circumstances, a crossover interaction was obtained, such that increased induction of negative affect produced a different direction of effect than did increased load, the case for process distinctiveness would be supported.

To take another example by returning to the alleged distinction between intergroup and interpersonal contexts, levels of social status, power, and interdependence are seen as affecting intergroup behavior (Brown & Turner, 1981). If these factors (status, power, and interdependence) similarly affect interpersonal relations, that would support process uniformity for the two allegedly distinct settings. By contrast, if they produce opposing directions of effects in the two settings, that would support process distinctiveness.

In an attempt to assess Moscovici's (1980) claim that distinct processes underlie numerical majority and minority influence, Kruglanski and Mackie (1990) reviewed the relevant literature. They not only invoked the first two methods, ecological validity and difference in experimental effects, as evidence

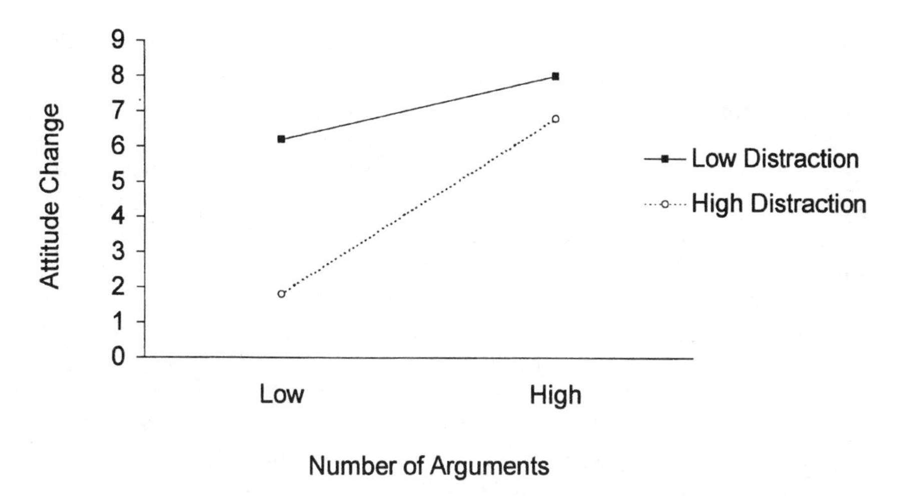

Figure 3.1. A Hypothetical Interaction Effect Between Distraction and Number of Arguments That Need Not Be Interpreted as Reflecting Distinct Distraction Processes When Arguments Are Few Versus Many

for process distinctiveness but also invoked the presence or absence of statistical interactions to thereby form an ordered scale of "strength of evidence." Specifically, they argued that an interaction between variable "X" and minority or majority group status would offer the strongest evidence for process distinctiveness.[8] Thus, absence of even a correlational relation and presence of an interaction between majority/minority status and another variable formed the weak- and strong-evidence endpoints of their scale.

In their interpretive logic, however, they failed to note the differential diagnostic power of ordinal (non-crossover) and disordinal (crossover) interactions. Aspects of measurement scales or differences in the magnitude by which subjective experience is altered by "equivalent" increments of a manipulated variable can create ordinal interactions that do not require the postulation of distinct underlying processes for their explanation. Noncrossover interactions therefore cannot be viewed as diagnostic of discriminative process validity.[9]

For example, imagine an experiment that manipulated distraction (high versus low) and the number of persuasive arguments, with 10 in the "low" and 20 in the "high" condition, respectively. Assume further that, in the context of two main effects, a significant ordinal interaction was obtained such that distraction strongly affected persuasion in the low but not the high argument-number condition (see Figure 3.1).

Given their stated reasoning, Kruglanski and Mackie (1990) would take this interaction as evidence of process distinctiveness under low and high numbers of arguments because distraction affected the two argument conditions differently, causing a larger decrease in persuasion under few arguments. One can also argue, however, that there is no difference in the underlying process that mediates the effect of number of arguments on attitude. Instead, the "effect" is due simply to the diminishing return of increased stimulus intensity and as such is similar to Weber's, Fechner's, or Steven's laws regarding the perception of physical stimuli. For these psychophysical laws, the essential idea is that an equivalent increase in subjective stimulus intensity requires ever increasing absolute intensities of the physical stimulus. Increases in the value of a physical stimulus therefore are more "impactful" in terms of their effect on perception when the intensity of the stimulus is already low and therefore near the absolute threshold.

Returning to the preceding example, assume that the number of arguments corresponds to a stimulus intensity dimension—strength of persuasive message. Further assume that the effect of high distraction is to halve the number of arguments processed, irrespective of whether the number of arguments is high or low. Taking these assumptions and tying them to the psychophysical principle stated above, under the high argument-number condition the difference between the 20 and 10 arguments respectively processed by participants subjected to low (zero) and high distraction is not as impactful (in terms of its effect of decreasing persuasion) as is the decrease from 10 to 5 arguments that will be produced by these same levels of distraction in the low number-of-arguments condition.

Direct Examination of Process and Process Distinctiveness

Correlational Analysis of Process

Stronger approaches to establishment of process distinctiveness will link the process both to antecedent and to consequent effects. Specifically, one can examine (a) the relation between key independent variables and process events (as dependent variables) and (b) the relation between the alleged process and the key dependent measures of interest. In line with our discussion of discriminative nomological validity, measures of relevant and irrelevant rival concepts are needed for the components of (a) and (b). Although better than no measurement of the alleged process, this procedure is weak in its failure to provide any evidence confirming the postulated temporal ordering of the three ingredients—independent variable, process, and final dependent variable.

Combined Correlational and Experimental Analysis

Statistical mediational assessment (Baron & Kenny, 1986; Judd & Kenny, 1981) is another approach now routinely used (Sigall & Mills, 1998) to assess mediational processes. The well-established tendency among two interacting groups for the numerically smaller one to exhibit stronger in-group identification and favoritism has been attributed to greater self-focus of the smaller group (Mullen, Brown, & Smith, 1992). Demonstration of process mediation with statistical mediational procedures (in addition to showing that manipulated numerosity affects the magnitude of bias) requires that self-focus be affected by numerosity, that self-focus be correlated with bias, and that when the effect of numerosity on self-focus is controlled (via covariance or regression analysis), its effect on bias disappears.

A problem here, as with the previously discussed procedures, is that it does not rule out other possible mediators (Sigall & Mills, 1998). Moreover, as with the preceding procedures, statistical mediational assessment cannot provide direct evidence of a causal connection between the alleged mediator and the key dependent variable in that it does not establish the temporal ordering implied by the alleged process explanation. The situational manipulation designed to assess the process may *simultaneously* affect both the key dependent measure and the alleged measure of process, leading one to mistakenly impute a process role for the latter. For example, cooperation and competition, conceptually and operationally defined by the structure of outcomes, may elicit different motives. Cooperation may mean, to participants, "talk to each other." Competition may mean "focus on the task" (cf. Bettencourt, Brewer, Croak, & Miller, 1992). On first thought, one might assume that these differences in meanings mediate the effects of cooperation and competition on a key measure (e.g., attraction). Instead, both the differences in meaning (talking vs. task focus) and the differences in liking or attraction toward one's coactors may be simultaneous effects that are consequences of (unmeasured) differences in categorization caused by the manipulation of goal structures. Cooperation may induce a superordinate one-group perception, whereas competition causes a two-group or multiple-group perception (Gaertner, Dovidio, Anastasio, Bachman, & Rust, 1993). Such categorization effects may simultaneously affect focus (interpersonal vs. task) and degree of liking, and if unmeasured, they will never be diagnosed as the critical underlying process event.

Experimental Analysis of Process

The approach described in the preceding section experimentally examines the effect of the antecedent situation on the process. It can be further strengthened by applying experimentation to all steps of the causal chain. This requires examination of three experimental components: the effect of the antecedent situation on the process, the effect of the antecedent situation on the key dependent variable, and a direct manipulation of the process so as to experimentally (rather than correlationally) examine its effect on the key dependent variable. Both nomological and process distinctiveness will be confirmed by application of the principles for trait validation to all three components of the chain.

Word, Zanna, and Cooper (1974) provide an instructive illustration. In Study 1, white interviewers met with confederates posing as applicants for a job. The race of the applicant was manipulated (black vs. white). The dependent measures included nonverbal behaviors (e.g., physical distance between the interviewer and the interviewee, eye contact with the confederate), the length of the interview, and the speech error rate of the subject (e.g., stutters, sentence incompletions, repetitions). In the second study, all subjects were white applicants who were interviewed by confederates. Some of the dependent measures used in Study 1, namely, "immediacy behaviors," were manipulated. In the immediate condition, the interviewer sat closer to the subject, made fewer speech errors, and conducted a longer interview, by comparison with the nonimmediate condition. Dependent variables were (a) judges' ratings of the subject's interview performance, (b) judges' ratings of reciprocated immediacy behaviors by the subjects, and (c) subjects' ratings of their postinterview mood state and attitudes toward the interviewer (confederate).

The results showed that black applicants were treated with less immediacy than whites (Study 1) and that this differential treatment impeded the performance and negatively affected the attitudes of job applicants (Study 2). In terms of our analysis of the needed components, the strength of Word et al.'s (1974) article lies in their completion of parts 1 and 3. Specifically, they test the effect of the antecedent situation (i.e., race) on the process (i.e., immediacy behaviors) (part 1), and they directly manipulate the process to examine its effect on the key dependent measure (i.e., applicant performance) (part 3). Thus, the research has strong analytical features. Nevertheless, an assessment of the effect of the antecedent situation (viz., race) on the key dependent variable (performance) is missing. Of greater importance, there are no assessments of rival process measures, other than those assessing immediacy. For instance, one unassessed rival process explanation is that black applicants may elicit more anxiety in white interviewers than do white applicants (Stephan & Stephan, 1985). In turn, the anxiety of the white interviewers not only may depress black performance (had performance of black and white applicants been measured) but also may have produced the effects found on the measures of immediacy. Moreover, even though the differential effects of the manipulation of immediacy in Study 2 may indeed have the effects experimentally shown by Word et al. (1974), these effects may be ancillary. They may be sufficient to produce the effects observed but not be a necessary ingredient of the underlying process. That is, even if immediacy behaviors were controlled, it is possible that differential anxiety on the part of the interviewers (had it been manipulated in Study 2 instead of immediacy) would have produced the same effects found in Word et al's. version of

Study 2. Clearly, it is important to measure and manipulate rival process variables to assess their comparative explanatory value.

Meta-Analytic Process Analysis

The use of meta-analytic synthesis of experimental procedures that incorporate the approaches described in the three preceding procedures will provide the strongest evidence on process uniformity or distinctiveness (see Driskell and Mullen, 1990, for a meta-analytic approximation of statistical mediational assessment). As has been suggested previously for the preceding three approaches, however, a most important addition to this meta-analytic amalgam is the inclusion (for comparative purposes) of measures that assess rival, as well as the hypothesized explanatory, processes. At the same time, it is important to note that differential reliability and validity of the measures used to assess rival explanatory processes will contribute to differential statistical confirmation of their explanatory strength, as assessed by statistical mediation procedures (Hunter & Schmidt, 1990). In turn, this can lead to erroneous inferences concerning the relative explanatory power of the (rival) processes they are assumed to tap.

EXTENSION OF DISCRIMINATIVE VALIDITY TO THE LEVEL OF THEORIES

In the preceding sections, we dealt with discriminative validity in terms of both constructs and processes. This analysis can now be extended to a more abstract or superordinate level—the discriminative validity of rival theoretical accounts. Are two theoretical explanations distinct, or do they amount to a compounding of multiple instances of concept capture?

In the arena of persuasion, the two theories that have dominated research during the last decade, the elaboration likelihood model (ELM) (Petty & Cacioppo, 1986) and heuristic-systematic model (HSM) (Chaiken, 1980, 1987), provide a useful illustration. Both hypothesize a dual-process feature in which there are two qualitatively distinct routes to persuasion, the central or systematic route for the processing of message arguments and the peripheral or heuristic route for the processing of cue information. Attitudes formed through the first route are longer lasting and more resistant to change than are those formed through the second. Here we ask if the distinctions between the two—other than the fact that models have been named, respectively, by selecting labels from opposing endpoints of a continuum that describes thoroughness of message or informational processing—are in any way theoretically important.

The proponents of each approach note differences between the two models. There are three issues on which either one or the other proponent attempts to delineate differences that distinguish the models from each other. For instance, Petty (1994) states that the ELM views heuristic processing as a subcategory of peripheral processing. Eagly and Chaiken (1993) concur, seeing heuristic processing as more narrow and refined in scope than peripheral processing because the heuristic processing mode implies that individuals use learned knowledge structures, simple schemas, or rules to make judgments. Unlike the ELM's treatment of peripheral processing, the HSM emphasizes underlying principles of cognitive processing that affect heuristic processing (Eagly & Chaiken, 1993). Some of these principles include availability, referring to the storage of a structure in memory for potential subsequent use (Higgins, King, & Mavin, 1982); accessibility, referring to the activation of a structure from memory (e.g., Chaiken & Eagly, 1983; Roskos-Ewoldsen & Fazio, 1992); the processing goal (e.g., motivation induced by concerns about accuracy, defense, or impression management); and reliability or strength, as manipulated for instance as the likelihood of receiving good advice from a good friend (Chaiken, 1987).

Second, Petty (1994) explains that the ELM hypothesizes a trade-off between the central and peripheral routes of processing, whereas the HSM allows for the possibility of increased impact of both systematic and heuristic processing as elaboration likelihood is enhanced. Thus, the ELM more clearly views the processing as consisting of a continuum, ranging from central to peripheral processing, than does the HSM. Petty does not believe, however, that the central and peripheral routes are mutually exclusive. In fact, bringing convergence rather than divergence to the two models, he states that these processes often co-occur and potentially may have a collective impact on attitudes. The trade-off hypothesis simply indicates that a variable has a decreased likelihood of affecting attitudes via a peripheral process as the elaboration likelihood increases. Echoing this point of view, Eagly and Chaiken (1993) mention a "concurrent processing assumption" (p. 328) that allows the HSM to propose "that heuristic and systematic processing can exert both independent (i.e., additive) and interdependent (i.e., interactive) effects on judgment" (p. 328).

Third, with respect to motivation, both the ELM and the HSM assume that people are inclined to adopt correct attitudes. Unlike the ELM, however, the HSM goes on to specify other possible motivations that people might have. These include "defense motivation" (e.g., Giner-Sorolla & Chaiken, 1997) and "impression motivation" (e.g., Chen, Shechter, & Chaiken, 1996). As such, the HSM hypothesizes a broader scope of potential motivations (Eagly & Chaiken, 1993). Finally, in discussing a "sufficiency threshold," the HSM hypothesizes a mediator to explain when certain variables will enhance the processing and elaboration that is seen in the central or systematic routes. The ELM lacks this theoretical mediator (Eagly & Chaiken, 1993).

As interesting and important as these distinctions may be, however, we know of no research that validates the distinctiveness of the two theories. Evidence concerning substantive arenas in which one or the other of the two theories makes (subsequently confirmed) predictions that are not made by the other, while constituting theoretical advance, does not invalidate the other theory or provide discriminative theoretical validity in a strong sense. The latter requires instead evidence in support of one of two competing predictions generated by each theory. Obviously, the alleged theoretical distinctiveness will acquire meaning only when such evidence is produced.[10]

CONCLUSION AND RECOMMENDATIONS

We have argued here that science evidences a needless and dysfunctional proliferation of conceptual baggage, whether considered at the abstract level of theory development or constrained to the more specific level of individual concepts. Although Don Campbell's interests were primarily methodological and epistemological, his substantive research was concerned with issues in social psychology. Consequently, it is appropriate that the specific illustrative examples of questionable discriminative validity that we have presented were taken from subareas of social psychology. At the same time, however, one can ask whether, in addressing an issue that seems to be an appropriate area of concern for those in social psychology, we are guilty of having generalized a local problem into a worldview by implying that it characterizes psychology or social science in general.

In large part, we drew our examples from areas of social psychology because we are more familiar with research in it; however, we believe that other areas are no less likely to exhibit these same problems. Sometimes, instances of the general issue arise much like differences between languages for the name of an object. Thus, when distinct scientific disciplines encounter evidence of the same phenomenological experience among a set of persons, they may assign distinctive labels to that which they observe. For instance, psychiatry and neurology repeatedly use different terms to describe fundamentally similar manifestations of language disorders. Specifically, in semantic disorders of aphasia, *driveling* is used in psychiatry, whereas *jargon augmentation* is used in neurology to refer to speech that is devoid of meaningful content yet characterized by tightly linked associations that are accompanied by a preservation of syntax. In nominal disorders, *word approximation* is used in psychiatry and *verbal paraphasia* is used in neurology to refer to instances in which words are used without reference to their precise meaning. In phonemic disorders, *clanging* is used in psychiatry and *literal phonemic paraphasia* is used in neurology to refer to situations in which associations between words occur on the basis of their sound, rather than their meaning.

Similar problems arise in the field of mental illness. That is, discriminative validity is an issue in differential diagnosis of psychopathology. For example, in defining categories of depression, the *Diagnostic and Statistical Manual of Mental Disorders* (third revision; *DSM-III*) differentiates major depression from dysthymia, even though their respective symptoms overlap to a substantial degree (van Praag, 1993). Mild cases of major depression are virtually indistinguishable from dysthymia. Similarly, it is unclear that severe dysthymia can be distinguished from major depressive episodes.

Taking another example, modern psychopathological nomenclature used in the *DSM-IV* (and the *DSM-III* as well) avoids use of the term *neurosis*—a term with a long history of use that started in the 18th century with respect to conditions such as hysteria, as well as types of depression. (Today, these same symptoms are considered to be anxiety disorders.) Later, Freud used the term *neurosis* to conceptualize and describe these, as well as other, related anxiety symptoms and traits. Despite the long-standing use of the terms *neurosis* and *neurotic* to characterize persons who suffer from fundamentally similar symptomatology indicative of anxiety (including heart palpitations, malaise, and neurasthenia), the term *neurosis* has been abandoned for use with anxiety disorders in the *DSM-IV.* It is now replaced with a plethora of discrete diagnostic categories: distinct *phobias*, including social, simple, and agoraphobia; *panic disorder*, subtyped as occurring with or without agoraphobia; *obsessive-compulsive disorder*; and *post-traumatic stress disorder.* In assessing the need for such differentiation, one must ask whether the underlying processes that account for the development of distinct phobias and the conceptually related anxiety disorders mentioned above do indeed differ. Similarly, one must ask whether the optimal approaches for their treatment will differ. Given van Praag's (1993) documentation of countless other instances of such dubious differentiation among diagnostic categories in psychopathology, we have little reason to think that the problem of compromised discriminative construct validity is constrained to social psychology.

At the same time, the preceding discussion calls for some remedial recommendations. For whatever form of discriminative validity is of concern—trait, nomological, process, or theoretical—measures of rival, potentially similar, and potentially dissimilar concepts are needed within the context of individual studies. To provide evidence of the discriminative construct validity of the primary dependent measure, the results of the study should show that the manipulation of independent variables affects the measure of interest but not the theoretically similar, but allegedly distinct, secondary measures. Likewise, for experimental social psychology, manipulations of independent variables not only should be confirmed with relevant manipulation checks that assess the presence of the subjective state intended to be induced by the manipulation but also should be accompanied by other manipulation check measures that assess rival states judged to be theoretically unrelated to the experimental manipulation. In an analogous manner, when dealing with discriminative process validity, researchers should include measurements of theoretically similar but rival process variables to show that when such variables serve as covariates, they do not affect the primary dependent measure to the same extent as the hypothesized process variable of interest. Such changes in practice need to become normative. For this to happen, they must be espoused, if not routinely required, by dissertation advisers, editors, and professional associations.

NOTES

1. Campbell and Fiske (1959) went on to argue that when trait A is assessed with any one type of measure, its average correlation with four other types of measures of A should exceed the correlation found among traits A, B, C, D, and E as measured with any single type of measure. Here, however (without meaning to diminish the importance of discriminative trait validity as evidenced by within-measure differences in the magnitude of within-trait versus cross-trait correlations), we question the general principle that is implicit in this specific stipulation. The issue here is the comparative importance of trait versus state variables. If traits are conceptualized as individual difference measures, as opposed to situational (or acute, or temporary) inductions of high or low levels of a state, then this latter requirement assumes that social psychological variables (situational variance) are less powerful or important than personality variables (individual difference variance). Taking instead the perspective of experimental social psychology, both trait and method can be conceptualized as consequences of experimenter action. That is, both can be operationalized as situational manipulations. From this perspective, then, why should the former "on average" be a stronger source of variance than the latter, when they both are the same thing—a consequence of situational variation? Even while retaining the conceptual distinction between personality and situation, however, there seems to be little logical or empirical justification for assuming that personality is a more potent source of explained variance than is situation. For instance, a striking feature of the work related to Milgram's obedience studies was the failure to find personality moderators of the basic situational effect that Milgram studied (Brown, 1986, p. 5).

2. We do not mean to imply here that they thought it was indeed sufficient. See, for instance, the next paragraph.

3. Specifically, Cook and Campbell (1979) discussed four separate types of validity: internal, external, statistical conclusion, and construct (specifically, construct validity of causes or effects).

4. We recognize that to whatever degree we are correct in identifying overlap between *cognitive dissonance* and *unbalanced triads*, Festinger's theory of cognitive dissonance nevertheless was broader than balance theory in that it elaborated specific antecedent conditions for producing the dissonant or unbalanced state (cf. Zajonc, 1968) that had

not been identified within balance theory. Moreover, it also specified alternative modes for reducing dissonance or unbalance.

5. The following discussion does not attempt to evaluate Kruglanski's lay epistemic theory. Instead, it is constrained solely to a consideration of individual measures (trait validation) of the need for closure. There is a substantial body of empirical research on issues related to the broader theory. (For recent discussions and summaries, see Cratylus, 1995; DeGrada, Kruglanski, Mannetti, Pierro, & Webster, 1996; Jamieson & Zanna, 1989; Kruglanski & Webster, 1996.) In numerous studies, need for closure has been experimentally manipulated as an independent variable. Research has examined its relation to the correspondence bias (Webster, 1993), impression primacy effects (Webster & Kruglanski, 1994), persuasion (Kruglanski, Webster, & Klem, 1993), and stereotypical judgments (Dijksterhuis, van Knippenberg, Kruglanski, & Schaper, 1996).

6. Curiously, however, there was little relation between the NFCS and both the Personal Need for Structure Scale ($r = .24$) and the Personal Fear of Invalidity Scale ($r = -.21$), even though (a) several items from each of these scales (particularly the PNS Scale) had been adopted for use on the NFCS and (b) all three scales are supposed to be based on lay epistemic theory (Kruglanski, 1989).

7. The sex of participants in this unpublished study is not reported. Meta-analytic evidence shows that females respond to failure by turning it inward and expressing self-disparagement, rather than responding with hostility toward others (Bettencourt & Miller, 1996). Do these outcomes reflect results obtained (primarily) with female participants?

8. Paralleling our own ordered criteria (but omitting the subsequent approaches we present), Kruglanski and Mackie (1990) noted that necessary covariation (experimental evidence) provides weaker confirmatory evidence for process distinctiveness, with natural covariation (correlational evidence) the weakest evidence. After imposing their logical (or intuitive) analysis, they concluded that all but one of the 21 variables that they considered, at best, only naturally covary with minority/majority source status (or have no relationship at all). They argued, therefore, that process uniformity underlies minority and majority influence.

9. It is important to note that the point we make here could not affect their conclusion because their application of a logical analysis of the "likely effects" of relevant variables on minority/majority influence failed to yield a single instance of an interaction among the 21 variables that they considered.

10. Although not bearing directly on the issue of discriminative theoretical validity, Kruglanski and Thompson (1999) question whether the two types of process invoked by both theories should be considered as distinct processes. They contend instead that both messages and cues should be subsumed under the broader category of persuasive evidence. Thus, they question the discriminative process validity of the two modes/routes of persuasion by proposing that once differences on persuasively relevant informational parameters are controlled, cue-based and message-based persuasion should be affected similarly by relevant processing variables (e.g., motivation and cognitive capacity). They offer evidence for this contention in a series of studies. The advantage of this view in terms of parsimony is obvious.

REFERENCES

Ammons, C. H., & Ammons, R. B. (1962). *The Quick Test (QT).* Missoula, MT: Psychological Test Specialists.

Aronson, E. (1969). The theory of cognitive dissonance: A current perspective. In L. Berkowitz (Ed.), *Advances in experimental social psychology* (Vol. 4, pp. 1-34). New York: Academic Press.

Bacon, F. (1853). *The physical and metaphysical works of Lord Bacon* (J. Dewey, Trans.). London: Bohn. (Original work published 1620)

Bagby, R. M., Parker, J.D.A., & Bury, A. S. (1990). A comparative citation analysis of attribution theory and the theory of cognitive dissonance. *Personality and Social Psychology Bulletin, 16*(2), 274-283.

Baker, R. C., & Guttfreund, D. G. (1993). The effects of written autobiographical recollection induction procedures on mood. *Journal of Clinical Psychology, 49*(4), 563-568.

Baron, R. M., & Kenny, D. A. (1986). The moderator-mediator variable distinction in social psychological research: Conceptual, strategic, and statistical considerations. *Journal of Personality and Social Psychology, 51,* 1173-1182.

Bettencourt, B. A., Brewer, M. B., Croak, M. R., & Miller, N. (1992). Cooperation and the reduction of intergroup bias: The role of reward structure and social orientation. *Journal of Experimental Social Psychology, 28,* 301-319.

Bettencourt, B. A., & Miller, N. (1996). Gender differences in aggression as a function of provocation: A meta-analysis. *Psychological Bulletin, 119,* 422-447.

Bieri, J. (1966). Cognitive complexity and personality development. In O. J. Harvey (Ed.), *Experience, structure and adaptability* (pp. 13-37). New York: Springer.

Bogen, J. E. (1975). Some educational aspects of hemispheric specialization. *UCLA Educator, 17,* 24-32.

Brown, R. (1986). *Social psychology* (2nd ed.). New York: Free Press.

Brown, R. J., & Turner, J. C. (1981). Interpersonal and intergroup behavior. In J. C. Turner & H. Giles (Eds.), *Intergroup behaviour* (pp. 33-65). Oxford, UK: Blackwell.

Byrne, D. (1971). *The attraction paradigm.* New York: Academic Press.

Cacioppo, J. T., & Petty, R. E. (1982). The need for cognition. *Journal of Personality and Social Psychology, 42,* 116-131.

Campbell, D. T. (1960). Recommendations for APA test standards regarding construct, trait, or discriminant validity. *American Psychologist, 15,* 546-553.

Campbell, D. T. (1986). Science's social system of validity-enhancing collective belief change and the problems of the social sciences. In D. W. Fiske & R. A. Shweder (Eds.), *Metatheory in social science: Pluralism and subjectivities* (pp. 108-135). Chicago: University of Chicago Press.

Campbell, D. T. (1988). Convergent and discriminant validation by the multitrait-multimethod matrix. In E. S. Overman (Ed.), *Methodology and epistemology for social science: Selected papers* (pp. 37-61). Chicago: University of Chicago Press.

Campbell, D. T., & Fiske, D. W. (1959). Convergent and discriminant validation by the multitrait-multimethod matrix. *Psychological Bulletin, 56*(2), 81-105.

Campbell, D. T., Miller, N., Lubetsky, J. & O'Connell, E. J. (1964). Varieties of projection in trait attribution. *Psychological Monographs, 78*(15, Whole No. 592).

Cartwright, D., & Harary, D. (1956). Structural balance: A generalization of Heider's theory. *Psychological Review, 63*, 277-293.

Chaiken, S. (1980). Heuristic versus systematic information processing and the use of source versus message cues in persuasion. *Journal of Personality and Social Psychology, 39*, 752-756.

Chaiken, S. (1987). The heuristic model of persuasion. In M. P. Zanna, J. M. Olson, & C. P. Herman (Eds.), *Social influence: The Ontario symposium* (Vol. 5, pp. 3-39). Hillsdale, NJ: Lawrence Erlbaum.

Chaiken, S., & Eagly, A. H. (1983). Communication modality as a determinant of persuasion: The role of communicator salience. *Journal of Personality and Social Psychology, 45*, 241-256.

Chen, S., Shechter, D., & Chaiken, S. (1996). Getting the truth or getting along: Accuracy- versus impression-motivated heuristic and systematic processing. *Journal of Personality and Social Psychology, 71*, 262-275.

Conte, H. R., Weiner, M. B., & Plutchik, R. (1982). Measuring death anxiety: Conceptual, psychometric, and factor-analytic aspects. *Journal of Personality and Social Psychology, 43*, 775-785.

Cook, T. D., & Campbell, D. T. (1979). *Quasi-experimentation: Design and analysis issues for field settings.* Chicago: Rand McNally.

Cratylus. (1995). De Nederlandse Need for Closure schaal [The Netherlands Need for Closure Scale]. *Nederlands Tijdschrift Voor de Psychologie, 50*, 231-232.

Cronbach, L. J. (1951). Coefficient alpha and the internal structure of tests. *Psychometrika, 16*, 297-334.

Cronbach, L. J. (1955). Processes affecting scores on "understanding of others" and "assumed similarity." *Psychological Bulletin, 52*, 177-193.

Cronbach, L. J., & Meehl, P. E. (1955). Construct validity in psychological tests. *Psychological Bulletin, 52*, 281-302.

Crowne, D. P., & Marlowe, D. (1964). *The approval motive.* New York: John Wiley.

DeGrada, E., Kruglanski, A. W., Mannetti, L., Pierro, A., & Webster, D. M. (1996). Un'analisi strutterale comparativa delle versioni USA e italiana della scala di "Bisogno di chiusura cognitiva" di Webster e Kruglanski [A comparative structural analysis of the U.S. and Italian versions of the "Need for Closure Scale" of Webster and Kruglanski]. *Testing, Psicometria, metodologia, 3*, 5-18.

Dijksterhuis, A., van Knippenberg, A., Kruglanski, A. W., & Schaper, C. (1996). Motivated social cognition: Need for closure effects on memory and judgment. *Journal of Experimental Social Psychology, 32*, 254-270.

Driskell, J. E., & Mullen, B. (1990). Status, expectations, and behavior: A meta-analytic review and test of the theory. *Personality and Social Psychology Bulletin, 16*, 541-553.

Duncan, O. T. (1984). *Notes on social measurement: Historical and critical.* New York: Russell Sage Foundation.

Eagly, A. H., & Chaiken, S. (1993). *The psychology of attitudes.* Fort Worth, TX: Harcourt Brace Jovanovich.

Eysenck, H. J. (1954). *The psychology of politics.* New York: Praeger.

Festinger, L. (1957). *A theory of cognitive dissonance.* Stanford, CA: Stanford University Press.

Festinger, L., & Carlsmith, J. M. (1959). Cognitive consequences of forced compliance. *Journal of Abnormal and Social Psychology, 58,* 203-210.

Forgas, J. P. (1995). Mood and judgment: The affect infusion model (AIM). *Psychological Bulletin, 117,* 39-66.

Forgas, J. P., & Fiedler, K. (1996). Us and them: Mood effects on intergroup discrimination. *Journal of Personality and Social Psychology, 70,* 28-40.

Freud, S. (1937). *The ego and mechanisms of defense.* London: Hogarth.

Gaertner, S. L., Dovidio, J. F., Anastasio, P. A., Bachman, B. A., & Rust, M. C. (1993). The common ingroup identity model: Recategorization and the reduction of intergroup bias. In W. Stroebe & M. Hewstone (Eds.), *European review of social psychology* (Vol. 4, pp. 1-26). New York: John Wiley.

Gilbert, D. T., & Hixon, J. G. (1991). The trouble of thinking: Activation and application of stereotypes. *Journal of Personality and Social Psychology, 60,* 509-517.

Gilbert, D. T., & Osborn, R. E. (1989). Thinking backwards: Some curable and incurable consequences of cognitive business. *Journal of Personality and Social Psychology, 57,* 940-949.

Giner-Sorolla, R., & Chaiken, S. (1997). Selective use of heuristic and systematic processing under defense motivation. *Personality and Social Psychology Bulletin, 23,* 84-97.

Greenberg, J., Pyszczynski, T., & Solomon, S. (1986). The causes and consequences of the need for self-esteem: A terror management theory. In R. F. Baumeister (Ed.), *Public self and private self* (pp. 189-212). New York: Springer-Verlag.

Greenberg, J., Pyszczynski, T., & Solomon, S., Rosenblatt, A., Veeder, M., Kirkland, S., & Lyon, D. (1990). Evidence for terror management theory II: The effects of mortality salience reactions to those who threaten or bolster the cultural worldview. *Journal of Personality and Social Psychology, 58,* 308-318.

Greenberg, J., Pyszczynski, T., Solomon, S., Simon, L., & Breus, M. (1994). The role of consciousness and accessibility of death-related thoughts in mortality salience effects. *Journal of Personality and Social Psychology, 67,* 627-637.

Greenberg, J., Simon, L., Harmon-Jones, E., Solomon, S., Pyszczynski, T., & Lyon, D. (1995). Testing alternative explanations for mortality salience effects: Terror management, value accessibility, or worrisome thoughts? *European Journal of Social Psychology, 45,* 417-433.

Greenberg, J., Simon, L., Pyszczynski, T., & Solomon, S. (1996). *Are mortality salience effects specific to reminders of mortality? Actual and imagined failure versus mortality salience.* Unpublished manuscript, University of Arizona, Tucson.

Greenberg, J., Solomon, S., & Pyszczynski, T. (1997). Terror management theory of self-esteem and cultural worldviews: Empirical assessments and conceptual refinements. In M. E. P. Zanna (Ed.), *Advances in experimental social psychology* (Vol. 29, pp. 61-139). San Diego: Academic Press.

Gross, S. R., & Miller, N. (1997). The "golden section" and bias in perceptions of consensus. *Personality and Social Psychology Review, 1*(3), 241-271.

Harrington, H., & Miller, N. (1993). Do group motives differ from individual motives? Considerations regarding process distinctiveness. In M. A. Hogg & D. Abrams (Eds.), *Group motivation: Social psychological perspectives* (pp. 149-172). New York: Harvester Wheatsheaf.

Higgins, E. T., King, G. A., & Mavin, G. H. (1982). Individual construct accessibility and subjective impressions and recall. *Journal of Personality and Social Psychology, 43*, 35-47.

Hunter, J. E., & Schmidt, F. L. (1990). *Methods of meta-analysis: Correcting error and bias in research findings.* Newbury Park, CA: Sage.

Jamieson, D. W., & Zanna, M. P. (1989). Need for structure in attitude formation and expression. In A. Pratkanis, S. Breckler, & A. G. Greenwald (Eds.), *Attitude structure and function* (pp. 46-68). Hillsdale, NJ: Lawrence Erlbaum.

Jordan, N. (1953). Behavioral forces that are a function of attitude and of cognitive organization. *Human Relations, 6*, 273-287.

Judd, C. M., & Kenny, D. A. (1981). *Estimating the effects of social interventions.* New York: Cambridge University Press.

Kreuger, J., & Clement, R. W. (1997). Estimates of social consensus by majorities and minorities: The case for social projection. *Personality and Social Psychology Review, 1*, 299-313.

Kruglanski, A. W. (1989). *Lay epistemics and human knowledge: Cognitive and motivational bases.* New York: Plenum.

Kruglanski, A. W. (1990a). Lay epistemic theory in social-cognitive psychology. *Psychological Inquiry, 1*, 181-197.

Kruglanski, A. W. (1990b). Motivations for judging and knowing: Implications for causal attribution. In E. T. Higgins & R. M. Sorrentino (Eds.), *The handbook of motivation and cognition: Foundations of social behavior* (Vol. 2, pp. 333-368). New York: Guilford.

Kruglanski, A. W., Atash, M. N., DeGrada, E., Mannetti, L., Pierro, A., & Webster, D. M. (1997). Psychological theory testing versus psychometric nay-saying: Comment on Neuberg et al.'s (1997) critique of the need for closure scale. *Journal of Personality and Social Psychology, 73,* 1005-1016.

Kruglanski, A. W., & Mackie, D. M. (1990). Majority and minority influence: A judgmental process analysis. In W. Stroebe & M. Hewstone (Eds.), *European review of social psychology* (Vol. 1, pp. 229-261). Chichester, UK: Wiley.

Kruglanski, A. W., & Thompson, E. P. (1999). Persuasion by a single route: A view from the unimodel. *Psychological Inquiry, 10*(2), 83-109.

Kruglanski, A. W., & Webster, D. M. (1996). Motivated closing of the mind: "Seizing" and "freezing." *Psychological Review, 103*, 263-283.

Kruglanski, A. W., Webster, D. M., & Klem, A. (1993). Motivated resistance to openness to persuasion in the presence or absence of prior information. *Journal of Personality and Social Psychology, 65*, 861-876.

Kuder, G. F., & Richardson, M. W. (1937). The theory of the estimation of test reliability. *Psychometrika, 2*, 151-160.

Marco, C. A., & Suls, J. (1993). Daily stress and the trajectory of mood: Spillover, response assimilation, contrast, and chronic negativity. *Journal of Personality and Social Psychology, 64*, 1053-1063.

Miller, N., & Pollock, V. E. (1994). Meta-analysis and some science-compromising problems in social psychology. In W. R. Shadish & S. Fuller (Eds.), *The social psychology of science* (pp. 230-261). New York: Guilford.

Moscovici, S. (1980). Toward a theory of conversion behavior. In L. Berkowitz (Ed.), *Advances in experimental psychology* (Vol. 13, pp. 209-239). New York: Academic Press.

Mullen, B., Brown, R., & Smith, C. (1992). Intergroup bias as a function of salience, relevance, and status: An integration. *European Journal of Social Psychology, 22*, 103-122.

Mullen, B., & Hu, L.-T. (1988). Social projection as a function of cognitive mechanisms: Two meta-analytic integrations. *British Journal of Social Psychology, 130*, 333-356.

Neuberg, S. L., Judice, T. N., & West, S. G. (1997). What the Need for Closure Scale measures and what it does not: Toward differentiating among related epistemic motives. *Journal of Personality and Social Psychology, 72*, 1396-1412.

Neuberg, S. L., & Newsom, J. (1993). Individual differences in chronic motivation to simplify: Personal need for structure and social-cognitive processing. *Journal of Personality and Social Psychology, 65*, 113-131.

Neuberg, S. L., West, S. G., Judice, T. N., & Thompson, M. M. (1997). On dimensionality, discriminant validity, and the role of psychometric analyses in personality theory and measurement: Reply to Kruglanski et al.'s (1997) defense of the Need for Closure Scale. *Journal of Personality and Social Psychology, 73*, 1017-1029.

Pacini, R., Muir, F., & Epstein, S. (1998). Depressive realism from the perspective of cognitive-experimental self-theory. *Journal of Personality and Social Psychology, 74*, 1056-1068.

Petty, R. E. (1994). Two routes to persuasion: State of the art. In G. d'Ydewalle, P. Eelen, & P. Berteleson (Eds.), *International perspectives on psychological science* (Vol. 2, pp. 229-247). Hillsdale, NJ: Lawrence Erlbaum.

Petty, R. E., & Cacioppo, J. T. (1986). The elaboration likelihood model of persuasion. In L. Berkowitz (Ed.), *Advances in experimental social psychology* (Vol. 19, pp. 123-205). San Diego: Academic Press.

Repetti, R. L. (1993). Short-term effects of occupational stressors on daily mood and health complaints. *Health Psychology, 12*, 125-131.

Richardson, M. W., & Kuder, G. F. (1939). The calculation of test-reliability coefficients based upon the method of rational equivalence. *Journal of Educational Psychology, 30*, 681-687.

Rokeach, M. (1960). *The open and closed mind.* New York: Basic Books.

Rosenblatt, A., Greenberg, J., Solomon, S., Pyszczynski, T., & Lyon, D. (1989). Evidence for terror management theory I: The effects of mortality salience on reactions to those who violate or uphold cultural values. *Journal of Personality and Social Psychology, 57*, 681-690.

Rosenthal, R. (1994). On being one's own case study: Experimenter effects in behavioral research—30 years later. In W. R. Shadish & S. Fuller (Eds.), *The social psychology of science* (pp. 214-229). New York: Guilford.

Roskos-Ewoldsen, D. R., & Fazio, R. H. (1992). The accessibility of source likability as a determinant of persuasion. *Personality and Social Psychology Bulletin, 18*, 19-25.

Ross, L., Greene, D., & House, P. (1977). The "false consensus effect": An egocentric bias in social perception and attributional processes. *Journal of Experimental Social Psychology, 13*, 279-301.

Sanford, R. N., Adorno, E., Frenkel-Brunswik, E., & Levinson, D. J. (1950). The measurement of implicit antidemocratic trends. In E. Adorno, E. Frenkel-Brunswik, D. J. Levinson, & R. N. Sanford (Eds.), *The authoritarian personality* (pp. 222-279). New York: Harper & Row.

Scher, S. J., & Cooper, J. (1989). Motivational basis of dissonance: The singular role of behavioral consequences. *Journal of Personality and Social Psychology, 56*, 899-906.

Schopler, J., & Insko, A. (1992). The discontinuity effect in interpersonal and intergroup relations: Generality and mediation. In W. Stroebe & M. Hewstone (Eds.), *European review of social psychology* (Vol. 3, pp. 121-152). New York: John Wiley & Sons.

Sears, D. O. (1983). The person-positivity bias. *Journal of Personality and Social Psychology, 44,* 233-250.

Sigall, H., & Mills, J. (1998). Measures of independent variables and mediators are useful in social psychology experiments: But are they necessary? *Personality and Social Psychology Review, 2*, 218-226.

Solomon, S., Greenberg, J., & Pyszczynski, T. (1991). A terror management theory of social behavior: The psychological functions of self-esteem and cultural worldviews. In M. E. P. Zanna (Ed.), *Advances in experimental social psychology* (Vol. 23, pp. 91-159). San Diego: Academic Press.

Steele, C. M., Spencer, S. J., & Lynch, M. (1993). Self-image resilience and dissonance: The role of affirmational resources. *Journal of Personality and Social Psychology, 64*, 885-896.

Stephan, W. G., & Stephan, C. W. (1985). Intergroup anxiety. *Journal of Social Issues, 41,* 157-175.

Stephan, W. G., & Stephan, C. W. (1996). *Intergroup relations.* Dubuque, IA: Brown & Benchmark.

Tellegen, A. (1982). *Brief manual for the multidimensional personality questionnaire.* Unpublished manuscript.

Thompson, M. M., Naccarato, M. E., & Parker, K. E. (1989, June). *Assessing cognitive need: The development of the Personal Need for Structure and Personal Fear of Invalidity scales.* Paper presented at the annual meeting of the Canadian Psychological Association, Halifax, Nova Scotia, Canada.

Thompson, M. M., Naccarato, M. E., Parker, K. E., & Moskowitz, G. (1993). *The development and validation of the Personal Need for Structure (PNS) and Personal Fear of Invalidity (PFI) measures.* Unpublished manuscript, New York University.

Travers, R. M. W. (1941). A study in judging the opinions of groups. *Archives of Psychology* (Whole No. 266).

Urban, L. M., & Miller, N. (1998). A theoretical analysis of crossed categorization effects: A meta-analysis. *Journal of Personality and Social Psychology, 74*, 894-908.

van Praag, H. M. (1993). *"Make-believes" in psychiatry or the perils of progress.* New York: Brunner/Mazel.

Wallen, R. (1943). Individuals' estimates of group opinion. *Journal of Social Psychology, 17*, 269-274.
Webster, D. M. (1993). Motivated augmentation and reduction of the overattribution bias. *Journal of Personality and Social Psychology, 65*, 261-271.
Webster, D. M., & Kruglanski, A. W. (1994). Individual differences in need for cognitive closure. *Journal of Personality and Social Psychology, 67*, 1049-1062.
Wegner, D. M. (1994). Ironic processes of mental control. *Psychological Review, 101*, 34-52.
Wilder, D. A. (1993). The role of anxiety in facilitating stereotypic judgments of outgroup behavior. In D. M. Mackie & D. L. Hamilton (Eds.), *Affect, cognition, and stereotyping: Interactive processes in group perception* (pp. 87-109). San Diego: Academic Press.
Wilder, D. A., & Shapiro, P. (1989). Role of competition-induced anxiety in limiting the beneficial impact of positive behavior by an outgroup member. *Journal of Personality and Social Psychology, 56*, 60-69.
Wilder, D. A., & Shapiro, P. (1991). Facilitation of outgroup stereotypes by enhanced ingroup identity. *Journal of Experimental Social Psychology, 27*, 431-452.
Wilder, D. A., & Simon, A. F. (1996). Incidental and integral affect as triggers of stereotyping. In R. M. Sorrentino & E. T. Higgins (Eds.), *Motivation and cognition: Vol. 3. The interpersonal context* (pp. 397-422). New York: Guilford.
Word, C. O., Zanna, M. P., & Cooper, J. (1974). The nonverbal mediation of self-fulfilling prophecies in interracial interaction. *Journal of Experimental Social Psychology, 10*, 109-120.
Zajonc, R. B. (1968). Cognitive theories in social psychology. In G. Lindzey & E. Aronson (Eds.), *The handbook of social psychology* (Vol. 1, pp. 320-412). Reading, MA: Addison-Wesley.

CHAPTER 4

Statistical Conclusion Validity for Intervention Research

A Significant (p < .05) Problem

Mark W. Lipsey

When conducting evaluative research on the effectiveness of intervention, a defining moment comes when the statistical analysis reveals whether the differences between treatment and control groups on the key outcome variables are statistically significant. If statistical significance is attained, the researcher has a warrant to draw positive conclusions about the impact of the intervention on the conditions it was attempting to change. Although such conclusions may be qualified by uncertainties associated with measurement, design, or other methodological issues, the court of scientific opinion nonetheless permits an argument to be made in support of the conclusion that the intervention was effective. If the methods are strong and convincing, the discussion may even move to matters of how large the effects were and their practical importance within the domain of application.

If statistical significance is not attained, however—even if it falls short of the conventional $p < .05$ level by a hair—any claim that the evidence shows intervention effects is judged to be specious. Indeed, the researcher who continued

---

AUTHOR'S NOTE: The data on which this chapter is based were collected with the support of grants from the National Institute of Mental Health (MH39958, MH42694, and MH51701) and the Russell Sage Foundation.

to press such a claim would generally be viewed as naive, at best, and, if too vigorous, maybe as biased enough to be trying to cheat on the rules of the game.

Significance testing, therefore, marks a fork in the road for intervention research that has very pronounced implications for the interpretation of the effects of the intervention under investigation. As with any decision-making procedure, of course, there is a possibility of error. An analysis might show statistical significance when, in fact, there were no meaningful intervention effects, or fail to reach significance when there were. The validity of the statistical conclusion about the relationship of the independent and dependent variables is what Cook and Campbell (1979) called *statistical conclusion validity.* They described the situation as follows:

> Covariation is a necessary condition for inferring cause, and practicing scientists begin by asking of their data: "Are the presumed independent and dependent variables related?" Therefore, it is useful to consider the particular reasons why we can draw false conclusions about covariation. We shall call these reasons (which are threats to valid inference-making) threats to *statistical conclusion validity*, for conclusions about covariation are made on the basis of statistical evidence. (p. 37)

Statistical conclusion validity thus was among the four types of validity Don Campbell discussed as relevant to experimental and quasi-experimental design in field settings. Among the four, however, it has received the least attention in his writings. On that basis, one might conclude that Campbell thought it was relatively unimportant or unproblematic. Indeed, within the historical context of his seminal volumes on quasi-experimental design (Campbell & Stanley, 1966; Cook & Campbell, 1979), there was little reason to believe that statistical conclusion validity was as troublesome in intervention research as internal validity or external validity.

Within recent decades, however, evidence has mounted that the statistical conclusion validity of much intervention research is very questionable and does not justify complacency. The most dramatic indications of serious problems have come from the findings of the increasing volume of meta-analysis of experimental and quasi-experimental intervention research. One purpose of this chapter is to present some of that meta-analytic evidence and thus add another voice to the chorus provided by other meta-analysts who have addressed this issue (e.g., Rosenthal & Rubin, 1985; Schmidt, 1992). In addition, the results of selected meta-analyses will be used to explore some statistically straightforward approaches to improving the statistical conclusion validity of intervention research.

There is perhaps no fuller tribute I can make to Don Campbell than my certainty that his response to a critical probing of a concept he popularized would be welcoming, even enthusiastic. As I heard him say on more than one

TABLE 4.1 The Probability of Correct and Incorrect Conclusions in Statistical Significance Testing for Treatment vs. Control Group Differences

	Population Circumstances	
Results of Significance Test on Sample Data	*Treatment and Control Means Differ (Null Hypothesis False)*	*Treatment and Control Means Do Not Differ (Null Hypothesis True)*
Significant difference (null hypothesis rejected)	Correct conclusion $(1 - \beta)$	Type I Error(α)
No significant difference (null hypothesis not rejected	Type II Error (β)	Correct conclusion $(1 - \alpha)$

occasion to someone voicing a critical view, "Pursue that—we need it for the dialogue!" For Don Campbell, knowledge, always tentative, was something hard won through the exertions of a "disputatious community of scholars."

ERRORS IN STATISTICAL CONCLUSIONS

The nature of the statistical conclusions that can be reached through statistical significance testing and the possible errors in those conclusions are well defined and can be represented in the familiar form shown in Table 4.1 (I assume a simple comparison between a treatment and a control group). A statistical test at a predetermined confidence level (usually .95, i.e., alpha = .05) either serves to reject the null hypothesis that the difference between the groups on their mean scores for a given dependent variable is zero, or it does not. The decision on this matter represents the statistical conclusion whose validity is of concern.

Unknown to the researcher (else why do the research?), the difference between the means for the entire population that the research sample is assumed to represent, compared with and without treatment, may actually be zero, or it may not. Any discrepancy between the mean difference observed on the sample and that which would be observed on the population if its scores could be obtained and compared under treatment and control conditions is assumed to come from sampling and measurement error. Statistical tests are based on assumptions about the nature of the sampling error and attempt to estimate the range of observed effects most probable if the true difference is assumed actually to be zero. If the probability of the observed effect occurring under this null hypothesis is sufficiently small (e.g., less than 1 chance in 20), the observed difference is assumed to most likely represent a "real" effect of treatment and not simply the "luck of the draw."

Although a review of these concepts helps set the stage for discussion, they are quite familiar to researchers. The application of this commonplace statistical inference framework to applied intervention research, however, has some important implications that are not well recognized and are worth highlighting. Consider, for instance, the risk of drawing an erroneous conclusion from significance testing. The probability of a Type I error is quite circumscribed in conventional statistical testing. If the critical assumptions of the statistical test are met, it is determined by the confidence level the researcher selects. Conventionally, this is 95%, permitting a .05 probability of a Type I error.

The risk of a Type II error, on the other hand, cannot be constrained by adopting some appropriate statistical criterion. It is chiefly a function of the magnitude of the population-level difference between the means (the effect size), the amount of measurement error in the observations upon which those means are based, and the size of the sample on which the difference is observed in the research study. The population-level effect is, of course, unknown and, in any event, not amenable to manipulation by the researcher. Measurement error may be unknown and, in any event, difficult to diminish. Sample size can, in principle, be easily manipulated by the researcher, though in practice there are often constraints on the availability of subjects, access to them, or the resources for including them in the study.

The risk of Type II error, then, is difficult for a researcher to control and, indeed, difficult to even appraise because it depends, in part, on the unknown population effect size. As a result, the probability of Type II error can easily be quite high in an intervention study, say 50% or higher (Hunter, Schmidt, & Jackson, 1982; Lipsey, 1990). This may come as a surprise to many researchers who, by focusing on Type I error when stipulating an alpha level for significance testing, believe they have ensured a high level of statistical conclusion validity. In the process, they may overlook situations in which the risk of Type II error is excessive in relation to the purposes of the study.

Another important aspect of statistical inference is that in any given intervention study, only *one* of the two forms of error is possible. That is, the statistical conclusion is at risk of *either* Type I error or Type II error, but not both. This is because the likelihood of each of these errors is a conditional probability that depends on the population circumstances. If the null hypothesis of no difference between the means is true in the population, then the only statistical conclusion error that can be made is a Type I error. On the other hand, if the null hypothesis is false in the population, only Type II error is possible. Thus, in circumstances in which the null hypothesis is false, statistical conclusion validity is entirely a function of the probability of Type II error, which, as noted above, may be relatively unconstrained and quite large if sample size and/or population effect size are unfavorable.

This aspect of statistical inference is especially threatening to applied intervention research, which is generally conducted to test the effectiveness of a practical treatment believed to have potential benefits. In this circumstance, the research hypothesis (in contrast to the statistical null hypothesis) is that the population effect is not zero; furthermore, establishing that this is the case may be important. If the treatment is indeed effective, then the only threat to statistical conclusion validity is a Type II error. If the risk of Type II error is large, therefore, the researcher can easily fail to find statistical significance for an important effect.

In evaluation research and related applied intervention research, there are often stakeholders who have invested considerable psychological, political, and financial resources in the belief that the treatment is effective. For the researcher to fail to detect meaningful effects in such circumstances because of Type II error is a grievous error that may have serious ramifications for program personnel, sponsors, beneficiaries, and society. It is little comfort to the affected parties that, technically, failure to reject the null hypothesis does not affirm that it is true, only that it cannot be assumed false. Research purportedly designed to assess intervention effects that then reports that the effects were not statistically significant will be widely interpreted as showing "no effects" irrespective of any disclaimers based on the nuances of statistical inference that the researcher might make.

In short, there is good reason to believe that statistical conclusion validity could often be weak in intervention research and that the consequences could be serious. The major threat to statistical conclusion validity, however, is not the familiar risk of Type I error that researchers routinely constrain to low levels by setting alpha equal to .05 or less when they do significance testing. The threat is from Type II error, the risk of which easily can be large even in otherwise well-designed intervention research. This is basically a problem of statistical power. Statistical power is the probability of rejecting the null hypothesis when, in fact, it is false; as such, it is defined as $1 - \beta$, where β is the probability of a Type II error (just as α is the probability of a Type I error; see Table 4.1). The issue of statistical conclusion validity with which intervention researchers must be most concerned, therefore, has to do with studies that have too little statistical power to detect intervention effects of a practically meaningful magnitude.

It follows that one important form of meta-evaluation of intervention research is to examine the Type II error rates that occur in practice to ensure that researchers are achieving reasonable levels of statistical conclusion validity. A good part of the problem of statistical conclusion validity, however, stems from the difficulty of assessing how good it is in typical bodies of intervention research. As noted, Type I and Type II error rates are defined around the population values of treatment effects, in particular, whether differences between treatment and control means are zero or meaningfully greater than zero. Population-level effects, of course, not only are unobserved but also are the very things intervention research is attempting to investigate. Without a god's-eye view of population effect sizes, therefore, it is almost impossible to make any empirical assessment of Type II error rates and typical levels of statistical conclusion validity.

It is at this point that the large cumulation of meta-analysis research is helpful. It may not provide a god's-eye view on "true" intervention effects, but it does yield estimates of population-level effects in many areas of intervention research that can be used to assess the statistical conclusion validity of that research. In what follows, an attempt is made to do just that.

META-ANALYSIS ESTIMATES OF POPULATION-LEVEL EFFECT SIZES

In a meta-analysis of intervention research, effect size statistics are computed for each study to represent the difference between treatment and control group means on the dependent variables of interest. Because the dependent variables in different studies rarely are measured on the same scale, the differences are standardized so that effect size statistics can be meaningfully aggregated and compared in various ways within and across studies. In particular, the effect size statistic used for this purpose is defined as $ES = (M_t - M_c)/s_p$, where M_t and M_c are the means of the treatment and control group, respectively, and s_p is the pooled standard deviation of those groups.

Each selected study contributes one or more observed effect sizes to a meta-analysis. By averaging over a number of individual studies in an intervention area, we presumably get a better indication of the population value which that type of intervention study is attempting to estimate than can be derived from any one study alone. The tactic I will use to examine statistical conclusion validity is based on this presumption. In a given meta-analysis of a set of studies relating to one type of intervention, I will take the mean effect size over all the studies in the set as an estimate of the population value and then examine the statistical conclusions of the individual studies with regard to how "correct" they are in relation to that estimated population value.

With this procedure, I can approximate the conditional probability matrix in Table 4.1. The mean effect sizes values will be used to categorize sets of studies according to whether the null hypothesis is true or false in the respective populations. Then the proportions of individual studies rejecting and not rejecting the null hypotheses will be examined in each instance so that some empirical estimate of Type I and Type II error rates can be made for samples of actual intervention research.

If the effect size estimates in the individual studies in a meta-analysis differ from the population value only by sampling error, and the pooled sample size across all the studies in each meta-analysis is relatively large, the above procedure will provide a very good assessment of statistical conclusion validity. In most instances, we cannot be sure these conditions are met. The pooled sample sizes for the meta-analyses of interest are generally large enough to give reasonable estimates, but there is good reason to believe that effect size estimates vary across studies by more than sampling error. Differences in method, procedure, specific nature of the samples, and other such factors also contribute to between-study variance in the effect size estimates. Although homogeneity tests are available with which meta-analysts can test for variance greater than sampling error, they are not used or reported in many meta-analyses.

Explorations of statistical conclusion validity using meta-analysis results, therefore, must be viewed as approximate and interpreted with caution. Variation across studies that derives from sources other than sampling error may represent nuisance variance of a sort that acts like sampling error to obscure actual intervention effects. It can also represent real differences in those intervention effects that are associated with undocumented treatment variation, dosage differences, and the like, and, to that extent, may not represent error variance at all. Nonetheless, given the difficulty of configuring any empirical examination of the statistical conclusion validity of actual intervention studies, it will be instructive to examine even crude indicators.

The meta-analysis data that will be used for this purpose come from a pool of more than 300 intervention meta-analyses my colleagues and I have collected in recent years (Lipsey & Wilson, 1993). These cover areas of psychological, educational, and behavioral interventions (e.g., psychotherapy, computer-based instruction, parent-effectiveness training, treatment programs for juvenile delinquents, smoking cessation programs, pain management interventions). For a subset of the meta-analyses in this collection, information relating to statistical significance and/or statistical power (e.g., sample size) is reported.

EMPIRICAL ESTIMATES OF TYPE I AND TYPE II ERROR RATES

Among the meta-analyses described above, 60 reported the proportion of the studies in the meta-analysis for which statistical significance was found. Another 81 meta-analyses provided enough information about the sample sizes of the included studies and the across-study variance in observed effect sizes so that the proportion of studies that were statistically significant could be estimated fairly well. Taking the mean effect size (ES) over all the studies included in each of these 141 meta-analyses as an approximation of the population effect size on the dependent variables represented in that intervention area, we can ask how often the individual studies found statistical significance when that population effect size was estimated to be zero (null hypothesis true) and when it was estimated to be nonzero (null hypothesis false).

Here, however, we encounter a problem. The null hypothesis is almost never literally true in the sense that the population effect is exactly zero. Similarly, the mean effect sizes found in meta-analysis are rarely exactly zero. For practical considerations of statistical conclusion validity, what is important is not that there be a high probability of correctly detecting *any* departure from a zero effect, no matter how trivial. The proper goal of intervention research is to detect any effects of meaningful magnitude relative to the nature of the intervention and the conditions it addresses.

To further pursue an assessment of statistical conclusion validity, therefore, I must adopt some threshold effect size that can be expected to be practically meaningful in typical intervention research. Practical significance, however, is largely a function of the nature of the outcome construct and its operationalization, the importance of changes in the target conditions, and a host of other such matters that cannot easily be generalized across intervention areas. As a result, there is, at best, only a loose relationship between the magnitude of statistical effects and their practical significance (Abelson, 1985; Lipsey, 1990).

To estimate statistical conclusion error rates, therefore, a somewhat arbitrary criterion, or perhaps several, must be set for the statistical effect size that intervention research should be expected to detect. For this purpose, I turn to effect size displays such as the binomial effect size display (BESD) (Rosenthal & Rubin, 1982) and those described by Cohen (1988). These represent standardized mean difference effect size values (and other effect size statistics) in terms of the proportions of treatment and control groups above an arbitrary "success" threshold.

Figure 4.1, for instance, is based on the indicator of overlap between distributions that Cohen (1988) called U3. It shows two normal distributions of dependent variable scores in which the mean for the treatment population is higher than that for the control population. Larger differences between the means, of course, produce greater separation between the distributions and correspond to larger effect sizes. Suppose, for instance, that we set a threshold value at the mean of the control group (μ_c in Figure 4.1) so that 50% of that group has scores below that value and 50% above. We can now ask what effect size value would correspond to, say, an increase of 5 percentage points in the number of persons with scores above that level (i.e., 45% below the threshold and 55% above). That effect size, as shown in the tabular summary in Figure 4.1, is a little more than .10 (i.e., one-tenth of a standard deviation difference between the mean of the treat-

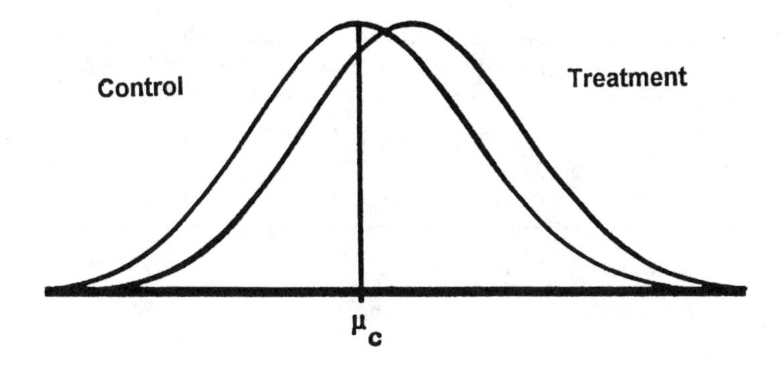

Effect Size	% of Tx Population Above μ_c	Increase in "Success" Rate
.10	54	8%
.20	58	16%
.30	62	24%
.40	66	32%
.50	69	38%

Figure 4.1. The Proportion of the Treatment Population Above a Success Threshold Set at the Mean (μ_c) of the Control Distribution for Different Effect Size Values

ment distribution and that of the control distribution). Note that an increase from 50% to 55% in the proportion of "successes" in this instance is actually a 10% improvement over baseline (.05/.50). If the "success" threshold is set further out in the control group distribution, so that a smaller proportion of that group is presumed to be successful prior to intervention, the result is to decrease the effect size value associated with a given percentage point success differential.

As a generality, an effect size that represents a 10% increase in the proportion of persons above a success threshold seems like it could be at least marginally meaningful in many intervention contexts. Given these relationships between effect size and success rate differentials, therefore, one might argue that effect size values of .10 and larger could easily be of practical significance. For purposes of estimating the error rates for statistical conclusions in intervention research, therefore, I will consider effect sizes of .09 and under as small enough to be virtually "null" and deem statistical significance in such cases to be a Type I error. For effect sizes of .10 and larger, divided into intervals according to how much larger, I will judge failure to find statistical significance as a Type II error.

TABLE 4.2 Mean Proportions of Studies With Significant and Nonsignificant Findings Within Meta-Analyses Showing Different Mean Effect Sizes

	Range of Mean Effect Sizes for Meta-Analyses				
	–.09 to .09	*.10 to .19*	*.20 to .49*	*.50 to .99*	*1.00 to 1.60*
Proportion significant	**.22**	.35	.46	.57	.68
Proportion nonsignificant	.78	**.65**	**.54**	**.43**	**.32**
Number of meta-analyses	5	12	55	56	13
Number of studies	286	367	2,606	2,933	522
Mean subject *n* per study	213	294	137	88	47

Although this is admittedly arbitrary, it affects the consideration of error only for very small estimated population effects. The matter of primary interest in this discussion is relatively large population effects, which intervention research should be expected to detect in any event.

Table 4.2 shows the estimated error rates (in boldface type) under this procedure for the 141 meta-analyses reporting information permitting the proportion of individual studies reaching statistical significance to be estimated. The mean effect size found in each of these 141 meta-analyses was used to estimate the population effect size for the respective intervention area and classify the studies included in the meta-analysis into the effect size ranges shown in Table 4.2. These meta-analyses included 6,714 individual intervention studies that either attained statistical significance or did not. For each effect size range, the respective proportions are shown.

In the smallest mean effect size category (between –.09 and +.09), assumed here to represent virtually null relationships, the estimated Type I error rate is .22. This describes situations in which statistical significance was found for what are assumed to be very small population effect sizes. This proportion is considerably higher than the .05 level expected from conventional alpha levels. Of course, though small, many of the effect sizes in this range are not actually zero, in which case the statistical testing in the individual studies is not actually in error in rejecting the null hypothesis. Nevertheless, given the mean sample sizes reported, the high proportion of these studies finding statistical significance is somewhat surprising. This finding may be an artifact of using meta-analysis data, which tend to overrepresent published studies because they are easier to locate. Published studies, in turn, tend to overrepresent statistically significant findings (Begg, 1994; Greenwald, 1975).

More interesting for present purposes are the estimated Type II error rates for different effect size ranges. For mean effect sizes in the .20-.49 range, the Type II error rate was estimated at .54; when the mean effect sizes are in the range of .50-.99, the Type II error rate was estimated at .43. These are strikingly high error rates. Intervention researchers could attain about the same accuracy simply by flipping a coin and declaring a significant effect on heads and nonsignificance on tails. Furthermore, these are effect size ranges within which many intervention studies apparently produce effects. As Table 4.2 shows, 5,539 (or 82%) of the 6,714 individual studies described there were found in meta-analyses reporting mean effect sizes between .20 and .99.

For mean effects greater than 1.00—that is, with more than a standard deviation's separation between treatment and control group means—the Type II error rate was estimated at 32%. That is, even with the meta-analysis showing a quite large intervention effect, on average, for these studies, nearly one third of them failed to find statistically significant differences.

HOW MIGHT STATISTICAL CONCLUSION VALIDITY BE IMPROVED?

At this point, it is rather obvious that statistical conclusion validity could well be a serious problem in practical intervention research. The proportion of studies attaining statistical significance in intervention areas where meta-analytic results show meaningful mean effects are well below any reasonable standard. The implication of this finding is that researchers often fail to find statistical significance when the intervention effects they are investigating may, in fact, be large; that is, they are often making Type II errors in their statistical conclusions. Indeed, in many cases, the conclusions about the significance of effects drawn from statistical testing appear to be no more accurate than a coin flip. In evaluation research and other such situations where the interventions under study represent appreciable social investments directed at important social problems, a Type II error makes a genuinely effective program appear to be a failure or, at best, of uncertain value.

Needless to say, this is not a satisfactory situation, but what can be done about it? Some have called for an end to statistical significance testing (e.g., Hunter, 1997), and the analysis here certainly supports the view that it can be very problematic. That in itself offers no solution, however. Experimental and quasi-experimental data do not speak for themselves; they must be interpreted with regard to what conclusions they support and how convincingly. With or without significance testing, intervention researchers need a credible procedure for assessing an observed difference between the means of a treatment and a control group and judging whether this difference represents a meaningful difference in the two groups.

A more conventional view is that intervention researchers must simply improve the design of their studies so that they have adequate statistical power and, hence, lower Type II error rates. To be sure, there is much that can be done to improve on present design practices, not all of which require larger sample sizes than may be available to the researcher (Lipsey, 1990, 1998). The problems of statistical conclusion validity identified in this analysis, however, have been around a long time (Cohen, 1962, 1994; Sedlmeier & Gigerenzer, 1989), and there is little indication that researchers have reacted by giving more attention to statistical power in their intervention designs. Part of the difficulty is that designing intervention research with sufficient power is a rather complex endeavor. To be done well, it requires a depth of knowledge about the characteristics of the research samples, independent and dependent variables, and magnitude of likely population effect sizes that is not often available to the researcher.

At least in applied areas where the consequences of Type II error can be very serious, the best approach may be to consider alternatives to statistical significance testing that provide more accurate statistical conclusions when applied to the same data. Ideally, such alternatives would be straightforward to apply and would not require new and unfamiliar statistical concepts and procedures.

The aspect of conventional significance testing that is most responsible for the problems of statistical conclusion validity shown above is the favored status of the null hypothesis. Null hypothesis testing presumes that the population effect size is zero and permits a contrary statistical conclusion only when that presumption can be shown, on the basis of the observed data, to be very unlikely. This is a rather conservative procedure for many circumstances of intervention research. It amounts to a presumption of guilt (failure) unless proven innocent.

Several straightforward ways to draw statistical conclusions on a less conservative basis while still staying within the framework of conventional statistical theory are worth considering. I will examine three alternatives to null hypothesis testing as a basis for drawing statistical conclusions and apply each to the meta-analysis data above to assess its ability to improve statistical conclusion validity for applied intervention research.

1. Point Estimates of Effect Size

In any intervention study based on sample data, the single best estimate of the magnitude of the population effect from a statistical standpoint is the observed effect size, that is, the effect size value estimated from the sample data. One approach to statistical conclusions, therefore, would be to compute the observed effect size and appraise it directly. If it is nonzero, the conclusion that the intervention has an effect is considered to be supported. Of course, the observed effect size will rarely be exactly zero, so this procedure will yield a very high rate of positive conclusions and a corresponding high rate of Type I errors.

TABLE 4.3 Mean Proportions of Studies Within Meta-Analyses Above and Below Different Effect Size Values

	Range of Mean Effect Sizes for Meta-Analyses				
	–.09 to .09	*.10 to .19*	*.20 to .49*	*.50 to .99*	*1.00 to 1.60*
Over ES = .00	.53	.73	.77	.84	.88
Under ES = .00	.47	.27	.23	.16	.12
Over ES = .05	.48	.67	.74	.83	.87
Under ES = .05	.52	.33	.26	.17	.13
Over ES = .10	.43	.59	.71	.81	.86
Under ES = .10	.57	.41	.29	.19	.14
Over ES = .20	.34	.43	.63	.77	.84
Under ES = .20	.66	.57	.37	.23	.16

A better version of this approach would be to first determine the threshold effect size value that distinguishes meaningfully large effects from trivial ones in the intervention context at issue. The conclusion about the "significance" of the observed effect size would then be made according to whether it was smaller or larger than that criterion value. Determining what constitutes a meaningful effect is a task that could well be salutary for intervention researchers, and reasonable approaches are available (Jacobson & Truax, 1991; Lipsey, 1990; Sechrest & Yeaton, 1982). Absent more discriminating criteria, conventions might be adopted that are based on the proportions of control and treatment populations above arbitrary "success" thresholds, as shown in Figure 4.1. For instance, an effect size of .20, corresponding to a 16% increase in the number of "successes," might be taken as the presumptive threshold for concluding that an effect has been found.

Table 4.3 shows how the studies included in the 141 meta-analyses would be classified in a decision scheme using effect size values of .00, .05, .10, and .20 as threshold values with which to compare the observed effect size in an individual study. In this scheme, the statistical conclusion that an effect has been detected is justified by an observed effect size equal to or larger than the designated threshold level but not by one that falls below that level.

A comparison of Table 4.3 with Table 4.2 shows first that the estimated Type I error rates, already high in Table 4.2, rise even higher under this threshold decision scheme. This is hardly surprising because the nature of this scheme is rather liberal in the standards required to support a claim that an effect has been detected.

Because the meta-analyses indicate that most intervention effects are actually positive and in the range of .10 and above, however, this threshold scheme for drawing statistical conclusions does result in considerably lower Type II error rates than those shown in Table 4.2 for conventional significance testing. In most instances, therefore, an intervention researcher would be more accurate in drawing statistical conclusions by directly judging the observed effect size against one of these criterion values than by conducting a significance test. Despite this, the error rates shown in Table 4.3 are still rather large, especially when the threshold for the observed effect size is not made trivially close to zero. For example, when population effect sizes are assumed to be in the .20 to .99 range, where the vast majority of meta-analysis results appear, comparison of the effects observed in individual studies with the modest ES = .20 threshold value still produces Type II error rates of more than 20%.

2. Relaxed Alpha

Using point estimates of effect sizes as the basis for statistical conclusions has the advantage of focusing attention on the observed effect size and the question of what effect magnitude is meaningful in the intervention context. It does not take into account, however, the degree of statistical uncertainty in the observed effect size. Thus, an observed effect size of .40 based on a sample of 20 subjects would have the same standing with regard to the statistical conclusions as one of .40 based on 200 subjects, despite its greater instability.

Within conventional significance testing, the simplest strategy for taking sampling error into consideration while lowering Type II error rates is to relax the alpha level at which significance is tested. Thus, in situations where statistical power would otherwise be marginal for effects of meaningful magnitude, Type II error rates are improved by tolerating more risk of Type I error. For instance, if statistical testing is done at, say, alpha = .20 instead of .05, statistical significance becomes "easier" to attain and Type II error rates will be lower when population effects are indeed present. Put another way, this approach accepts .80 confidence regarding Type I error as sufficient, given the greater protection against Type II error this might afford.

Table 4.4 shows the Type I and Type II error rates estimated for the studies in the 141 meta-analyses when different alpha levels are used for significance testing. These values are derived from calculations based on the observed effect sizes and mean sample sizes reported for these meta-analyses. At alpha = .05, this procedure produces the same error rates as shown in Table 4.2, as expected because Table 4.2 represents conventional statistical testing at the conventional alpha level.

At alpha levels of .10 through .30, however, Type I error rates increase and Type II error rates decrease, as required by statistical theory. What is surprising

TABLE 4.4 Mean Proportions of Studies Within Meta-Analyses Significant and Nonsignificant at Different Alpha Levels

	Range of Mean Effect Sizes for Meta-Analyses				
	–.09 to .09	*.10 to .19*	*.20 to .49*	*.50 to .99*	*1.00 to 1.60*
Significant at .05	.22	.35	.46	.57	.68
Nonsignificant at .05	.78	.65	.54	.43	.32
Significant at .10	.25	.41	.52	.63	.73
Nonsignificant at .10	.75	.59	.48	.37	.27
Significant at .15	.28	.44	.55	.66	.75
Nonsignificant at .15	.72	.56	.45	.34	.25
Significant at .20	.30	.47	.58	.69	.77
Nonsignificant at .20	.70	.53	.42	.31	.23
Significant at .25	.32	.50	.60	.71	.78
Nonsignificant at .25	.68	.50	.40	.29	.22
Significant at .30	.34	.52	.62	.72	.79
Nonsignificant at .30	.66	.48	.38	.28	.21

is how little change there is. Even with significance testing at alpha = .30, which is uncomfortably far from conventional levels, Type II error rates for effect sizes in the .20 to .99 range are still nearly 30% or higher. Although attractive in principle, therefore, increasing alpha to as much as six times the conventional .05 level simply does not make enough difference in the error rates to be very promising.

3. "Goodness of Fit"

A common suggestion for improving the presentation of statistical conclusions is for researchers to report confidence intervals around the observed effect size rather than whether or not significance was attained (Cohen, 1994; Loftus, 1996; Reichardt & Gollob, 1997). If a 95% confidence interval includes zero, of course, this is equivalent to a finding of nonsignificance. Nonetheless, the confidence interval has the advantage of showing the range of nonzero effect sizes that are also consistent with the observed effect size and expected likelihood of sampling error. When rather large and potentially meaningful effect sizes are included in the confidence interval along with zero, there is some justification for concluding that some of those values are at least as likely to characterize the population effect (Rosenthal & Rubin, 1994).

One straightforward approach to modifying conventional statistical significance testing to yield improved Type II error rates in intervention research, therefore, is to focus on the full range of effect size values encompassed in the confidence interval. For instance, researchers might determine in advance what minimal effect size value represents a meaningful treatment effect, as was suggested earlier in the discussion of point estimates of effect size. This threshold value, of course, must be based on a credible argument about what constitutes a "significant" effect in the intervention context. Alternatively, a defensible statistical convention might be serviceable for setting a threshold if no other basis can be established. Using the overlapping distributions framework shown in Figure 4.1, for instance, an effect size of .20 might be defended as a default threshold value on the grounds that it represents a nontrivial 16% increase in the success rate.

Given an appropriate threshold value, this approach to drawing statistical conclusions is very straightforward. It requires only that a conventional .95 confidence interval be put around the observed effect size. If that confidence interval includes zero, of course, the effect has not reached statistical significance by the usual standards and the associated uncertainty of the effect size estimate must be recognized. If the confidence interval also includes the specified threshold value, however, the hypothesis that a meaningful intervention effect has occurred is not rejected. Thus, the null hypothesis may not be rejected, but neither is the hypothesis that the intervention has produced an effect as large as the predetermined criterion value.

Of course, if the confidence interval includes zero but *not* the specified criterion effect size value, the statistical evidence clearly would not support a claim of meaningful intervention effects. Finally, if the confidence interval includes or exceeds the criterion value and does not include zero, the null hypothesis is rejected by conventional criteria; this outcome produces the least degree of uncertainty about the magnitude of effects.

This procedure applies a "goodness of fit" logic to observed effects. The critical question is not whether the statistical evidence is clearly inconsistent with the null hypothesis but whether it is consistent with an appropriately formulated non-null hypothesis. Additional discussion of this logic can be found in Abelson (1997), Murphy and Myors (1988), and Rogers, Howard, and Vessey (1993).

Table 4.5 applies this "goodness of fit" procedure to the data from the 141 meta-analyses. The descriptive data about observed effect sizes and sample sizes from the meta-analyses were used to estimate the proportion of the individual studies for which the 95% confidence intervals would include various specified effect size values. When that criterion value was within the confidence interval, the statistical conclusion was that an effect had been detected. When that criterion value was outside the confidence interval, the conclusion was that no effect had been detected.

TABLE 4.5 Mean Proportions of Studies Within Meta-Analyses With .95 Confidence Intervals That Include Different Effect Size Values

	Range of Mean Effect Sizes for Meta-Analyses				
	–.09 to .09	*.10 to .19*	*.20 to .49*	*.50 to .99*	*1.00 to 1.60*
Confidence interval includes ≥ .20	.67	.79	.86	.92	.94
Confidence interval below .20	.33	.21	.14	.08	.06
Confidence interval includes ≥ .25	.64	.74	.84	.91	.94
Confidence interval below .25	.36	.26	.16	.09	.06
Confidence interval includes ≥ .40	.52	.54	.76	.87	.92
Confidence interval below .40	.48	.46	.24	.13	.08
Confidence interval includes ≥ .50	.44	.38	.69	.84	.91
Confidence interval below .50	.56	.62	.31	.16	.09

Comparison with the previous tables indicates that this procedure yields Type I error rates that are generally higher than the other approaches examined. In the critical cases where the population effects are assumed to be .20 or greater, however, this procedure results in lower Type II error rates than those other approaches. Indeed, in that effect size range, the Type II error rates are generally below 20%. Although still well short of the 5% error rate conventional for Type I error, 20% is still a great improvement over the Type II rates shown earlier in Table 4.2. Even when the criterion level for a "significant" effect size is set as high as .50, the Type II error rates under this procedure are still much better than those shown for conventional statistical significance testing.

Of the variants of significance testing statistics examined here, therefore, the application of confidence intervals under a "goodness of fit" logic seems the most promising. This approach provides conventional significance testing information by revealing whether the confidence interval includes zero. At the same time, however, it gives the researcher the opportunity to draw statistical conclusions about the likelihood that the intervention effect is meaningfully large and, hence, make a more balanced judgment about whether a "significant" effect has been found.

CONCLUSION

Intervention researchers simply cannot take statistical conclusion validity for granted. As this chapter has argued, there is ample reason to believe that the heavy use of conventional significance testing in intervention research has resulted in unacceptable levels of statistical conclusion error, particularly Type II error. The meta-analysis findings reviewed above provide an empirical perspective on statistical conclusion validity that is otherwise hard to obtain. What they reveal is quite unsettling. The statistical conclusions drawn from conventional significance testing appear to be wrong at least as often as they are right. When the consequence is to undermine intervention programs that, in fact, are effective, research becomes a source of misinformation rather than enlightenment.

The various suggestions for alternative approaches made above all provide more accurate statistical conclusions than conventional procedures. The most important concept that the better of these approaches involve is specification of the population effect size the research is attempting to detect. Conventional null hypothesis testing recognizes only two categories—zero and not zero—with a built-in bias toward zero. The key to improving the statistical conclusion validity of intervention research is for researchers to develop a better understanding of what statistical effect size is meaningful in an intervention context. Although this is not an easy determination to make, it is far from impossible, and much good advice is already available (e.g., Jacobson & Truax, 1991; Lipsey, 1990; Sechrest & Yeaton, 1982). With a meaningful effect size criterion value specified, the researcher is in a much stronger position to design research with adequate power and to adapt statistical significance testing to give that criterion value the same consideration as the null hypothesis.

Among Don Campbell's many contributions to the study of intervention was the development of the concepts of internal validity, external validity, construct validity of the treatment, and statistical conclusion validity as major criteria for assessing the quality of experimental and quasi-experimental research. Of these, he gave the fullest expression and discussion to internal validity. The conclusion of this chapter, however, is that there are very significant threats to the statistical conclusion validity of intervention research that warrant at least equal attention. Don, I'm sure, would agree.

REFERENCES

Abelson, R. P. (1985). A variance explanation paradox: When a little is a lot. *Psychological Bulletin*, *97*, 129-133.

Abelson, R. P. (1997). On the surprising longevity of flogged horses: Why there is a case for the significance test. *Psychological Science*, *8*(1), 12-15.

Begg, C. B. (1994). Publication bias. In H. Cooper & L. V. Hedges (Eds.), *The handbook of research synthesis* (pp. 399-409). New York: Russell Sage Foundation.

Campbell, D. T., & Stanley, J. C. (1966). *Experimental and quasi-experimental designs for research.* Chicago: Rand McNally.

Cohen, J. (1962). The statistical power of abnormal-social psychological research: A review. *Journal of Abnormal and Social Psychology, 65*, 145-153.

Cohen, J. (1988). *Statistical power analysis for the behavioral sciences* (2nd ed.). Hillsdale, NJ: Lawrence Erlbaum.

Cohen, J. (1994). The world is round ($p < .05$). *American Psychologist, 49(12), 997-1003.*

Cook, T. D., & Campbell, D. T. (1979). *Quasi-experimentation: Design and analysis for field settings.* Chicago: Rand McNally.

Greenwald, A. G. (1975). Consequences of prejudice against the null hypothesis. *Psychological Bulletin, 82*(1), 1-20.

Hunter, J. E. (1997). Needed: A ban on the significance test. *Psychological Science, 8*(1), 3-7.

Hunter, J. E., Schmidt, F. L., & Jackson, G. B. (1982). *Meta-analysis: Cumulating research findings across studies.* Beverly Hills, CA: Sage.

Jacobson, N. S., & Truax, P. (1991). Clinical significance: A statistical approach to defining meaningful change in psychotherapy research. *Journal of Consulting and Clinical Psychology, 59*, 12-19.

Lipsey, M. W. (1990). *Design sensitivity: Statistical power for experimental research.* Newbury Park, CA: Sage.

Lipsey, M. W. (1998). Design sensitivity: Statistical power for applied experimental research. In L. Bickman & D. J. Rog (Eds.), *Handbook of applied social research methods* (pp. 39-68). Thousand Oaks, CA: Sage.

Lipsey, M. W., & Wilson, D. B. (1993). The efficacy of psychological, educational, and behavioral treatment: Confirmation from meta-analysis. *American Psychologist, 48*(12), 1181-1209.

Loftus, G. R. (1996). Psychology will be a much better science when we change the way we analyze data. *Current Directions in Psychological Science, 5*(6), 161-171.

Murphy, K. R., & Myors, B. (1988). *Statistical power analysis: A simple and general model for traditional and modern hypothesis tests.* Mahwah, NJ: Lawrence Erlbaum.

Reichardt, C. S., & Gollob, H. F. (1997). When confidence intervals should be used instead of statistical tests, and vice versa. In L. L. Harlow, S. A. Mulaik, & J. H. Steiger (Eds.), *What if there were no significance tests?* (pp. 259-284). Hillsdale, NJ: Lawrence Erlbaum.

Rogers, J. L., Howard, K. I., & Vessey, J. T. (1993). Using significance tests to evaluate equivalence between two experimental groups. *Psychological Bulletin, 113*(3), 553-565.

Rosenthal, R., & Rubin, D. B. (1982). A simple, general purpose display of magnitude of experimental effect. *Journal of Educational Psychology, 74*, 166-169.

Rosenthal, R., & Rubin, D. B. (1994). The counternull value of an effect size: A new statistic. *Psychological Science, 5*(6), 329-334.

Rosenthal, R., & Rubin, D. B. (1985). Statistical analysis: Summarizing evidence versus establishing facts. *Psychological Bulletin*, *97*, 527-529.
Schmidt, F. L. (1992). What do data really mean? Research findings, meta-analysis, and cumulative knowledge in psychology. *American Psychologist*, *47*(10), 1173-1181.
Sechrest, L., & Yeaton, W. H. (1982). Magnitudes of experimental effects in social science research. *Evaluation Review*, *6*, 579-600.
Sedlmeier, P., & Gigerenzer, G. (1989). Do studies of statistical power have an effect on the power of studies? *Psychological Bulletin*, *105*(2), 309-316.

CHAPTER 5

Effect Sizes in Behavioral and Biomedical Research

Estimation and Interpretation

Robert Rosenthal

PROLOGUE

When you're working in controversial areas and you're young and you're at the University of North Dakota in the 1950s, it pays to have senior mentors and advisers. I was lucky enough to have such mentors when they were sorely needed, and one of them was Don Campbell.[1] Beginning December 1, 1958, Don Campbell was my mentor at a distance. In a letter he wrote that day, he agreed to participate in a symposium I was organizing for the American Psychological Association to be held the following summer in Cincinnati. (Other members of that symposium on "The Problem of Experimenter Bias" were Martin Orne, Walter Reitman, and Hank Riecken.) Don took me under his wing in the organization of that symposium, because he accurately diagnosed me as a naive backwoods psychologist teaching at the time at the University of North Dakota.

From then on, Don inspired me, educated me, honored me, and supported me. He saw me through some tough times when most fellow psychologists thought I was crazy for believing that psychological experimenters could unintentionally influence their research subjects to respond in accordance with the experimenters' expectations. Later, he intervened on my behalf when some highly placed critics tried to stop the publication of the book describing the research with Lenore Jacobson on teachers' expectation effects in classrooms. Throughout my career, Don helped me and supported me.

In more recent years, I have on occasion been asked to write letters about Don. I told him that those requests so honored me that I would add it to my vita that I had been asked! I take pride, in primitive identification with Don, that we both spent time studying at the University of California, Berkeley (he a lot and I a little), that we both taught at Ohio State (he a lot and I a little), that we both published in less than fully refereed journals (he a little and I a lot), and that the topic of the present chapter on effect sizes, which is central to the enterprise of cumulating psychology, can be seen to be relevant to the very center of Don's thinking, not only about the methodology of psychology but about all the sciences:

> A crucial part of the egalitarian antiauthoritarian ideology of the seventeenth-century "new science" was the ideal that each member of the scientific community could replicate a demonstration for himself. . . . Each scientist was to be allowed to inspect the apparatus and try out the shared recipe. . . . A healthy community of truth seekers can flourish where such replication is possible. It becomes precarious where it is not. (Campbell, 1988, p. 516)

INTRODUCTION

As psychology has matured as a social and behavioral science, there seems to have been a gradual but accelerating increase in attention to, computation of, and reporting of the effect sizes obtained in social and behavioral research. Although there is not yet a well-worked-out history of this development, contributing factors may have included the increasing sophistication about power analysis among behavioral researchers, the rapid ascendancy of the meta-analytic enterprise, and a very long-term increase in dissatisfaction with null hypothesis decision procedures (e.g., Cohen, 1965, 1988, 1990, 1994; Cooper, 1981; Cooper & Hedges, 1994; Glass, McGaw, & Smith, 1981; Rozeboom, 1960).

The purposes of this chapter are (a) to provide an overview of types of effect size indicators, (b) to provide simple procedures for helping us understand the practical meaning of effect size estimates, and (c) to suggest that for simplicity, versatility, and consistency, it is hard to improve upon the use of Pearson's product moment correlation as the all-around index of effect size for use in individual studies and in the meta-analytic context (Rosenthal, Rosnow, & Rubin, in press).

EFFECT SIZE AND SIGNIFICANCE TESTS

When behavioral and biomedical researchers speak of "the results" of research, they are still most often referring to the statistical significance (*p* values) of the results, somewhat less often to the effect size estimates associated with those *p* values, and still less often to both the *p* value and the effect size. To make explicit the relationship between these two kinds of results, we can write the following prose equation (Cohen, 1965; Rosenthal & Rosnow, 1991):

Significance Test = Effect Size × Study Size.

Any particular test of significance can be obtained by one or more definitions of effect size multiplied by one or more definitions of study size. For example, if we are interested in $\chi^2_{(1)}$ as a test of significance, we can write:

$$\chi^2_{(1)} = \phi^2 \times N$$

where $\chi^2_{(1)}$ is a χ^2 on 1 degree of freedom (e.g., from a 2 × 2 table of counts), where ϕ^2 is the squared Pearson product moment correlation (the effect size) between membership in the row category (scored 1 or 0) and membership in the column category (scored 1 or 0), and where N (the study size) is the total number of sampling units, for example, found in the cells of the 2 × 2 table.

If we are interested in t as a test of significance, we have a choice of many equations (Rosenthal, 1991; 1994b), of which two are

$$t = \frac{r}{\sqrt{1-r^2}} \times \sqrt{df}$$

$$t = d \times \sqrt{df}\big/2$$

where r is the point biserial Pearson r between group membership (scored 1 or 0) and obtained score, d is the difference between means divided by the pooled standard deviation σ, and df is the degrees of freedom, usually $N - 2$.

TWO IMPORTANT FAMILIES OF EFFECT SIZES

Two of the most important families of effect sizes are the r family and the d family.

The *r* Family

The r family includes the Pearson product-moment correlation in any of its popular incarnations with labels:

r when both variables are continuous,

ϕ when both variables are dichotomous,

r_{pb} when one variable is continuous and one variable is dichotomous, and

ρ when both variables are in ranked form.

The r family also includes Z_r, the Fisher transformation of r, and the various squared indices of r and r-like quantities including r^2, ω^2 (omega squared), ε^2 (epsilon squared), and eta^2. Because squared indices of effect size lose their directionality (is the treatment helping or hurting, is the correlation positive or negative), they are of little use in scientific work for which information on directionality is essential. Another reason to avoid the use of the squared indices of effect size is that the practical magnitude of these indices is likely to be seriously misinterpreted as much smaller than is really the case—as illustrated a little further on when discussing the Physicians' Aspirin Study (Rosenthal, 1990; Rosenthal & Rosnow, 1991; Rosenthal & Rubin, 1979, 1982).

The *d* Family

The three central members of the d family are Cohen's d, Hedges's g, and Glass's Δ; all three employ the same numerator, the difference between the means of the groups being compared (i.e., $M_1 - M_2$). The denominators of these three indices differ, however:

$$\text{Cohen's } d = \frac{M_1 - M_2}{\sigma}$$

$$\text{Hedges's } g = \frac{M_1 - M_2}{S}$$

$$\text{Glass's } \Delta = \frac{M_1 - M_2}{S_{control}}$$

where σ is the square root of the pooled variance computed from the two groups that is,

$$\sigma = \sqrt{\Sigma(X - M)^2 / n},$$

where S is the square root of the pooled unbiased estimate of the variance

$$S = \sqrt{\Sigma(X - M)^2 / (n-1)},$$

and where S_{control} is like the S in the denominator of Hedges's g but computed only for the control group.

The d family of effect sizes also includes such other indices of differences as the raw difference in proportions d^1 (Fleiss, 1994) and the difference between two proportions after each has been transformed to (a) radians (Cohen's h, Case 1; Cohen, 1988, p. 200), (b) probits, or (c) logits (Glass et al., 1981).

EFFECT SIZES FOR THE ONE-SAMPLE CASE

The effect size estimates discussed so far have applied to situations in which we wanted to index the magnitude of a linear relationship between two variables by means of a correlation or by means of a comparison between the means of two conditions, say by d, Δ, or g. There are situations, however, in which there is only a single sample in our experiment, perhaps with each sampling unit exposed to two different experimental conditions. For example, teachers' favorableness of nonverbal behavior is recorded toward children for whom they hold more versus less favorable expectations. One test of significance of this effect of teachers' expectations on their nonverbal behavior could be the t for correlated observations.

Two equations illustrating our basic relationship between significance tests and effect sizes for this one-sample case are (Rosenthal, 1994b)

$$t = \frac{r}{\sqrt{1-r^2}} \times \sqrt{df}$$

$$t = d \times \sqrt{df}$$

The first of these equations shows that an r index can be used in the one-sample case, and the second equation shows that a d index can be used in the one-sample case. It should be noted, however, that the r index is identical in the one-sample and two-sample cases, whereas the d index is quite different in the one-sample and two-sample cases (Cohen, 1988; Rosenthal, 1994b).

Dichotomous Data

When the data are dichotomous rather than continuous, a number of d family indices are available including Cohen's g, Cohen's h (Case 2), and a newer index, Π. The index Cohen's g is simply the difference between an observed proportion and .50. For example, the magnitude of an electoral victory is given directly by g. If .60 of the electorate voted for the winner, then $g = .60 - .50 = .10$. Such an effect size might be regarded as enormous in the case of an election result but might be regarded as far less noteworthy as the result of a true-false test! The index Cohen's h (Case 2) is the difference between an observed proportion and a theoretically expected proportion after each of these proportions has been transformed to radians (an arcsin transformation). For example, in a multiple-choice examination in which one of four alternatives is correct and the position of the correct alternative has been assigned at random, guessing alone should yield an accuracy rate of .25. If the actual performance on this examination were found to be .75, we would compute h by transforming the actual (.75) and the expected (.25) proportions by means of 2 arcsin $\sqrt{P}$ yielding

$$h = 2 \text{ arcsin } \sqrt{.75} - 2 \text{ arcsin } \sqrt{.25} = 2.09 - 1.05 = 1.04.$$

The reason for employing the arcsin transformation is to make the hs comparable.

Differences between raw proportions are not all comparable, for example, with respect to statistical power. Thus, a difference between proportions of .95 and .90 yields an h of .19, whereas a difference between proportions of .55 and .50 yields an h of only .10 (Cohen, 1988).

The one-sample effect size index, Π, is expressed as the proportion of correct guesses if there had been only two choices to choose between. When there are more than two choices, Π converts the proportion of hits to the proportion of hits made if there had been only two equally likely choices:

$$\Pi = \frac{P(k-1)}{P(k-2)+1}$$

when P is the raw proportion of hits and k is the number of alternative choices available. The standard error of Π is

$$SE_{(\Pi)} = \frac{1}{\sqrt{N}} \left[\frac{\Pi(1-\Pi)}{P(1-P)} \right]$$

This index would be especially valuable in evaluating performance on a multiple-choice type of examination in which the number of alternatives varied from item to item. The index allows us to summarize the overall performance so that we could compare performances on tests made up of varying numbers of alternatives per item. Further details can be found in Rosenthal and Rubin (1989, 1991) and in Schaffer (1991).

EFFECT SIZES FOR COMPARING EFFECT SIZES

The Two-Sample Case

Sometimes the basic research question concerns the difference between two effect sizes. For example, a developmental psychologist may hypothesize that two cognitive performance measures will be more highly correlated in preschool children than in fifth graders. The degree to which the hypothesis is supported will depend on the difference between the correlations obtained from preschoolers and fifth graders, $r_1 - r_2$. Cohen's q is just such an index, one in which each r is transformed to Fisher's Z_r before the difference is computed, so that

$$\text{Cohen's } q = Z_{r1} - Z_{r2}$$

The One-Sample Case

Cohen's q also can be employed when an obtained effect size is to be compared to a theoretical value of r. In this case, we simply take the difference between the Z_r associated with our observed sample and the Z_r associated with our theoretical value of r (Cohen, 1988).

COMPARING THE *r* AND *d* FAMILIES

It seems natural to employ r-type effect size estimators when the original effect size estimates are reported in r-type indices, such as in meta-analyses of validity coefficients for test instruments (e.g., Hunter & Schmidt, 1990). Similarly, it seems natural to employ d-type effect size estimates when the original studies have compared two groups so that the difference between their means and their within group Ss or σs are available. In meta-analytic work, however, it is often the case that the effect size estimates will be a mixture of r-type and d-type indices.

Because r- and d-type estimates can be converted readily to one another, obtaining both types of estimates will involve no hardship. It will be necessary, however, to make a decision in meta-analytic work to convert all effect size estimates to just one particular index, usually to r or Z_r for the r family, or to Hedges's g (or Cohen's d) for the d family. Although any of these effect size estimates can be employed, there are some reasons to view r as the more generally useful effect size estimate.

Generality of Interpretation

If our data came to us as rs, it would not make much sense to convert rs to ds because the concept of a mean difference index makes little sense in describing a linear relationship over a great many values of the independent variable. On the other hand, given a d-type effect size estimate, r makes perfectly good sense in its point biserial form (i.e., just two levels of the independent variable).

Suppose that our theory calls for us to employ at least three levels of our independent variable because we predict a quadratic trend in the relationship between level of arousal and subsequent performance. The magnitude of the effect associated with our quadratic trend contrast is quite naturally indexed by r- but not so naturally indexed by d-type indices.

Consistency of Meaning in the One-Sample Case

The r-type index requires no computational adjustment in going from the two-sample or multisample case to the one-sample case. As noted above, r is identically related to t for the two-sample and for the one-sample case. That is not the case for the d-type indices, however. For example, the definition of the size of the study changes by a factor of two in going from a t test for two samples to a t test for one sample.

Simplicity of Interpretation

Finally, r is more simply interpreted in terms of practical importance than are the usual d-type indices, such as Hedges's g or Cohen's d. The details are given in the following section.[2]

THE INTERPRETATION OF EFFECT SIZES

Despite the growing awareness of the importance of estimating effect sizes, there is a problem in evaluating various effect size estimators from the point of view of practical usefulness (Cooper, 1981). Rosenthal and Rubin (1979, 1982) found that neither experienced behavioral researchers nor experienced statisticians had a good intuitive feel for the practical meaning of common effect size estimators and that this was particularly true for such squared indices as r^2, ω^2, ε^2, and similar estimates.

The Physicians' Aspirin Study

At a special meeting held on December 19, 1987, it was decided to end, prematurely, a randomized double-blind experiment on the effects of aspirin on reducing heart attacks (Steering Committee of the Physicians Health Study Research Group, 1988). The reason for this unusual termination of such an experiment was that it had become so clear that aspirin prevented heart attacks (and deaths from heart attacks) that it would be unethical to continue to give half the physician research subjects a placebo. What was the magnitude of the experimental effect that was so dramatic as to call for the termination of this research? Was $r^2 = .80$ or .60, so that the corresponding rs would have been .89 or .77? Was $r^2 = .40$ or .20, so that the corresponding rs would have been .63 or .45? No, none of these. Actually, r^2 was .00 or, to four decimal places, .0011, with a corresponding r of .034. The decision to end the aspirin experiment was an ethical necessity—it saved lives. Most social and behavioral scientists are surprised that life-saving interventions can be associated with effect sizes as small as rs of .034 and r^2s of .0011.

TABLE 5.1 Effects of Aspirin on Heart Attacks Among 22,071 Physicians

	Heart Attack	*No Heart Attack*	*Total*
I. Raw Counts in Four Conditions			
Aspirin	104	10,933	11,037
Placebo	189	10,845	11,034
Total	293	21,778	22,071
II. Percentages of Patients			
Aspirin	0.94	99.06	100
Placebo	1.71	98.29	100
Total	1.33	98.67	100
III. Binomial Effect Size Display			
Aspirin	48.3	51.7	100
Placebo	51.7	48.3	100
Total	100	100	200

The Binomial Effect Size Display

Table 5.1 shows the results of the aspirin study in terms of raw counts and percentages, and as a binomial effect size display (BESD). This display is a way of showing the practical importance of any effect indexed by a correlation coefficient.

The correlation is shown to be a simple difference in outcome rates between the experimental and the control groups in this standard table, which always adds up to column totals of 100 and row totals of 100 (Rosenthal & Rubin, 1982). We obtain the BESD from any obtained effect size r by computing the treatment condition success rate as .50 plus $r/2$ and the control condition success rate as .50 minus $r/2$. Thus, an r of .20 yields a treatment success rate of $.50 + .20/2 = .60$ and a control success rate of $.50 - .20/2 = .40$ or a BESD of

	Success	*Failure*	Σ
Treatment	60	40	100
Control	40	60	100
Σ	100	100	200

TABLE 5.2 Other Examples of Binomial Effect Size Displays

	Problem	*No Problem*	*Total*
I. Vietnam Service and Alcohol Problems ($r = .07$)			
Vietnam veteran	53.5	46.5	100
Non-Vietnam veteran	46.5	53.5	100
Total	100	100	200

	Death	*Survival*	*Total*
II. AZT in the Treatment of AIDS ($r = .23$)			
AZT	38.5	61.5	100
Placebo	61.5	38.5	100
Total	100	100	200

	Less Benefit	*Greater Benefit*	*Total*
III. Benefits of Psychotherapy ($r = .39$)[a]			
Psychotherapy	30.5	69.5	100
Control	69.5	30.5	100
Total	100	100	200

a. The analogous r for 464 studies of interpersonal expectancy effects was .30 (Rosenthal, 1994a).

Had we been given the BESD to examine before knowing r, we could easily have calculated it mentally for ourselves; r is simply the difference between the success rates of the experimental versus the control group (.60 – .40 = .20).

This type of result seen in the Physicians' Aspirin Study is not at all unusual in biomedical research. Some years earlier, on October 29, 1981, the National Heart, Lung, and Blood Institute discontinued its placebo-controlled study of propranolol because results were so favorable to the treatment that it would be unethical to continue withholding the life-saving drug from the control patients. The effect size r was .04, and the leading digits of the r^2 were .00! As behavioral researchers we are not used to thinking of rs of .04 as reflecting effect sizes of practical importance, but when we think of an r of .04 as reflecting a 4% decrease in heart attacks, the interpretation given r in a binomial effect size display, the r does not appear to be quite so small.

Additional Results

Table 5.2 gives three further examples of binomial effect size displays. In a study of 4,462 Army veterans of the Vietnam War era (1965-1971), the correlation between having served in Vietnam (rather than elsewhere) and having suffered from alcohol abuse or dependence was .07 (Centers for Disease Control Vietnam Experience Study, 1988). The top display in Table 5.2 shows that the difference between the problem rates of 53.5 and 46.5 per 100 is equal to the correlation coefficient of .07.

The center display in Table 5.2 shows the results of a study of the effects of AZT on the survival of 282 patients suffering from AIDS or AIDS-related complex (ARC) (Barnes, 1986). This result of a correlation of .23 between survival and receiving AZT (an r^2 of .054) was so dramatic as to lead to premature termination of the clinical trial on the ethical grounds that it would be improper to continue to give placebos to the control group patients.

The bottom display in Table 5.2 shows the results of a famous meta-analysis of psychotherapy outcome studies reported by Smith, Glass, and Miller (1980). Interestingly for behavioral researchers, the magnitude of the effect of psychotherapy was substantially greater than the effects of a good many breakthrough medical interventions. Table 5.3 shows the effect sizes obtained in a convenience sample of 19 different studies; 8 of the studies, employing dependent variables of convulsions, AIDS events, alcohol problems, heart attacks, and death, were associated with effect size *r*s of less than .10. One desirable result of our consideration of these biomedical effect size estimates is to make those of us working in the social and behavioral sciences less pessimistic about the magnitude and importance of our research results (Rosenthal, 1990, 1995).

OTHER EFFECT SIZE ESTIMATES FOR 2 × 2 TABLES OF COUNTS: THE BIOMEDICAL CONTEXT

The effect size index, *r*, can be applied readily to any 2 × 2 table of counts. Three other indices of effect size have been found useful in biomedical contexts: (a) relative risk, (b) odds ratio, and (c) risk difference. All three are illustrated for several hypothetical outcomes in Table 5.4. Each study compared a control condition to a treatment condition with two possible outcomes: not surviving or surviving.

Relative Risk

Relative risk is defined as the ratio of the proportion of the control patients at risk (not surviving) divided by the proportion of the treated patients at risk. With the cells of the 2 × 2 table of counts labeled A, B, C, and D from upper left to lower right (as shown in Table 5.4), relative risk (RR) is defined as

$$RR = \frac{A}{A+B} \bigg/ \frac{C}{C+D}$$

TABLE 5.3 Effect Sizes of Various Independent Variables

Independent Variable	*Dependent Variable*	*r*	r^2
Aspirin[a]	Heart attacks	.03	.00
Beta carotene[b]	Death	.03	.00
Streptokinase[c]	Death	.03	.00
Propranolol[d]	Death	.04	.00
Magnesium[e]	Convulsions	.07	.00
Vietnam veteran status[f]	Alcohol problems	.07	.00
Garlic[g]	Death	.09	.01
Indinavir[h]	Serious AIDS events	.09	.01
Testosterone[i]	Adult delinquency	.12	.01
Compulsory hospitalization versus treatment choice[j]	Alcohol problems	.13	.02
Cyclosporine[k]	Death	.15	.02
Ganzfeld perception[l]	Accuracy	.16	.03
Cisplatin and vinblastine[m]	Death	.18	.03
AZT for neonates[n]	HIV infection	.21	.04
Cholesterol-lowering regimen[o]	Coronary status	.22	.05
AZT[p]	Death	.23	.05
Treatment choice versus AA[q]	Alcohol problems	.27	.07
Psychotherapy[r]	Improvement	.39	.15
Compulsory hospitalization versus AA[s]	Alcohol problems	.40	.16

a. Steering Committee of the Physicians Health Study Research Group (1988).
b. Alpha-Tocopherol, Beta Carotene Cancer Prevention Study Group (1994).
c. Gruppo Italiano per lo Studio della Streptochinasi Nell'Infarto Miocardico (1986).
d. Kolata (1981).
e. Foreman (1995).
f. Centers for Disease Control Vietnam Experience Study (1988).
g. Goldfinger (1991).
h. Knox (1997).
I. Dabbs and Morris (1990).
j. Cromie (1991).
k. Canadian Multicentre Transplant Study Group (1983).
l. Chandler (1993).
m. Cromie (1990).
n. Altman (1994).
o. Roberts (1987).
p. Barnes (1986).
q. Cromie (1991).
r. Smith, Glass, and Miller (1980).
s. Cromie (1991).

TABLE 5.4 Three Examples of Four Effect Size Estimates

	Die	Live	*Relative Risk*	*Odds Ratio*	*Risk Difference*	*r*
Control	A	B	$\dfrac{A}{A+B} \Big/ \dfrac{C}{C+D}$	$\dfrac{A}{B} \Big/ \dfrac{C}{D}$	$\dfrac{A}{A+B} - \dfrac{C}{C+D}$	
Treatment	C	D				
Study I						
	Die	Live				
Control	10	990	10.00	10.09	.01	.06
Treatment	1	999				
Study II						
	Die	Live				
Control	10	10	10.00	19.00	.45	.50
Treatment	1	19				
Study III						
	Die	Live				
Control	10	0	10.00	∞	.90	.90
Treatment	1	9				

A limitation of this effect size estimate can be seen in Table 5.4. We examine the three study outcomes closely and ask ourselves the following: If we had to be in the control condition, would it matter to us whether we were in Study I, Study II, or Study III? We think most people would rather have been in Study I than II, and we think that virtually no one would have preferred to be a member of the control group in Study III. Despite the very important phenomenological differences among these three studies, Table 5.4 shows that all three relative risks are identical: 10.00. That feature may be a serious limitation to the value and informativeness of the relative risk index.

Odds Ratio

The odds ratio is defined as the ratio of the not-surviving control patients to the surviving control patients divided by the ratio of the not-surviving treated patients to the surviving treated patients. For cells as labeled in Table 5.4, the odds ratio (OR) is defined as

$$OR = \frac{A/B}{C/D}$$

The odds ratio behaves more as expected in Table 5.4 than does the relative risk, in that the odds ratio increases with our phenomenological discomfort as we go from the results of Study I to Study II to Study III. The high odds ratio for Study I, however, seems alarmist. Indeed, if the data showed

	Die	Live	
Control	10	999,990	10^6
Treated	1	999,999	10^6
	11	1,999,989	$2(10^6)$

so that an even smaller proportion of patients were at risk, the odds ratio would remain at 10.00, an even more alarmist result.

The odds ratio is also unattractive for Study III. Because all the controls die, perhaps we could forgive the infinite odds ratio; however, very different phenomenological results yield an identical odds ratio. If the data showed

	Die	Live	
Control	1,000,000	0	10^6
Treated	999,999	1	10^6
	1,999,999	1	$2(10^6)$

we would again have an infinite odds ratio, definitely an alarmist result. In this case even the problematic relative risk index would yield a phenomenologically more realistic result of 1.00.

Risk Difference

The risk difference is defined as the difference between the proportion of the control patients at risk and the proportion of the treated patients at risk. For cells as labeled in Table 5.4, the risk difference (RD) is defined as

$$RD = \frac{A}{A+B} - \frac{C}{C+D}$$

The last column of Table 5.4 shows the Pearson product-moment correlation (r) between the independent variable of treatment (scored 0, 1) and the dependent variable of outcome (scored 0, 1). Comparison of the risk differences with r in Table 5.4 (and elsewhere) shows that the risk difference index is never unreasonably far from the value of r. For that reason, the risk difference index may be the one least likely to be quite misleading under special circumstances, so we prefer it as our all-purpose index if we have to use one of the three indices under discussion. But even here we feel we can do better.

Standardizing the Three Risk Measures

We propose a simple adjustment that standardizes our measures of relative risk, odds ratio, and risk difference (Rosenthal & Rubin, 1998). We simply compute the correlation r between the treatment and outcome and display r in a binomial effect size display (BESD) as described above.

Table 5.5 shows the BESD for the three studies of Table 5.4. Although the tables of counts of Table 5.4 varied from Ns of 2,000, to 40, to 20, the corresponding BESDs of Table 5.5 all show the standard margins of 100, which is a design feature of the BESD. The computation of our new effect size indices is straightforward. We simply compute relative risks, odds ratios, and risk differences on our standardized tables (BESDs) to obtain standardized relative risks, standardized odds ratios, and standardized risk differences. The computation of these three indices is simplified because the A and D cells of a BESD always have the same value (as do the B and C cells). Thus, the computational equations simplify to A/C for standardized relative risk (*SRR*), to $(A/C)^2$ for standardized odds ratio (*SOR*), and to (A – C)/100 for standardized risk difference (SRD).

Table 5.5 shows the standardized relative risks increasing as they should in going from Study I to Study III. The standardized odds ratios also increase as they go from Study I to Study III but without the alarmist value for Study I and the infinite value for Study III. (A standardized odds ratio could go to infinity only if r were exactly 1.00, an unlikely event in behavioral or biomedical research.) The standardized risk difference is shown in Table 5.5 to be identical to r, which is an attractive feature emphasizing the interpretability of r as displayed in a BESD.

MINIMIZING ERRORS IN THINKING ABOUT EFFECT SIZES: THE COUNTERNULL VALUE OF AN EFFECT SIZE

The counternull value of an effect size was recently introduced as a new statistic (Rosenthal & Rubin, 1994). It is useful in virtually eliminating two common errors: (a) equating failure to reject the null with the estimation of the effect size as equal to zero and (b) equating rejection of a null hypothesis on the basis of a

TABLE 5.5 Standardized Outcomes of Table 5.4

	Die	Live	*Standardized Relative Risk (A/C)*	*Standardized Odds Ratio* $(A/C)^2$	*Standardized Risk Difference (r) (A – C)/100*
Control	A	C			
Treatment	C	A			
Study I					
	Die	Live			
Control	53	47	1.13	1.27	.06
Treatment	47	53			
Study II					
	Die	Live			
Control	75	25	3.00	9.00	.50
Treatment	25	75			
Study III					
	Die	Live			
Control	95	5	19.00	361.00	.90
Treatment	5	95			

significance test with having demonstrated a scientifically important effect. In most applications, the value of the counternull is simply twice the magnitude of the obtained effect size (e.g., d, g, Δ, Z_r). Thus, with $r = .10$ found to be nonsignificant, the counternull value of $r = .20$ is exactly as likely as the null value of $r = .00$. For any effect size with a symmetric reference distribution, such as the normal or any t distribution, the counternull value of an effect size can always be found by doubling the obtained effect size and subtracting the effect size expected under the null hypothesis (usually zero). Thus, if we found that a test of significance did not reach the chosen level (e.g., .05), the use of the counternull would keep us from concluding that the mean effect size was, therefore, probably zero. The counternull value of $2d$ or $2Z_r$ would be just as tenable a conclusion as concluding $d = 0$ or $Z_r = 0$.

The counternull is a kind of confidence interval conceptually related to the more traditional (e.g., 95%) confidence interval. As Cohen, with his customary wisdom, pointed out, the behavioral and medical sciences would be more advanced had we always routinely reported not only p values but effect size estimates with confidence intervals as well (Cohen, 1990, 1994).

CONCLUSION

In this chapter, we have considered major types of effect size indicators and some simple procedures for helping us understand the practical meaning of effect size estimates, and we have emphasized the simplicity, versatility, and consistency of Pearson's effect size indicator *r*. The hope is that these considerations will be useful as the social and behavioral sciences continue their increasing emphasis on effect size estimation and their decreasing emphasis on dichotomous null hypothesis decision procedures.

EPILOGUE

I am very grateful to Don for all he did for me for all those years to educate me, to inspire me, to honor me, and to support me. I know how he would want me, and all the others that he helped over the years, to thank him. He would want us to pass it on to the next generation of scholars.

NOTES

1. Two other mentors or "psychological sponsors" at that time were Harold Pepinsky—who, fittingly, had helped Pauline Pepinsky to develop the very concept of psychological sponsor—and Henry Riecken, who had anticipated so much of the work on the social psychology of the psychological experiment and who was responsible for the financial support of the National Science Foundation for the work I was doing in those early days.

2. Although there has not been space here to develop the idea, it should be noted that useful distinctions can be made among various types of *r*s, including r_{alerting}, r_{contrast}, $r_{\text{effect size}}$, and r_{BESD}. For details, see Rosenthal, Rosnow, and Rubin (in press).

REFERENCES

Alpha-Tocopherol, Beta Carotene Cancer Prevention Study Group. (1994). The effect of vitamin E and beta carotene on the incidence of lung cancer and other cancers in male smokers. *New England Journal of Medicine, 330*, 1029-1035.

Altman, L. K. (1994, February 21). In major finding, drug limits HIV infection in newborns. *New York Times*, pp. A1, A13.

Barnes, D. M. (1986). Promising results halt trial of anti-AIDS drug. *Science, 234*, 15-16.

Campbell, D. T. (1988). *Methodology and epistemology for social science: Selected papers* (E. S. Overman, Ed.). Chicago: University of Chicago Press.

Canadian Multicentre Transplant Study Group. (1983). A randomized clinical trial of cyclosporine in cadaveric renal transplantation. *New England Journal of Medicine, 309*, 809-815.

Centers for Disease Control Vietnam Experience Study. (1988). Health status of Vietnam veterans: 1. Psychosocial characteristics. *Journal of the American Medical Association, 259*, 2701-2707.

Chandler, D. L. (1993, February 15). Study finds evidence of ESP phenomenon. *Boston Globe*, pp. 1, 8.

Cohen, J. (1965). Some statistical issues in psychological research. In B. B. Wolman (Ed.), *Handbook of clinical psychology* (pp. 95-121). New York: McGraw-Hill.

Cohen, J. (1988). *Statistical power analysis for the behavioral sciences* (2nd ed.). Hillsdale, NJ: Lawrence Erlbaum.

Cohen, J. (1990). Things I have learned (so far). *American Psychologist, 45*, 1304-1312.

Cohen, J. (1994). The earth is round ($p < .05$). *American Psychologist, 49*, 997-1003.

Cooper, H. M. (1981). On the significance of effects and the effects of significance. *Journal of Personality and Social Psychology, 41*, 1013-1018.

Cooper, H., & Hedges, L. V. (Eds.). (1994). *Handbook of research synthesis.* New York: Russell Sage.

Cromie, W. J. (1990, October 5). Report: Drugs affect lung cancer survival. *Harvard Gazette*, pp. 1, 10.

Cromie, W. J. (1991, September 13). Study: Hospitalization recommended for problem drinkers. *Harvard Gazette*, pp. 3-4.

Dabbs, J. M., Jr., & Morris, R. (1990). Testosterone, social class, and antisocial behavior in a sample of 4,462 men. *Psychological Science, 1*, 209-211.

Fleiss, J. L. (1994). Measures of effect size for categorical data. In H. Cooper & L. V. Hedges (Eds.), *Handbook of research synthesis* (pp. 245-260). New York: Russell Sage.

Foreman, J. (1995, July 27). Medical notebook: A new confirmation for a pregnancy drug. *Boston Globe*.

Gruppo Italiano per lo Studio della Streptochinasi Nell'Infarto Miocardico. (1986, February 22). Effectiveness of intravenous thrombolitic treatment in acute myocardial infarction. *Lancet*, pp. 397-402.

Glass, G. V., McGaw, B., & Smith, M. L. (1981). *Meta-analysis in social research.* Beverly Hills, CA: Sage.

Goldfinger, S. E. (1991, August). Garlic: Good for what ails you. *Harvard Health Letter, 16*(10), 1-2.

Hunter, J. E., & Schmidt, F. L. (1990). *Methods of meta-analysis: Correcting error and bias in research findings.* Newbury Park, CA: Sage.

Kolata, G. B. (1981). Drug found to help heart attack survivors. *Science, 214*, 774-775.

Knox, R. A. (1997, February 25). AIDS trial terminated: 3-drug therapy hailed. *Boston Globe*, pp. A1, A16.

Roberts, L. (1987). Study bolsters case against cholesterol. *Science, 237*, 28-29.

Rosenthal, R. (1990). How are we doing in soft psychology? *American Psychologist, 45*, 775-777.

Rosenthal, R. (1991). *Meta-analytic procedures for social research* (rev. ed.). Newbury Park, CA: Sage.

Rosenthal, R. (1994a). Interpersonal expectancy effects: A 30-year perspective. *Current Directions in Psychological Science, 3*, 176-179.

Rosenthal, R. (1994b). Parametric measures of effect size. In H. Cooper & L. V. Hedges (Eds.), *Handbook of research synthesis* (pp. 231-244). New York: Russell Sage.

Rosenthal, R. (1995). Progress in clinical psychology: Is there any? *Clinical Psychology: Science and Practice, 2*, 133-150.

Rosenthal, R., & Rosnow, R. L. (1991). *Essentials of behavioral research: Methods and data analysis* (2nd ed.). New York: McGraw-Hill.

Rosenthal, R., Rosnow, R. L., & Rubin, D. B. (in press). *Contrasts and effect sizes in behavioral research: A correlational approach.* New York: Cambridge University Press.

Rosenthal, R., & Rubin, D. B. (1979). A note on percent variance explained as a measure of the importance of effects. *Journal of Applied Social Psychology, 9*, 395-396.

Rosenthal, R., & Rubin, D. B. (1982). A simple, general purpose display of magnitude of experimental effect. *Journal of Educational Psychology, 74*, 166-169.

Rosenthal, R., & Rubin, D. B. (1989). Effect size estimation for one-sample multiple-choice-type data: Design, analysis, and meta-analysis. *Psychological Bulletin, 106*, 332-337.

Rosenthal, R., & Rubin, D. B. (1991). Further issues in effect size estimation for one-sample multiple-choice-type data. *Psychological Bulletin, 109*, 351-352.

Rosenthal, R., & Rubin, D. B. (1994). The counternull value of an effect size: A new statistic. *Psychological Science, 5*, 329-334.

Rosenthal, R., & Rubin, D. B. (1998). *Some new effect sizes for tables of counts.* Unpublished manuscript, Harvard University.

Rozeboom, W. W. (1960). The fallacy of the null hypothesis significance test. *Psychological Bulletin, 57*, 416-428.

Schaffer, J. P. (1991). Comment on "Effect size estimation for one-sample multiple-choice-type data: Design, analysis, and meta-analysis" by Rosenthal and Rubin (1989). *Psychological Bulletin, 109*, 348-350.

Smith, M. L., Glass, G. V, & Miller, T. I. (1980). *The benefits of psychotherapy.* Baltimore: Johns Hopkins University Press.

Steering Committee of the Physicians Health Study Research Group. (1988). Preliminary report: Findings from the aspirin component of the ongoing physicians' health study. *New England Journal of Medicine, 318*, 262-264.

CHAPTER 6

Realism, Validity, and the Experimenting Society

Melvin M. Mark

In Cook and Campbell (1979), as in many conversations, Don Campbell identified himself as an "evolutionary critical realist" (p. 28). The first component of this label, "evolutionary," was an important aspect of much of Campbell's thought. The metaphor of natural selection was central both to his overall approach to human knowledge processes (e.g., Campbell, 1972, 1974, 1977) and to his view of program and policy evaluation (Campbell, 1984). The notion of evolution is also a useful frame through which to view Campbell's own work: His later writings can be viewed as having evolved from earlier ones, forged through the selection mechanisms of a great mind drawing observations from practice and scholarship. For example, in his "contagious cross-fertilization" model (Campbell, 1984), Campbell offered a complex vision of program evaluation in which the funding, evaluation, and dissemination of social initiatives are intertwined in a way that mirrors evolutionary processes; this model is

---

AUTHOR'S NOTE: Address correspondence to Melvin Mark, Department of Psychology, Penn State, University Park, PA 16802, or via e-mail at M5M@psu.edu. I thank Gary Henry and George Julnes, who, as my coauthors on a book presenting a realist theory of evaluation (Mark, Henry, & Julnes, in press), have stimulated and contributed to many of the thoughts presented in this chapter. In addition, thanks go to Chip Reichardt, whose prior work on validity helped lead to some of the ideas presented herein and who provided valuable comments on a prior draft. Finally, I appreciate the insightful comments of an anonymous reviewer and thank Len Bickman for his support and for his efforts in bringing this volume to fruition.

clearly a more advanced life-form, so to speak, than Campbell's earlier, simpler notions of an experimenting society (e.g., Campbell, 1969c).

In this chapter, I speculate about some of the ways that Campbell's seminal work about research and about social experimentation might continue to evolve. In this speculative process, I rely heavily on the other component of Campbell's self-identified label, that is, on the "critical realist" part of "evolutionary critical realist." The realist lens is an appropriate one for examining Campbell's notions of validity and selected aspects of his vision of an experimenting society. As illustrated in the citation to Cook and Campbell (1979), Campbell described himself as a critical realist for much of his career. Moreover, fundamental concepts from critical realism informed Campbell's conception of validity and in turn influenced his view of social experimentation. As suggested by his own "fish-scale model" of interdisciplinary scholarship (Campbell, 1969a), Campbell mined the intersections of several traditional scholarly domains, one of which was realist philosophy of science. He imported large nuggets of insights from realist philosophy into his work on validity and social experimentation. In this chapter, I overview some of the payoff I believe may come from continuing to work in that vein.

CRITICAL REALISM AND CAMPBELL: A "*READER'S DIGEST*" VERSION

Campbell, like other realists, believed in an external world such that all we see, hear, and otherwise perceive is not simply subjective construction: "Evolutionary epistemology has in it an unproven assumption of a real world external to the organism, with which the organism is in dialectic interaction" (Campbell, 1984, p. 30). The assumption of a real world external to the organism seems uncontroversial to most people, but its acceptance is not a given in a world containing radical constructivists. Nor is this a trivial assumption. Indeed, as Bhaskar, one of the key figures in contemporary realism, points out, the scientific process—and indeed any belief in the possibility of critiquing and improving our understanding of the world—depends on this assumption:

> For it is only if the working scientist possesses the concept of an ontological realm, distinct from his current claims to knowledge of it, that he can philosophically think about the possibility of a rational criticism of these claims. To be a fallibilist about knowledge, it is necessary to be a realist about things. (Bhaskar, 1978, p. 43)

Campbell most assuredly was a fallibilist about knowledge. Fallibilism is a central tenet of *critical* realism (as opposed to some other forms of realism, particularly naive realism, which is the name given to the belief that we perceive the world directly as it exists—that our eyes simply see what is there). Fallibilism was also one of the cornerstones of Campbell's work. Fallibilism permeated Campbell's general approach to knowledge processes, as evident throughout his writing on this subject: "All knowing is highly presumptive, involving presumptions not directly or logically justifiable" (Campbell, 1969b, p. 66). Fallibilism was also central to most of Campbell's work on research methods, including his writings on multiple operationalization, triangulation, the multitrait-multimethod matrix, plausible rival hypotheses, validity threats, quasi-experimentation, and evaluation. Because he took seriously the critical realist assumption that all our methods are fallible, Campbell developed techniques and stances designed to move us toward more defensible and more valid inferences.

Although it is less strongly highlighted and, I think, far less well remembered, Campbell also sounded a fallibilist theme in his writings on the experimenting society. For example, in "Reforms as Experiments" (1969c), one of his early papers on the topic, Campbell described a set of designs (interrupted times series, regression-discontinuity, and the randomized experiment) and illustrated their application to the evaluation of social programs and policies. He also addressed the fallibility of such evaluations, under the heading "Multiple Replication in Enactment":

> Too many social scientists expect single experiments to settle issues once and for all. This may be a mistaken generalization from the history of great crucial experiments in physics and chemistry. In actuality the significant experiments in the physical sciences are replicated thousands of times. . . . Because we social scientists have less ability to achieve "experimental isolation," because we have good reasons to expect our treatment effects to interact significantly with a wide variety of social factors many of which we have not yet mapped, we have much greater needs for replication experiments than do the physical sciences.
>
> The implications are clear. We should not only do hard-headed reality testing in the initial pilot testing and choosing of which reform to make general law; but once it has been decided that the reform is to be adopted as standard practice in all administrative units, we should experimentally evaluate it in each of its implementations. (Campbell, 1969c, pp. 427-428)

In short, Campbell was a critical realist. He clearly held to what arguably are the core beliefs of critical realism, that (a) there is an external world and (b) our knowledge of it is imperfect. As summarized in Cook and Campbell (1979), his "perspective is realist because it assumes that causal relations exist outside of the human mind, and it is critical realist because it assumes that these valid causal relationships cannot be perceived with total accuracy by our imperfect sensory and intellective capacities" (p. 29). Although the fallibilism of critical realism permeated Campbell's work, some other realist concepts were not so strongly represented. Moreover, realist scholarship has continued since the time of Campbell's most influential work. Thus, there are notions from critical realism about which we can ask: How would Campbell's views—and, given his widespread influence, the views of many if not most social scientists—about validity and about the experimenting society be different if they incorporated more fully some additional themes of contemporary realism?

In the sections that follow, I attempt to address this question. The reader should be mindful of several caveats. First, there are many "varieties of realism" (Harré, 1986). Realist scholarship is filled with debate and disagreement, which I do not try to summarize here. Second, some of the people I cite use other, more specific labels than "critical realist," or vary over time in their apparent adherence to realism. Third, I am employing a selective and somewhat eclectic sampling of contemporary realist theorists. My goal is not to review realism comprehensively, but instead to use selected realist themes to re-examine Campbell's notions of validity and the experimenting society. Fourth, when I refer to Campbell's work on validity, I am referring to his work on the validity associated with experiments, quasi-experiments, and other cause-probing methods. Finally, when I refer to "the experimenting society," I do not mean to imply a particular single model of social interventions. Instead, I mean to refer broadly to social programs and policies and their evaluation.

REALISM AND THE NATURE OF VALIDITY

Campbell's work on the validity of experiments and quasi-experiments is one of his most notable achievements. His contributions to the conceptualization of validity, as exemplified by his distinction between internal and external validity, are in themselves deserving of our gratitude and careful study. Campbell's notions of validity are also intrinsically intertwined with his seminal work on research design, including quasi-experimentation. This linkage is evident in the opening sentence of Campbell and Stanley (1966): "In this chapter we shall examine the validity of 16 experimental designs against 12 common threats to valid inference" (p. 1). It seems somewhat ironic, then, that the concept of validity is not explicitly defined in that work. Internal validity and external validity are each defined:

> *Internal validity* is the basic minimum without which any experiment is uninterpretable: Did in fact the experimental treatment make a difference in this specific experimental instance? *External validity* asks the question of *generalizability*: To what populations, settings, treatment variables, and measurement variables can this effect be generalized? (Campbell & Stanley, 1966, p. 5)

But validity per se is not defined. Although Campbell and Stanley tell us what questions internal and external validity address, they do not, so to speak, define for us what it means to answer these questions validly.

Unlike Campbell and Stanley, Cook and Campbell (1979) do provide an explicit definition of validity per se: "We shall use the concepts *validity* and *invalidity* to refer to the best available approximation to the truth or falsity of propositions" (p. 37). On first reading, this definition may seem puzzling. Why is validity defined in terms of *approximations* of the truth, instead of in terms of truth itself? And what does truth mean, anyway? If validity is defined in terms of its relationship to the truth, but the truth is not defined, how satisfactory is the definition of validity? In addition, the particular phrasing of the Cook and Campbell definition of validity may seem troubling: What if the "best available approximation" to the truth is *way* off? Ancient astronomers were far from the truth in claiming that the sun revolved around the earth, but was this not the best approximation available to them at the time?

Some clarification of these issues comes by recognizing the relationship between Cook and Campbell's definition of validity and the work of Popper, a critical realist whose specific brand of realism is sometimes called "convergent realism." Campbell often cited Popper approvingly, especially in his emphasis on the process of winnowing out inadequate theories. According to Popper (1972), the goal of scientific theories is to approximate the truth, where truth means correspondence between a statement (a theory, or proposition, or even an effect-size estimate) and the real state of the world. Campbell agreed that the *goal* of research should be truth as defined in terms of correspondence between our inferences and the actual state of reality (Campbell, 1977). At the same time, he acknowledged that epistemological limits—limits on human knowing—make it impossible for us to know whether we have found the truth as defined by correspondence (Campbell, 1991). Cook and Campbell's definition of validity thus appears to parallel—perhaps by virtue of being influenced by—Popper and other critical realists, in acknowledging that mere mortals simply cannot know if we have some statement right; instead, in Campbellian terms, all we can do is to strive to rule out plausible rival hypotheses. Both Popper and Campbell indicate that humans can hold out at best for approximations of truth. From this perspective, the Cook and Campbell definition of validity looks better than it might at first glance.

There are, however, several logical problems lurking underfoot. I will deal with only one, summarized nicely by Aronson, Harré, and Way (1995): "Correspondence is like pregnancy: it does not admit of degrees. . . . How can something fit the world better or worse if it is false, that is, does not correspond to [the facts about the real state of the world]?" (p. 115). This may not seem to be such a problem if you are dealing with a single prediction or finding, such as the average effect size for a treatment. In this simple case, if the truth is, say, "5," then an estimate of "6" is a better approximation than an estimate of "8."

Things are more complex for theories—including program theories—that make multiple predictions. In this regard, echoing Aronson et al. (1995), one can reasonably ask: If two theories are both wrong, in terms of not fully corresponding to the real state of affairs, how can one be better at being wrong than the other? The answer that Popper and others appear to have adopted is that the better theory is the one with a better ratio of true observations to false ones. Unfortunately, this seemingly sensible approach has severe logical failings. As a simple example, this approach makes the "truth" of a theory dependent on the number of predictions it makes, not just on the number of correct predictions it makes. For instance, a theory that makes 500 correct predictions but 150 incorrect ones would be judged as less true than a theory that makes 5 correct predictions and 1 incorrect one. Even more troubling (and complex) reversals of our intuitions exist, along with other logical failings, as summarized by Aronson et al. (1995, pp. 115-124).

Unfortunately, philosophers have been more successful in identifying the shortcomings of various ways of defining truth than in creating unassailable definitions. Still, realist philosophers continue to struggle with how to define truth. One recent noteworthy effort comes from Aronson et al. (1995) in their book *Realism Rescued*. Their work focuses on "type-hierarchies," a kind of branching model that Aronson and colleagues believe can be used to characterize scientific theorizing. As a simple example, one could have a type-hierarchy with "animal" at the top, branching down to "vertebrates" and "invertebrates," with "mammal" being one of the branches out of "vertebrates," and so on. Aronson and colleagues argue that such type-hierarchies are at the core of scientific theories. They also contend that there are natural kinds in the real world that are structured in the same sort of hierarchical relationship as in the type-hierarchies of our theories. Based on these assumptions, Aronson et al. define the *verisimilitude* of a theory as the extent to which the theoretical system models that part of the real world to which the theory applies. Aronson et al. even offer a specific index of verisimilitude based on Tversky's (1977) index of similarity. Truth, then, according to Aronson et al. (1995), is "a limiting case of verisimilitude. The truth of a theory occurs when the chunk of the hierarchy picked out by the theory exactly resembles the actual hierarchy" (p. 10).

This formulation may seem unexceptional, and it probably falls short of a completely satisfactory definition of truth—which, is after all, a rather tall order. Nevertheless, the Aronson et al. framework is noteworthy for several reasons. First, it points out that there can be more to realism than such general assumptions as that there is an external world, imperfectly known. Realist scholars continue to work on such issues as how we can define truth and whether we should try. Second, Aronson et al's. claim about the relationship between verisimilitude and truth may better illuminate what thoughtful scholars of research methods, such as Cook and Campbell, may have had in mind when they defined validity in terms of approximations of the truth. At least, it may give the rest of us another, perhaps better, way of thinking about such definitions.

Third, the type-hierarchy approach used by Aronson et al. raises some interesting questions about validity errors, as discussed in a subsequent section on mechanisms. Finally, Aronson et al's. work stimulates an interesting question about whether validity is—or should be thought of as—an attribute of our results (or, stated differently, our conclusions or inferences), or whether validity is an attribute of our methods.

To see how Aronson et al.'s framework raises the latter question, consider its translation from the more encompassing theories for which it was developed, to the cause-probing of a single experiment or quasi-experiment. An experiment can provide an estimate of the average effect size of the treatment on the outcome measure. Assume, as realists do, that there actually is some average effect-size estimate in that chunk of the world picked out by the study. Verisimilitude then can be defined as the extent to which the study's effect-size estimate matches the real average effect size. Truth is the limiting case in which the two match exactly. Verisimilitude and truth, then, are defined, but where is validity?

Is validity, as used by Campbell and his colleagues, equivalent to what Aronson et al. and many other philosophers mean by verisimilitude? Or is validity a separate, procedural concept? That is, could the term *validity* be used to refer to the attributes of research procedures, and the terms *verisimilitude* and *truth* to refer to the accuracy of the conclusions and inferences we draw? This would not be totally inconsistent with the use of the term *validity* in logic. Philosophers generally use the term *validity* to refer to an argument in which, *if* the premises were true, the conclusion would be true. In logic, validity does not refer to the truth-value of a statement. One can think of validity in logic, instead, as a procedural attribute. A valid argument "follows the rules" that will give a true conclusion *if* the premises are true.

A procedural view of truth itself is embodied in the work of Putnam (1981, 1990), a contemporary American realist. Putnam (1981) presented a conception of truth that he sees as an alternative to the view of truth as correspondence. He summarizes that position in a preface to a later book:

> To claim of any statement that it is true, that is, that it is true in its place, in its context, in its conceptual scheme, is, roughly, to claim that *it could be justified were epistemic conditions good enough*. If we were to allow ourselves the fiction of "ideal" epistemic conditions. . . , one can express this by saying that a true statement is one that could be justified were epistemic conditions ideal. (Putnam, 1990, p. vi)

Put simply, epistemic conditions refer to the circumstances and ways in which we can "put nature to the test." Putnam illustrates the concept of ideal epistemic conditions in terms of the assertion that "there is a chair in my study." In this case, the ideal epistemic conditions would be met if one could be in Putnam's study, with adequate lighting, with nothing obscuring one's vision, with no visual or mental deficiencies, and so on, to observe whether or not a chair is there. (Although this example may sound like it, Putnam's position is decidedly not naive realism, for he does not suggest that ideal epistemic conditions are generally—or ever—met.)

In thinking about validity, we might borrow Putnam's concept of ideal epistemic conditions. In the case of program evaluations and other social research, the ideal epistemic conditions would presumably be ones in which there were no validity threats influencing the results. That is, validity could be conceptualized as an index of the degree to which the actual epistemic situation approximates or deviates from the ideal epistemic situation. The more numerous and more powerful the validity threats that are present, the lower the validity—that is, the worse the epistemic conditions. Even the best epistemic conditions actually available for social research, however, are not truly ideal (Reichardt & Mark, 1998). Even the optimal experiment can give us the wrong answer, if only by chance. In applied social research, the best available epistemic conditions (process) do not necessarily provide truthful conclusions (outcomes).

So what conclusions, if any, are to be drawn from all of this? One is that the conception of validity in the work of Campbell and his associates is perhaps less clear than it might be. Of course, such a comment could be applied to many, many others. Nor is this comment meant to detract from the major accomplishment of Campbell and his collaborators. Indeed, one might argue that, by slightly sidestepping intractable issues and getting on to important business, Campbell and his colleagues followed the better, and more productive, route. Nevertheless, a second possible, tentative conclusion is that social researchers might benefit from a more complex model of validity and related concepts. Following Cronbach, it has often been said that validity is an attribute of inferences. As researchers, we often turn around and discuss the validity of research designs. Although this may simply be a kind of useful shorthand, perhaps instead it is diagnostic that we need a more differentiated set of concepts to apply.

One possibility would be apply the concepts of verisimilitude and truth, as derived from Aronson et al. (1995), to our *inferences,* while applying the concept of validity to our *methods.* Cook and Campbell's definition might then be revised something along these lines: "We shall use the concepts *validity* and *invalidity* to refer to shortcomings in research procedures that reduce the likelihood that one's research results will approximate truth." Verisimilitude and truth could also be defined, perhaps along the lines of Aronson et al.

At the very least, this might clarify our conversations and debates. At the same time, I must acknowledge that any other practical consequences of an improved conceptualization of validity may be limited. For the practicing researcher, there may be little day-to-day impact of the precise way in which validity is defined, at least within certain limits. Studies are likely to be designed and carried out the same regardless of whether validity is defined in terms of the best available approximation of the truth, correspondence, degree of verisimilitude, or the ideal epistemic situation, and regardless of whether truth, verisimilitude, and validity are explicitly differentiated. (On the other hand, sharply different definitions of truth, validity, or similar concepts may be associated with differences in practice, as illustrated by the more divergent positions in the qualitative-quantitative debate, e.g., Smith & Heshusius, 1986.) If the implications of the proposed change are limited, this is probably in part because most researchers take a commonsense position. Researchers wisely do not require that all the puzzles of epistemology and ontology be solved before they go about their business.

Still, there may be some beneficial consequences of an improved model of validity, if only for the way we think about our work. As Hillary Putnam (1990) has written, "*Of course* philosophical problems are unsolvable; but as Stanley Cavell once remarked, 'there are better and worse ways of thinking about them.' " (p. 19). *Perhaps* a more differentiated system, including verisimilitude, truth, and validity, would be a better way of thinking.

REALISM, TYPES OF VALIDITY, AND THE EXPERIMENTING SOCIETY

It may be, then, that modest enhancements in our conceptualization of validity per se will not greatly affect practice. On the other hand, validity typologies, which effectively tell us what different *types* of validity we should worry about, probably have more impact on practice. Cronbach (1982) acknowledged this influence, for example, when he contended that the classic Campbell and Stanley (1966) framework, by describing internal validity as more important than external validity, led to possible errors in evaluation practice. In this section, I examine the validity framework of Campbell, primarily as presented in Campbell and Stanley (1966) and in Cook and Campbell (1979). I examine this framework from the lens of selected notions from contemporary realism and then consider possible implications for the experimenting society. By way of preview, the problem is not what the Campbellian framework does; rather, it is that many of us in applied social research, including evaluation, have assumed that this framework suffices for our needs and have not sought out or developed other frameworks that may also be important for our work.

Realist Claim: We Must Understand Structures as Well as Causal Relations

Many contemporary realists assume that the world can be described in terms of underlying structure as well as in terms of causal relations. A focus on underlying structure involves attention not only to the composition of things but also to classification, often through the development of a taxonomy or category system. This focus, of course, is not unique to those who publicly wave the realist banner, but it is a major theme in 20th century realism. For example, according to Bhaskar (1978), "Science. . . is concerned with both taxonomic and explanatory knowledge: with what kind of things are there are, as well as how the things behave. It attempts to express the former in real definitions of the natural kinds and the latter in statements of causal laws" (p. 20).

The explanatory function that Bhaskar mentions, which is addressed in more detail later, can be seen primarily as involving the discovery of causal powers. The taxonomic work of science that Bhaskar also identifies is exemplified in such historical examples as the development of the periodic table of elements, the identification of the composition of water, the laying out of different genera and species of plants and animals, and the discovery of the structure of genes. Taxonomic work involves identifying the composition of things (e.g., what is water made of?) as well as the classification of cases into categories (e.g., which creatures are mammals?). The taxonomic work of science or, more generally, the study of underlying structure is also evident in Aronson et al.'s (1995) emphasis on type-hierarchies. When these and other realists talk about the taxonomic work of science, they often refer to the concept of natural kinds. Natural kinds refer to clusters of entities that share common properties, with the properties being determined by the relevant laws or causal structure of the world (cf. Aronson et al., 1995, p. 39; Boyd, 1991, p. 129).

Elsewhere, Bhaskar (1978) says, "Science consists in a continuing dialectic between taxonomic and explanatory knowledge; between knowledge of what kinds of things there are and knowledge of how the things behave" (p. 211). That is, according to Bhaskar, the study of structure and the study of causal relations interact and support each other. For example, seeing that alternative chemical agents (e.g., acids vs. bases) have different effects contributed to the development of a classification scheme for these agents. In turn, having a reasonable classification scheme facilitated additional study of the effects of each type of agent. Understanding of structure underlies and supports understanding of causal powers, and vice versa.

At this point, I should acknowledge that there is some risk in applying contemporary realists' notions about underlying structure directly to the social sciences. Many realist theorists take their examples from the physical sciences, and it is easier to see their applicability in that realm. For example, it seems far less controversial to apply the notion of natural kinds to items from the "natural world," such as carbon versus hydrogen, than to human constructions such as marriage or social programs. Even realists can wonder whether social structures, unlike the structures of the natural world, exist independently of humans' conceptions of them (e.g., Bhaskar, 1979, pp. 47-56).

We do not need to assume, however, that there is some single periodic chart to be discovered for social programs so that they can be decomposed into a specific single form of natural kinds the way natural elements are. Instead, I believe, there is a more commonsense position that still leads us to act as if the concept of natural kinds applies to social programs and other social entities. We can acknowledge that the structures of social programs and other social entities may have fuzzy boundaries and may shift over time and context. We can acknowledge that different levels of analysis, and even different structural schemes within a given level of analysis, can appropriately be applied; realism does not imply the belief of a "unique best taxonomy" (Aronson et al., 1995, p. 43; Hacking, 1991, p. 111). We can even acknowledge that our attempts at identifying structure will likely contain some features that are merely conventions and have no real relationship to underlying structure (Boyd, 1991). All that is required is that we believe that social entities, including social programs, have structure; that some local projects (or entire programs) fall into the same grouping while others fall into different groupings; that we can identify aspects of structure, although imperfectly, in a way that corresponds at least in part to the real underlying structure; and that knowing about structure will tell us something useful about the effects and merits of a program. In other words, we need only to assume that it is useful to act *as if* social programs can be classified in terms of natural kinds (Julnes & Mark, 1998; Mark, Henry, & Julnes, in press). Although in principle someone could disagree with this assumption, I wonder: How and why would a disbeliever, who instead believes in an undifferentiated social world, engage in the practice of evaluation?

Implications for a Validity Framework

In the Cook and Campbell validity framework, to the extent that questions of structure fit, they apparently fall within construct validity, defined as "the approximate validity with which we can make generalizations about higher-order constructs from research operations" (1979, p. 39). (In the Campbell and Stanley framework, with internal and external validity, underlying structure appears to have no place, except as construct validity is foreshadowed.) In the Cook and Campbell framework, construct validity focuses on the relationship between the specific research operations used in a study (either manipulations or measures) and the abstract labeling of them (cause or effect constructs).

It would be unfair to the validity frameworks of Campbell and Stanley and of Cook and Campbell (and to their creators) to criticize them for not giving more attention to structure. After all, Campbell and his associates' objective was to clarify the validity of cause-probing research, such as experiments and quasi-experiments. This was a sizable task, and generations of social science researchers have benefited from their work. Nevertheless, it may be a worthwhile exercise to ask how these validity frameworks would differ if Campbell and his collaborators had chosen to focus, equally or more, on the study of structure, as is emphasized by many realists. I certainly do not give a complete answer here, for this would require presenting in full such an alternative validity framework, but I suggest some areas of difference.

First, such a validity framework would highlight that the study of structure is not just an issue of avoiding mono-operational and mono-method bias and avoiding experimenter and subject effects (cf. Cook & Campbell, 1979, on construct validity). It would also make clear that the study of structure is not just a matter of the empirical pattern of relations between measures (or manipulations), as a misapplication of the classic Campbell and Fiske (1959) article to the broader issue of structure might seem to suggest. Instead, the study of structure also involves interaction among work on (a) underlying composition (e.g., what are the core features of integrated services models that differentiate them from other approaches to service delivery?); (b) the attributes of different program types, including treatment effects (e.g., what if any effects do integrated services provide that other systems do not?); and (c) the taxonomies that can used to describe structure (e.g., where in a classification system do integrated services models fit relative to other service delivery models? Are there actually different subtypes of integrated services models?).

Second, the study of structure accordingly will often be facilitated by methods other than the cause-probing methods of experimentation and quasi-experimentation. Instead, a validity framework that emphasized structure as much as causal relations would include, and encourage, research on composition and on classification, as well as theorizing about typologies. These additional kinds of research will raise validity issues that are somewhat different from those that arise in cause-probing research.

Third, if a new validity framework includes these other kinds of studies, in addition to experiments and quasi-experiments, it also needs explicit consideration of the validity associated with merged evidence from different kinds of studies. What validity issues arise, for instance, in integrating the findings from an earlier cluster analysis, used to infer program subtypes, into a quasi-experiment used to estimate treatment effects?

Implications for the Experimenting Society

Even if we never see a new validity framework that more strongly emphasizes structure, the study of structure could be addressed more in practice in the evaluation of social programs and policies. Campbell, however, appeared to be cautious about such a focus. He perhaps most explicitly addressed the issue of structure in evaluation in a 1986 chapter in which he sought to relabel internal and external validity for applied social science, including evaluation. Campbell (1986) suggested that internal validity be renamed as "local molar causal validity." "Local" was meant to imply that the causal relation is demonstrated in some particular setting and time. "Molar" was intended to imply that the treatment may be a complex package, rather than a singly, pure causal variable.

> The molar approach assumes that clinical practice, participant observation, and epidemiological studies already have accumulated some wisdom, suggesting treatments that are worth further testing as molar packages. If these packages turn out to have striking molar efficiency, we will, of course, be interested in further studies, both clinically and theoretically guided, that will help us to determine which of several major conjectured components is most responsible for the effect. (Campbell, 1986, p. 69)

Thus, Campbell presumes that clinical practice and other sources will provide interventions worthy of study. The study of structure, at least in the sense of decomposing the treatment, at best is to come later in Campbell's view, at least as expressed in Campbell (1986).

There are at least five possible counterarguments to Campbell's suggestion that the study of structure receive less emphasis than the study of treatment effects in evaluation. First, the treatment packages that evolve from service delivery practice (or from other sources) may in fact include different types of interventions; for example, different types of interventions may be offered at one site than at another or, within sites, to one client than another. In such cases, an experimental evaluation would average across the different effect sizes associated with the different interventions. This was one of the key points, more than two decades ago, of the House, Glass, McLean, and Walker (1978) critique of the Follow Through evaluation. Unfortunately, evaluators have not since solved this problem, and more often ignore it than consider it. Serious attention to structure could help researchers distinguish between the different intervention types that fall under the same program label.

Second, increased emphasis on structure will aid in attempts to generalize (Cook, 1993; Cronbach, 1982; Mark, 1986). For instance, attempts to diffuse a program should be facilitated by knowledge of the essential features of each program subtype.

Third, Campbell's view may have overstated the cost-effectiveness of subjecting novel programs to experimental tests. Just as most spontaneous genetic permutations fail in nature, perhaps many if not most interventions that arise spontaneously in practice also will be failures. Perhaps social problems could be alleviated more efficiently not only by testing those interventions that spontaneously arise in practice but also by placing more emphasis on the study of structure. This might result in more frequent application of apparently more effective interventions (metaphors of selective breeding and genetic testing seem to apply here but carry some undesirable connotations).

Fourth, it appears that even if Campbell's preferred sequencing is correct, in too few cases does evaluation actually reach the systematic study of structure. That is, let us assume that Campbell is correct that studies of treatment effect should be the horse, and the study of structure the cart that follows. In practice, evaluators seem almost always to be riding the horse and almost never to be building the cart. At the least, evaluators may need to get to the study of structure more often and more quickly than is currently the case.

Fifth and finally, increased attention to structure can be accomplished with, at best, modest investment of evaluation resources (see, e.g., Conrad & Buelow, 1990) and can even be a side effort of evaluations designed to study treatment effects (see, e.g., Julnes, 1995).

For these and perhaps other reasons, some evaluators have focused on structure. Indeed, the evaluability assessment (e.g., Wholey, 1987) and program theory (e.g., Bickman, 1990; Chen, 1990) movements in evaluation can be seen in part as highlighting structure. These literatures often encourage evaluators to attend to the composition of a program, in terms of its various components; however, they often emphasize causal linkages, that is, the mediators of a program's effects, far more than structure. Moreover, these literatures appear to have had limited impact on the study of structure in evaluation. They have not generally inspired evaluators to focus iteratively on the composition of a program (or a set of projects) and on the program's (or projects') effects. Nor have they inspired widespread theorizing about the taxonomies that might be used to classify interventions within one or more areas (Lipsey, 1997). The literatures on evaluability assessment and program theory, in short, may have heightened evaluators' awareness of program structure as an issue but have had less impact on the systematic study of structure in practice.

Despite this pessimistic assessment, there are important examples of evaluation work that emphasize structure. In fact, there may a modest trend whereby small but increasing numbers of evaluators are acknowledging the importance of structure in their empirical and conceptual work. For instance, Lipsey (1992, 1997) illustrates the value of attending to the structure of interventions in the context of a meta-analysis of treatments for juvenile delinquents. Lipsey provides several examples that demonstrate the utility of addressing program structure, including an analysis of the effectiveness of interventions for youths institutionalized in juvenile justice institutions. In this analysis, Lipsey grouped treatments into four classes based on the nature of the intervention (e.g., skills training, cognitive-behavioral, and multiservice versus group counseling and challenge programs). That is, Lipsey devised a tentative taxonomy of juvenile justice interventions. He also categorized treatments based on the degree of integrity of treatment implementation. Program type accounted for sizable variation in effect size, as did treatment integrity. In short, accuracy of statements about program impact was greatly increased even by relatively coarse attempts at differentiating programs based on their underlying structure. (For other important examples in which evaluators emphasize the study of structure, see, e.g., Conrad & Buelow, 1990; Trochim, 1989).

Lipsey (1997) has also recently made an argument that overlaps largely with the realist case for the study of underlying structure in evaluation. Lipsey argues for what he calls "social intervention theory," which would describe the common form and processes of successful (and unsuccessful) programs and policies. According to Lipsey, "What is remarkable is that so much program evaluation has been completed and reported since evaluation emerged as a distinct field of study about thirty years ago but so little effort has been invested in finding general patterns in those findings" (p. 9). Perhaps this state of affairs seems less remarkable when we consider, first, that our predominant validity framework focuses on single studies, rather than on theory, models, or collections of studies, and second, that it emphasizes the estimation of treatment effects rather than the study of structure or mechanisms. Again, it would be unfair to criticize Campbell and his collaborators for failing to do that which they did not set out to do. Perhaps it *is* fair to criticize the field of evaluation, and others who would be in service of the experimenting society, for clinging too closely to a validity framework for cause-probing research, and for not finding or developing one for structure-probing research.

In short, it can be argued from contemporary realist scholarship that the study of underlying structure of social programs should be an important task in the experimenting society. Evaluators and others should be aware that this task can be addressed through a number of methodological and conceptual techniques. For instance, Lipsey uses the metaphor of building social intervention theory with the thousands of bricks of individual evaluation studies. This is the approach Lipsey took in his own meta-analytic work on juvenile justice interventions. Evaluators should remember, though, that this is not the only way to study structure. Some research might focus directly on the composition of a program or set of programs. Other empirical and theoretical work might focus directly on taxonomies.

Realist Claim: Causal Relations Involve Mechanisms

Recall that Bhaskar referred to the interplay in science between taxonomic work on structure and explanatory work. In dealing with the explanatory work of science, many contemporary realists emphasize "underlying generative mechanisms" (e.g., Bhaskar, 1978, pp. 45-52). If an intervention has desired (or undesired) effects, presumably this arises because of some casual sequence that the intervention sets into motion. For example, many interventions appear to be based on an implicit, probably simplistic model that providing information (about HIV, or drugs, or whatever) will lead to attitude change, which will in turn lead to behavior change that will result in the desired outcomes. In at least some of its formulations, the realist notion of underlying generative mechanisms seems to fit well with modern social science techniques that study mediation, such as various approaches to structural equations modeling. In the parlance of evaluation theory, realists' notion of underlying generative mechanisms corresponds to the sort of linked causal models that program theorists emphasize.

In contrast, Campbell's validity framework emphasizes causal relations between cause and effect variables, rather than models that focus on the mediators between cause and effect variables (this exclusion of mediating variables in Campbell's validity framework is perhaps most explicit in Cook and Campbell's 1979 discussion of what they call "developmental sequences," p. 62). Why did Campbell, a critical realist, choose not to emphasize the mediational models that seem more compatible with the realist notion of underlying generative mechanisms?

At least three possible explanations exist. One is pragmatic: Campbell may have chosen to ignore mediational models because his central task, addressing the validity issues that arise in studies assessing cause-effect relationships, was sufficiently large. A second possible explanation is historical: Campbell initially developed the internal-external validity distinction in reaction to the Fisherian analysis of variance approach to research methods in the 1950s (Campbell, 1986). The experiment was a powerful research technology at the time Campbell forged the pivotal distinction between internal and external validity. Modern techniques for studying mediation, such as structural equations models, did not exist at the time. In addition, mediational models were probably less frequent in the positivist-influenced social sciences of that time—and in the philosophy of science of the day. Given the historical context in which it occurred, it is not surprising that Campbell's ground-breaking work did not focus on mediation.

A third explanation is more conceptual and would account for Campbell's choice not to emphasize mediation even in later discussions of validity types (e.g., Campbell, 1986): Campbell was an evolutionary critical realist who emphasized the identification of *manipulable solutions* to problems. As noted earlier, Campbell thought that practice would give us molar treatment packages that would be worthy of experimental or quasi-experimental tests designed to assess their merit. Concern about mediation presumably took a back seat in Campbell's thinking to the more immediate task of identifying successful interventions.

Perhaps, though, the question as I formulated it is not the real question of interest. Perhaps we should not ask why Campbell did not emphasize mediation. He did not, and instead he and his colleagues developed a valuable validity framework for research that tests cause-effect relationships. Perhaps we should instead ask whether researchers in the social sciences, including applied areas such as evaluation, need also to seek out or develop validity frameworks that give more attention to mediation.

Implications for a Validity Framework

Before we address possible changes, consider how mediation can be addressed within the validity frameworks to which Campbell contributed. There are at least two possible ways to see mediation in terms of the Cook and Campbell validity framework. First, mediation can be seen as an extension or at least a cousin of construct validity. If cause-and-effect constructs are correctly identified, this usually helps pin down the mediational process, at least to some extent.

Alternatively, and more satisfactorily, one could apply the entire Cook and Campbell framework sequentially to each paired linkage in a mediational chain. Consider, for example, the following mediational model:

intervention → attitude change → behavior change.

In this simple example, we could first apply Cook and Campbell's validity categories to the "intervention → attitude change" link, and then apply it to the "attitude change → behavior change" link. With this extension, one could make the case that, even though Campbell did not emphasize mediation, his validity framework can be applied readily to it.

Some complications can arise with this approach, however. Imagine, for example, that the attitude measure in the preceding mediational change actually consisted of two components, cognitive and affective. Perhaps the intervention impacts the cognitive component of attitude, and perhaps, in our independent test of the "attitude change → behavior change" link, we unknowingly instead manipulated the affective component of attitude. We might conclude that we have demonstrated mediation when this is not the case. Such problems can, of course, be addressed within the Cook and Campbell framework, in this case in the context of construct validity. Nevertheless, they highlight the need for some expansion and perhaps translation when moving from the Campbellian framework to mediation.

Perhaps more important is what is missing without a validity framework specifically focused on mediation and underlying mechanisms. Many, if not most, contemporary quantitative researchers probably think of structural equations models and similar statistical procedures for testing mediation (see, e.g., Kenny, Kashy, & Bolger, 1998) as *the* method for studying mediation. That, however, is not the case. Mediation (or, more generally, the study of causal process) can also be studied in other ways (Mark, 1986, 1990). For instance, a presumed mediator can be manipulated directly, with the researcher attempting either to block or to enhance its operation, or the researcher could test the differential predictions of one mediational model versus another, in terms of the conditions under which treatment effects will occur. Furthermore, statistical techniques are not the only way to trace causal links: Qualitative methods also can serve this purpose. The work of Campbell and Stanley and of Cook and Campbell not only clarified validity considerations but also awakened researchers to design options they might otherwise never have considered. If a similar validity framework were developed with an explicit focus on underlying generative mechanisms, it might similarly enhance research practice concerning the study of mediation.

An expanded validity framework should also provide a language or procedure for describing the errors in one's inferences about mechanisms. Using the language suggested earlier, a revised framework might describe not only the (procedural) validity problems associated with research procedures but also the ways in which verisimilitude can be high (approaching the truth) or low. One way to do this involves the use of type-hierarchies, as suggested by Aronson et al. (1995). Figure 6.1 presents a simplified, partial representation of a type-hierarchy describing some of the subcategories of mechanisms that may fall under the general category of "attitude change." For example, working down the leftmost branch, some attitude change presumably occurs because of mechanisms that involve cognitive consistency (e.g., the desire people have to be consistent in their opinions and behavior). Two subtypes of cognitive consistency are cognitive dissonance, which posits that inconsistency creates discomfort, which motivates people to change, and self-perception, a similar concept but without the psychological discomfort and motivation.

Using the type-hierarchy in Figure 6.1, we can see that different types of errors are possible in inferences about mechanisms. Although there is no inconsistency between Cook and Campbell's validity framework and these errors, the use of the type-hierarchy does highlight different kinds of problems:

- First, the identified mechanism may be at too general a level. For instance, one might conclude that the mechanism is attitude change when the more accurate conclusion really is self-perception. The overly general conclusion is not false, in a logical sense, and in some cases might not cause any problems. On the other hand, if too general a mechanism is identified, problems in generalization may sometimes arise. For instance, program developers at other sites may attempt to trigger attitude change through some other mechanism that may not be as successful.

- Second, the identified mechanism may be at too specific a level. For example, one might conclude that the mechanism is postdecision dissonance when the real mechanism is actually dissonance (of any kind). This error could cause program staff and future program planners to ignore alternative ways of stimulating the real mechanism, dissonance.

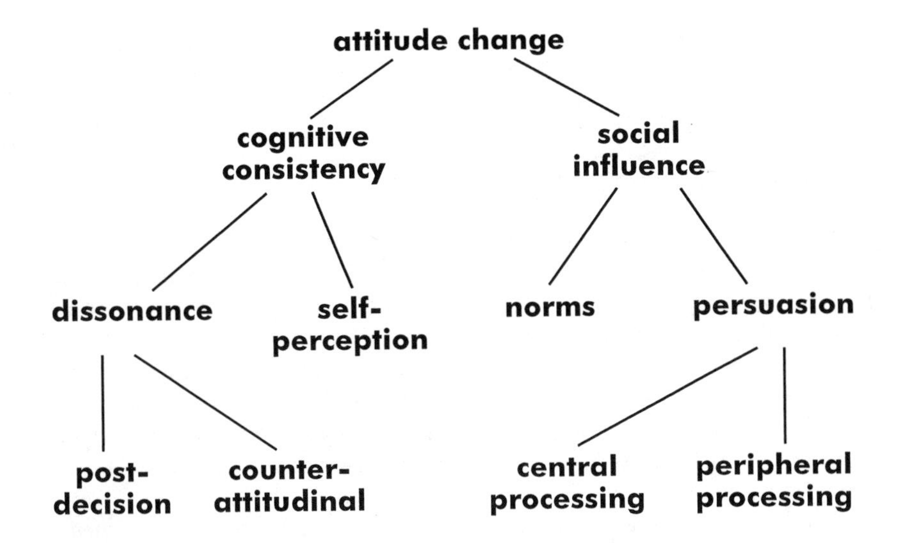

Figure 6.1. A Partial Type-Hierarchy of Mediational Processes Involving Attitude Change

- Third, the identified mechanism, although at the right level of specificity, may be the wrong mechanism. For example, this error would arise if the real mediator was dissonance and the evaluator concluded that it was self-perception. It is in this case that the Aronson et al. (1995) strategy of measuring verisimilitude is easiest to apply: If the real mediator is dissonance, self-perception is a closer wrong answer—that is, has higher verisimilitude—than, say, persuasion. Verisimilitude can be estimated roughly as the distance between two mechanisms in the type-hierarchy. Thus, lesser verisimilitude would occur if the actual mechanism is dissonance and the evaluator identified some mechanism that is not even within the attitude change type-hierarchy (e.g., a mechanism that fell under a different general category, such as compliance or obedience).

- Fourth, conclusions can be in error as to the operation of multiple mediators. For example, a program may be mediated by both dissonance and normative social influence, and the evaluator may fail to note the operation of both. Alternatively, the evaluator may identify two mediators when in fact there is only one.

- Finally, combinations of errors may occur. For example, one might select an overly general mechanism and also fail to recognize that another mediator is operating.

Note that all these types of error about mediators can occur in a study without internal validity problems, and even without any of the threats to construct validity specified by Cook and Campbell. Also note that this type-hierarchy approach can be applied more widely. For instance, similar errors can occur in the labeling of the cause (construct validity) when a "molar cause" is examined in an experiment.

Implications for the Experimenting Society

The study of underlying mechanisms has in fact received a good deal of attention in the practice of evaluation in recent years. Following the program theory movement, it has become common for evaluators to develop a graphic representation of the presumed causal flow of the program. Sometimes the sequential models are even tested. Pawson and Tilley (1997), drawing on realism, have proposed an evaluation theory that places the testing of causal mechanisms at the core of evaluation. Thus, it is not difficult to envision an evaluation practice that incorporates the realist theme of underlying generative mechanisms; we are part of the way there.

More difficult is a set of questions about how to balance the study of mediation with other concerns. How would an evaluation practice look if it included attention to structure as well as attention to mediation? When should we emphasize the study of structure, and when the study of mechanisms? When are both of these less important than other issues, such as the estimation of program effects to assess a program's merit and worth, as advocated by Campbell? Assuming the study of mediation is worthwhile in a particular case, how are we to conduct it if we cannot specify a tentative but plausible mediational model at the start of an evaluation?

These are issues for a theory of evaluation that can provide guidance about the types of activities evaluators should engage in under different circumstances. Such a theory should be informed by a perspective on the way in which evaluation findings can contribute to social betterment. Elsewhere, Gary Henry, George Julnes, and I have attempted to lay out such a theory (Mark, Henry, & Julnes, in press). One of our core premises is that evaluation can serve the cause of social betterment by contributing information and enlightenment to the democratic institutions and processes that are charged with creating, managing, overseeing, and improving social programs and policies. From this perspective, the answers to questions such as when to study structure and when to study mechanisms depend on an analysis of the likely contribution each could make, in a particular set of circumstances, to democratic efforts at social betterment. Although this answer may seem commonsensical, a review of the evaluation literature suggests it may not be so apparent. That is, many evaluation theorists and practitioners fail to consider the options that exist for inquiry and instead advocate carrying out the same sort of evaluation in virtually all conditions. Mark et al. (in press) suggest that instead one should recognize, first, that evaluation can serve different purposes (e.g., assessment of merit and worth, oversight and compliance, program improvement, and knowledge development) and, second, that evaluators can engage in different approaches to systematic inquiry in service of these alternative purposes. In short, if evaluation takes the study of mechanism and the study of structure seriously and, at the same time, recognizes the various ways that evaluation can contribute to social betterment, a complex framework is required for decision making about the planning of evaluations (see Mark et al., in press).

Realist Claim: Nature Is Embedded, Stratified, and Hierarchical

Contemporary realists propose that nature involves a complex layering of levels. Various conceptually overlapping terms have been used in advancing this point. For example, Pawson and Tilley (1997, p. 64) refer to "the stratified nature of social reality." By this they mean that human actions are embedded in broader systems of social rules and institutions, and that it is only within such contexts that human actions make sense. They give the example of a person signing a check. This act is meaningful only within the context of a social system with certain kinds of institutions, including a money and banking system, and certain kinds of rules, such as paying for one's groceries before taking them from the store. Individual actions are embedded within, and only interpretable within, a social structure. Bhaskar (1978) makes an even more general point, differentiating between different levels or "domains"; in short, he contends that unobservable entities and processes underlie actual experience. Aronson et al. (1995) point out that different type-hierarchies, involving different levels of analysis, may be applicable depending on one's goals and interests.

I will not examine these notions in detail, nor will I consider the distinctions among these and other realist writings on other related notions. Instead, I will briefly address one implication for evaluation practice, and one for our expanding validity framework, of the realist view of nature as hierarchical, stratified, and embedded.

Implications for the Experimenting Society

A key implication of this notion for evaluation is that there is not a single intrinsically correct level at which to study underlying mechanism or structure. For example, rather than examine some program's mediation at the psychological level of attitude changes, one might alternatively investigate a physiological level of neuropsychological change, or a sociological level of norms and roles. How, then, is one to select a level of analysis?

I believe that the answer, although perhaps challenging in practice, in principle is quite simple. For servants of the experimenting society, utility is the proper criterion for selecting a level of analysis for research on structure or mechanisms (Julnes & Mark, 1998; Mark, Henry, & Julnes, 1998). Imagine, for example, that assessing the role of attitude change as a mediator would likely help guide decision making about a program, but studying neuropsychological processes would not. In such a case, the choice seems obvious.

Implications for a Validity Framework

Most realists, from Popper to Aronson and colleagues, emphasize theories or models when they talk about the truth of the products of science. In contrast, although he certainly discussed theory, Campbell emphasized more the single cause-probing study. For instance, Cook and Campbell (1979) stated that "modern philosophers of science. . . have exaggerated the role of comprehensive theory in scientific advance and have made experimental evidence seem almost irrelevant" (p. 24). They went on to emphasize that the unpredicted or exploratory experimental result often has been key in advancing science.

Perhaps what is needed is an expanded framework that integrates these two extremes, by integrating single studies and theories, along with the concepts of validity, verisimilitude, and truth. At the lower level of single studies, validity, in a procedural sense, would predominate. Verisimilitude would not be emphasized at this level, because it is difficult to assess, even tentatively, for a single study, in that there is little or no converging evidence. To begin to assess verisimilitude (approximation to truth) and not just the quality of the research process, one needs Campbellian triangulation.

Moving up a step in the framework from single studies, we would come to aggregations or syntheses of studies, including meta-analyses and mixed-method syntheses. At this level, a combination of validity (in the procedural sense) and verisimilitude/truth judgment would apply. Elek-Fisk, Raymond, and Wortman (Chapter 2, this volume) demonstrate how the Campbellian view of validity can be applied to meta-analyses. In doing so, they also illustrate how a procedural approach to validity can be applied to aggregations of research. In addition, considerations of verisimilitude can be addressed at the level of multiple studies by assessing the degree of convergence across studies on some inference.

Moving further up to the theory/model level, the concepts of verisimilitude and truth come to apply even more, and the concept of validity even less. Even so, in such a multilevel model of validity, there would be vertical linkages, such that a theory/model would be linked to research syntheses, which would in turn be linked to individual studies.

The idea of such an interlinked system may help explain how fallibilists such as Campbell can maintain hope in the search for better answers. Campbell often cited Quine's reference to Neurath's metaphor of a boat needing repair while at sea:

> In science we are like sailors who must repair a rotting ship while it is afloat at sea. We depend on the relative soundness of all other planks while we replace a particularly weak one. Each of the planks we now depend on we will in turn have to replace. No one of them is a foundation, nor point of certainty, no one of them is incorrigible. (Campbell, 1969b, p. 43)

If we have the infrastructure of a theory or model, with notions of structure and mechanisms, there may be firmer footing as we examine the soundness of a plank from an individual study, and vice versa.

CONCLUSION

In this chapter, I have examined Campbell's work on the validity of experiments and quasi-experiments, as well as his view of the experimenting society, from the perspective of critical realism. This has led to several suggestions. First, the basic notion of validity might fruitfully be elaborated. One approach—deserving of further consideration, I believe—is to restrict *validity* to refer to procedural aspects (e.g., whether a design rules out history or not) and to use the terms *verisimilitude* and *truth* to refer to the accuracy of our inferences. Second, the concept of underlying structure might be dealt with in more detail, in an expansion of the Campbellian framework or in a complementary validity framework and also in our evaluation efforts in support of an experimenting society. This would involve an increased focus on composition and classification, on the development of taxonomies, and on the interaction between such work and the study of effects. Third, the study of underlying generative mechanisms, which has gained increasing attention in evaluation practice, might also fruitfully receive more attention in a complementary or expanded validity framework. As illustrated with a type-hierarchy involving attitude change, this could provide a language for better describing errors in our inferences, in addition to the more familiar framework for procedural validity threats. Fourth, another possible area for future development would be to extend the validity framework hierarchically, with theories and models linked to individual cause-probing studies.

These represent potential ways of continuing the work of Campbell on validity and on the experimenting society. They are all possible future evolutions of the validity framework devised by Campbell and his excellent collaborators, Stanley and Cook. If any of us aspire to improve on Campbell's work, we should keep our aspirations modest, relative to Campbell's own achievements. Earlier in this chapter, I quoted Putnam (1990): "*Of course* philosophical problems are unsolvable; but as Stanley Cavell once remarked, 'there are better and worse ways of thinking about them' " (p. 19). Methodological problems and issues of applying social science methods to alleviate social problems may or may not be unsolvable, but the work of Don Campbell has provided generations of social scientists with better ways of thinking about them.

REFERENCES

Aronson, J. L., Harré, R., & Way, E. C. (1995). *Realism rescued.* Chicago: Open Court.

Bhaskar, R. A. (1978). *A realist theory of science.* Atlantic Highlands, NJ: Humanities Press.

Bhaskar, R. A. (1979). *The possibility of naturalism.* Atlantic Highlands, NJ: Humanities Press.

Bickman, L. (Ed.). (1990). *Advances in program theory* (New Directions for Program Evaluation, no. 47). San Francisco: Jossey-Bass.

Boyd, R. (1991). Realism, anti-foundationalism, and the enthusiasm for natural kinds. *Philosophical Studies, 61*, 127-148.

Campbell, D. T. (1969a). Ethnocentrism of disciplines and the fish-scale model of omniscience. In M. Sherif & C. W. Sherif (Eds.), *Interdisciplinary relationships in the social sciences* (pp. 328-348). Chicago: Aldine.

Campbell, D. T. (1969b). A phenomenology of the other one: Corrigible, hypothetical, and critical. In T. Mischel (Ed.), *Human action: Conceptual and empirical issues* (pp. 41-69). New York: Academic Press.

Campbell, D. T. (1969c). Reforms as experiments. *American Psychologist, 24*, 409-429.

Campbell, D. T. (1972). On the genetics of altruism and the counter-hedonic components in human culture. *Journal of Social Issues, 28*, 21-37.

Campbell, D. T. (1974). Evolutionary epistemology. In P. A. Schilpp (Ed.), *The philosophy of Karl Popper* (pp. 413-463). La Salle, IL: Open Court.

Campbell, D. T. (1977). *Descriptive epistemology: Psychological, sociological, and evolutionary.* William James Lectures, Harvard University.

Campbell, D. T. (1984). Can we be scientific in applied social science? In R. F. Connor, D. G. Attman, & C. Jackson (Eds.), *Evaluation studies review annual* (Vol. 9, pp. 26-48). Beverly Hills, CA: Sage.

Campbell, D. T. (1986). Relabeling internal and external validity for applied social scientists. In W. M. K. Trochim (Ed.), *Advances in quasi-experimental design and analysis* (New Directions for Program Evaluation, no. 31, pp. 67-77). San Francisco: Jossey-Bass.

Campbell, D. T. (1991). Coherentist empiricism, hermeneutics, and the commensurability of paradigms. *International Journal of Educational Research, 15*, 587-597.

Campbell, D. T., & Fiske, D. W. (1959). Convergent and discriminant validation by the multitrait-multimethod matrix. *Psychological Bulletin, 56*, 81-105.

Campbell, D. T., & Stanley, J. C. (1966). *Experimental and quasi-experimental designs for research.* Chicago: Rand McNally.

Chen, H-t. (1990). *Theory-driven evaluations.* Thousand Oaks, CA: Sage.

Conrad, K. J., & Buelow, J. R. (1990). Developing and testing program classification and function theories. In L. Bickman (Ed.), *Advances in program theory* (New Directions for Program Evaluation, no. 47, pp. 73-91). San Francisco: Jossey-Bass.

Cook, T. D. (1993). A quasi-sampling theory of the generalization of causal relationships. In L. B. Sechrest & A. G. Scott (Eds.), *Understanding causes and generalizing about them* (New Directions for Program Evaluation, no. 57, pp. 39-82). San Francisco: Jossey-Bass.

Cook, T. D., & Campbell, D. T. (1979). *Quasi-experimentation: Design and analysis issues for field settings.* Chicago: Rand McNally.

Cronbach, L. J. (1982). *Designing evaluations of educational and social programs.* San Francisco: Jossey-Bass.

Hacking, I. (1991). A tradition of natural kinds. *Philosophical Studies, 61*, 109-126.

Harré, R. (1986). *Varieties of realism.* Oxford, UK: Blackwell.

House, E. R., Glass, G. V, McLean, L. D., & Walker, D. F. (1978). No simple answer: Critique of the Follow Through Evaluation. *Harvard Educational Review, 48*, 128-160.

Julnes, G. (1995, November). *Context-confirmatory methods for supporting disciplined induction in post-positivist inquiry.* Paper presented at the annual meeting of the American Evaluation Association, Vancouver.

Julnes, G. J., & Mark, M. M. (1998). Evaluation as sensemaking: Knowledge construction in a realist world. In G. Henry, G. W. Julnes, & M. M. Mark (Eds.), *Realist evaluation* (pp. 33-52). San Francisco: Jossey-Bass.

Kenny, D. A., Kashy, D. A., & Bolger, N. (1998). Data analysis in social psychology. In D. T. Gilbert, S. T. Fiske, & G. Lindzey (Eds.), *The handbook of social psychology* (4th ed., vol. 1, pp. 233-265). Boston: McGraw-Hill.

Lipsey, M. W. (1992). Juvenile delinquency treatment: A meta-analytic inquiry into the variability of effects. In T. D. Cook, H. Cooper, D. S. Cordray, H. Hartmann, L. V. Hedges, R. J. Light, T. A. Louis, & F. Mosteller (Eds.), *Meta-analysis for explanation: A casebook* (pp. 83-127). New York: Russell Sage Foundation.

Lipsey, M. W. (1997). What can you build with thousands of bricks? Musings on the cumulation of knowledge in program evaluation. In D. J. Rog & D. Fournier (Eds.), *Progress and future directions in evaluation: Perspectives on theory, practice, and methods* (New Directions for Evaluation, no. 76, pp. 7-23). San Francisco: Jossey-Bass.

Mark, M. M. (1986). Validity typologies and the logic and practice of quasi-experimentation. In W. M. K. Trochim (Ed.), *Advances in quasi-experimental design and analysis* (New Directions for Program Evaluation, no. 31, pp. 47-66). San Francisco: Jossey-Bass.

Mark, M. M. (1990). From program theory to tests of program theory. In L. Bickman (Ed.), *Advances in program theory* (New Directions for Program Evaluation, no. 47, pp. 37-51). San Francisco: Jossey-Bass.

Mark, M. M., Henry, G., & Julnes, G. J. (1998). A realist theory of evaluation practice. In G. Henry, G. W. Julnes, & M. M. Mark (Eds.), *Realist evaluation* (pp. 3-23). San Francisco: Jossey-Bass.

Mark, M. M., Henry, G., & Julnes, G. J. (in press). *Evaluation, making sense of social policies and programs: Description, classification, causal analysis, and values inquiry.* San Francisco: Jossey-Bass.

Pawson, R., & Tilley, N. (1997). *Realistic evaluation.* Thousand Oaks, CA: Sage.

Popper, K. (1972). *Objective knowledge: An evolutionary approach.* Oxford, UK: Oxford University Press.

Putnam, H. (1981). *Reason, truth, and history.* Cambridge, UK: Cambridge University Press.

Putnam, H. (1990). *Realism with a human face.* Cambridge, MA: Harvard University Press.

Reichardt, C. S., & Mark, M. M. (1998). Quasi-experimentation. In L. Bickman & D. J. Rog (Eds.), *Handbook of applied social research methods* (pp. 193-228). Thousand Oaks, CA: Sage.

Smith, J. K., & Heshusius, L. (1986). Closing down the conversation: The end of the quantitative-qualitative debate among educational inquirers. *Educational Researcher, 15*(1), 4-12.

Trochim, W. M. K. (1989). Pattern matching and program theory. *Evaluation and Program Planning, 12*, 355-366.

Tversky, A. (1977). Features of similarity. *Psychological Review, 84*, 327-352.

Wholey, J. (1987). Evaluability assessment: Developing program theory. In L. Bickman (Ed.), *Using program theory in evaluation* (New Directions for Program Evaluation, no. 33, pp. 77-92). San Francisco: Jossey-Bass.

PART II

Social Experiments

CHAPTER 7

Toward the Dream of the Experimenting Society

Robert G. St.Pierre
Michael J. Puma

One of Donald Campbell's most influential writings was his "methods for the experimenting society" paper (Campbell, 1971). Presented at the American Psychological Association meetings in 1971, this semi-utopian vision of a rational society in which social science would be used to inform policy making continues to be wonderfully appealing, even as we near its 30-year anniversary. In this seminal work, Campbell envisioned the experimenting society as

> one which will vigorously try out proposed solutions to recurrent problems, which will make hard-headed and multi-dimensional evaluations of the outcomes, and which will move on to try other alternatives when evaluation shows one reform to have been ineffective or harmful. (Campbell, 1971, p. 1)

One author's (St.Pierre) worn and tattered version of the 1971 experimenting society paper, edges brown with age, has a marginal comment in Campbell's handwriting that states "Needs much reworking. Suggestions, objections, citations all appreciated. DTC, Sept. 71." Although a revision of this paper eventually was published in a collection of Campbell's works (Campbell, 1988), we prefer the early, unpublished version, because it shows that Campbell himself was uncertain about whether an experimenting society was possible.

Campbell used the term "experimenting" in a broad sense, incorporating the idea that a rational society should be willing to experiment with, or try out, different solutions to social problems and subject these ideas to "hard-headed" evaluations. Here, hard-headed is used to describe a study that allows us to make the strongest possible causal attributions. Where the point is to assess the effectiveness of a social science program, this generally means implementing a study whose centerpiece is a randomized experiment, so as to do the best possible job of eliminating plausible rival hypotheses (explanations other than the treatment) for any observed program outcomes.

Campbell, however, never would have restricted us to a single research design or approach when trying to assess the impact of a program. Any and all methods that help to control or explain plausible rival hypotheses are useful. Because of their utility in controlling rival hypotheses, he was an advocate of randomized experiments where possible, and of strong quasi-experiments where randomized studies are not feasible.[1] The use of these alternative designs, however, is only part of a good evaluation study, and Campbell saw great value in using qualitative approaches to understand the program being evaluated, and to bolster our understanding of experimentally-derived findings, calling qualitative knowing absolutely essential as a prerequisite foundation for quantification (Campbell, 1984). We agree with this, meaning that our bias is for using randomized experiments as the best way of assessing the effectiveness of social programs, while also advocating the use of qualitative approaches as the best way of understanding what is going on in a social program.[2]

We do not intend this chapter as an examination of the relative merits of experimental versus non-experimental approaches, nor of qualitative versus quantitative data. In fact, the distinction is almost irrelevant if the point is to use the best research tool for the purpose at hand. We will only note that, when writing about the experimenting society, Campbell often phrased his questions in terms of finding out about the relative effectiveness of different approaches to solving a given social problem. If this is the question to be answered, then a study that includes a randomized experiment often is, in our view, the best way of getting an unbiased answer.

Mirroring the way he approached his own writing, Campbell urged evaluators to be critical and to continuously examine the relevance of the concept of the experimenting society. In this spirit, we ask: Have we realized Campbell's hopes for an experimenting society? If not, is it because his rationalism is incommensurate with today's complex global society? Is it a technical failure of not finding the right methodological tools? Is it because we have done a poor job of convincing the users of evaluation results that we have something important to contribute to their understanding of what works, for whom, in what circumstances, and why? In this chapter, we offer evidence about progress we have made and point to what we see as the most important problems blocking wider acceptance of the experimenting society.

THE EXPERIMENTING SOCIETY FOR LOW-INCOME FAMILIES

In the past 30 years, our society has taken important steps in designing, implementing, and evaluating programs to ameliorate social ills. Different approaches have been tried, and in many instances, policymakers and practitioners have relied on evaluation evidence to help them do a better job. Increasingly, experimental methods have been used to better understand the effectiveness of social programs. However, the process has not always lived up to Campbell's view of a systematic search for program innovation and improvement. Using our background in the evaluation of social and educational programs for low-income families, one of the areas where this nation has invested a huge amount of programmatic and evaluative effort, we examine the degree to which we have approximated Campbell's vision.

"Vigorously Try Out Proposed Solutions to Recurrent Problems"

Campbell saw the experimenting society as one that was constantly innovating, trying out new ideas, preferring action to inaction. Let us consider how this approach has been implemented with respect to programs for low-income families.

Despite three decades of persistent antipoverty efforts, the percentage of children living in poverty rose during the 1970s and 1980s and has remained stable during the 1990s (U.S. Department of Health and Human Services, 1998). Suggested causes include changes in family formation, with dramatic increases in the number of single- parent families; changing labor markets involving increased demand for technical skills, growing pressure to increase worker productivity, and increased competition in a global economy; and declines in social benefits to poor families (Duncan, 1991).

Through rhetoric and programmatic investments in antipoverty programs, our society has shown that we believe it is possible to intervene in the lives of families to disrupt or even break the cycle of poverty. Although there are arguments about whether the level of investment has been sufficient, there is no dispute that over the past 25 to 30 years, federal, state, and local governments, private foundations, and private industry have funded a broad array of interventions intended to interrupt the poverty cycle. These can be categorized into three major types:

- *Child-focused programs.* One approach to addressing the problems of poverty is to intervene in the lives of poor preschool children, in an attempt to improve the children's cognitive and social competence and prepare them to enter school on equal terms with more fortunate children. This approach is typified by Head Start, the major federal early childhood program for preschoolers, and other similar preschool programs. Such programs are typically part-day, part-year programs that deliver high-quality, intensive early childhood services to 3- to 5-year-olds. In some programs, these services include health and social welfare services as well as traditional cognitive and social services for children. (See, for example, Barnett & Boocock, 1998.)

- *Parenting programs.* These programs seek to affect children indirectly, by helping parents learn to care for their children in ways that will promote child development. Proponents of this approach (e.g., Missouri's Parents as Teachers or Arkansas's Home Instruction Program for Preschool Youngsters) believe that parents are their children's first and best teachers. Parents are expected to understand how children normally develop, to be able to detect developmental problems in their own children, and to be effective teachers so that their children can succeed. (See, for example, Clarke-Stewart, 1983, 1988.)
- *Adult-focused programs.* The third strategy for addressing the needs of low-income families focuses primarily on adults and, in particular, adult single parents. Welfare (i.e., the old Aid to Families With Dependent Children program, now Temporary Assistance for Needy Families, or TANF), welfare-to-work (e.g., the Job Training Partnership Act, the Job Opportunities and Basic Skills program, the Food Stamp E&T program, and an array of state-sponsored initiatives), and adult education all aim to move adults, particularly women with children, off welfare and into work and to increase their economic well-being. (See, for example, Gueron & Pauly, 1991.)

Certainly, different approaches to breaking the cycle of poverty for low-income parents have been tried. There has been a consistent long-term commitment to achieving the goal of self-sufficiency for these families, and both the public and private sectors have experimented with alternative strategies to achieving this goal. Consequently, at least within this one sector of public policy, trial-and-error experimentation has been the rule.

"Make Hard-Headed and Multi-Dimensional Evaluations of the Outcomes"

Over the years, our methodological tools for research design and statistical analysis have become increasingly sophisticated. We also have seen an increased reliance on program evaluation and monitoring to help determine whether programs are achieving their desired ends, and to find ways to make incremental improvements in program operations. In particular, federal agencies have called for randomized studies when they want to evaluate the effectiveness of their programs, and such studies typically are accorded greater weight in the decision-making process. All the intervention approaches described above have been studied, in part, by using randomized experiments.

In the 1970s and 1980s, Bob Boruch, a longtime proponent of the use of randomized experiments for determining the effectiveness of social programs, spent a substantial amount of time cataloging true experiments as a way to convince evaluators and decision makers of the feasibility of this approach (Boruch, 1974; Boruch & Wothke, 1985). In a recent conversation, Boruch (personal communication, 1998) acknowledged that he had moved away from this work, in large part because social experiments were becoming so numerous that tracking them required special expertise in each content field. For example, Greenberg and Shroder (1997) prepared a 500-page compendium of social experiments covering areas such as AFDC, food stamps, Medicaid, tax issues, electricity, health care, criminal justice, low-income children, youths, teen parents, unemployment, housing, substance abuse, mental health, the elderly, and homelessness.

As an example of the extent to which the social experiment is now seen as an important social science tool, consider the work done by the U.S. General Accounting Office (GAO), one of the major suppliers of information for congressional decision making. Up through the 1970s, the GAO relied primarily on case studies and small-scale reviews to provide information to congressional committees and individual members. When the Program Evaluation and Methodology Division was established in the late 1970s, however, the GAO began to conduct more systematic inquiries, and in a recent assessment of Head Start research, it criticized the Department of Health and Human Services for not conducting an experimental evaluation of the effectiveness of Head Start (U.S. General Accounting Office, 1997).

A preference for experimental evidence is not solely the province of researchers. For better or for worse, many congressional leaders have now become convinced of the utility of randomized trials for understanding the effectiveness of social programs and often include in legislation rudimentary specifications for high-quality evaluations. In addition to their normal reliance on testimony and anecdotes from program advocates, implementers, and participants, legislators and other program funders now routinely ask whether research and evaluation studies have been done on the program; in particular, they ask informed questions about the quality of the research design and the resulting validity of the evidence produced by such studies. The result is that studies based on strong research designs often are given greater weight in the policy-making process. For example, recent congressional testimony about the lack of effectiveness of one of Head Start's demonstration programs was accepted with little criticism, precisely because of the strength of the research design and of the study in general (St.Pierre, 1998). In the same hearing, committee leaders recognized that much of the research that had been presented as evidence of Head Start's effectiveness was based on studies of very small-model early childhood programs (Schweinhart, Barnes, & Weikart, 1993) and therefore called for a random assignment evaluation of project Head Start so that the "true effectiveness" of the program could be determined.

The approaches to dealing with the problems of low-income families have developed on the basis of real and perceived failures and successes, with evaluative evidence playing an important part in the process of adopting and discarding different solutions. Although evaluative evidence has been used for program assessment and improvement, and there is an increased appreciation for the strength of randomized designs, decisions have not always been as "hard-headed" as Campbell hoped for nearly 30 years ago. Reviews of research evidence about the effectiveness of child-focused, adult-focused, and parenting programs lead to the following conclusions:

- *Child-focused programs.* These programs have been found to help prepare children for school, and longer-term benefits have been detected for children entering public schools and beyond (Karweit, 1994). Research, however, has not found consistent evidence supporting the effectiveness of child-focused approaches. In general, program effects have been found to be highly associated with program intensity (i.e., the number of hours/days of instruction) as well as the extent and breadth of the services offered (Barnett, 1995; Yoshikawa, 1995).
- *Parenting programs.* These approaches have gained widespread popularity as another strategy for improving the lives of low-income families, but the research to date has shown that parenting programs alone are insufficient to improve children's developmental outcomes. On one hand, at least some important aspects of child development occur on their own timetable and cannot wait for the benefits of parenting programs to trickle down from parents to children; on the other hand, parenting programs do not typically provide interventions that are broad enough to address the range of issues confronting poor parents, all of which can affect child development (Ramey & Ramey, 1992).
- *Adult-focused programs.* A comprehensive review of the impact of welfare-to-work programs (Gueron & Pauly, 1991) concluded that although almost all led to small gains in earnings, most participants remained in poverty and continued to receive welfare benefits. Even mothers who obtained jobs through such programs frequently left or lost them because of a lack of transportation or child care, or because their jobs did not provide health benefits for their children. Similarly, reviews of adult basic education programs have concluded that adults' literacy skills and job opportunities have not been greatly increased by these programs, which are equivalent to instruction provided in Grades 1-8; by adult secondary education programs, which are equivalent to instruction provided in Grades 9-12; or by English as a Second Language (ESL) programs for individuals whose native language is other than English (Fischer & Cordray, 1995).

  Furthermore, little is known about child outcomes in families targeted by welfare-to-work programs, and some researchers (Ramey, Ramey, Gaines, & Blair,

> 1994) question the premise that adult education programs can benefit children at all, arguing that no studies have demonstrated that increasing parental job competence and self-esteem is sufficient to enhance short- or long-term outcomes for children. Again, these programs continue to proliferate despite mounting research evidence that at least some simply are not going to work as promised.

The above discussion shows that different strategies have been tried to improve conditions for low-income families; however, programs often are sustained despite strong research evidence that indicates a lack of effectiveness—well-meaning and hard-working people simply want to believe they are making a difference and resist abandoning their programs when evaluators bring them negative news. This is not surprising. We live in a society that is increasingly skeptical of science, especially given the confusing and conflicting stories in the popular press that claim great scientific breakthroughs that are then subsequently revised (e.g., cold fusion, cures for cancer), and grandiose promises often are made to get initial funding that then make it hard to admit failure.

The vision of a society willing to try different approaches to achieve desired ends seems to have been realized, but the process of sorting and selecting the best approaches to continue has not always matched Campbell's vision of rational decision making. In the current system, programs gain supporters and are difficult to dislodge even in the face of strong evidence that they don't work.

"Try Alternatives When Evaluation Shows a Reform to Have Been Ineffective"

As discussed above, there is compelling research evidence that single-focus approaches (i.e., those targeting only children, only parents, or only adults) have, at best, been only modestly successful, either individually or even when taken in combination (St.Pierre, Layzer, & Barnes, 1998). Early childhood education can increase children's cognitive development, but perhaps not as much as when parents also change their parenting skills. Parenting programs may sometimes improve parenting skills, but children's development does not show the hoped-for improvement. Neither of these programmatic approaches addresses outcomes such as parental economic self-sufficiency, and programs that do so directly have had quite modest success.

In response to these shortcomings, and in the spirit of Campbell's experimenting society, a new class of strategies called "two-generation" programs has gained popularity by recognizing the multigenerational, multidimensional aspects of family poverty and by trying to attack the problems associated with poverty from multiple directions. Hundreds of two-generation projects now exist across the nation, serving thousands of families, and funded by hundreds of millions of public and private dollars. Some examples of these programs include the Comprehensive Child Development Program, the Even Start Family Literacy

Program, New Chance (Smith, 1995), and the new large federal investment in the Early Head Start program (Love, 1998).

This is exactly the kind of self-critical, inventive approach that Campbell advocated for an experimenting society—when finding that a particular strategy is not working, take action to try something different. Two-generation programs have proliferated more because research has indicated the limitations of single-focus approaches rather than because research has demonstrated the benefits of two-generation programs. In fact, recent studies have indicated that federally funded two-generation programs, as initially implemented, are not proving to be a panacea (Quint, Bos, & Polit, 1997; St.Pierre, Layzer, Goodson, & Bernstein, 1997).

On the positive side, new ideas are being tested as envisioned by Campbell in his active, innovating, learning, evolutionary society. Where his vision has not matched reality is in his hope for the use of scientific principles (especially social experiments) to make decisions about which innovations to try, which approaches to continue and expand, and which approaches to abandon because they have been shown to be ineffective or harmful.

ACCEPTANCE OF THE EXPERIMENTING SOCIETY AT THE LOCAL LEVEL

One reason why innovation has been only partly guided by rational analysis, especially in the area of programs designed to improve the life condition of poor families with children, is frontline resistance to the evidence that evaluators seek to provide. Donald Campbell worried about problems that the experimenting society might cause for practitioners. In early writings, Campbell (1969) suggested that practitioners be protected by shifting "from the advocacy of a specific reform to the advocacy of the seriousness of the problem, and hence to the advocacy of persistence in alternative reform efforts should the first one fail" (p. 410). Campbell hoped that practitioners would align themselves more with the achievement of an ultimate goal (e.g., getting poor families out of poverty and ensuring the strong development of low-income children) than with the continuation of a particular approach or intervention. By the time he wrote *Methods for the Experimenting Society,* however, Campbell realized that this view may not always fit with the energy and zeal that is required to do a good job of starting and maintaining a social program.

It turns out that Campbell's worry was justified. Although the acceptance of, and reliance on, strong research designs for evaluating the effectiveness of social programs has become increasingly important at the federal level, this view has been difficult to replicate at the local level. Local practitioners face the problems of, on one hand, trying their best to design and implement a strong program, while on the other hand seeking to convince potential program supporters that their program "works" in order to maintain support and continued funding. As a consequence, most practitioners believe that their programs work and "see" improvements in the people they are trying to help. Campbell understood the importance of the "commonsense" information that program practitioners gather:

> We should recognize that participants and observers have been evaluating program innovations for centuries without the benefit of quantification or scientific method. This is common-sense knowledge which our scientific evidence should build upon and go beyond, not replace. . . . One should attempt to systematically tap all the qualitative common-sense program critiques and evaluations that have been generated among the program staff, program clients and their families, and community observers. . . . When such evaluations are contrary to the quantitative results, the quantitative results should be regarded as suspect until the reasons for the discrepancy are well understood. (Campbell, 1979, pp. 52-53)

Given Campbell's acknowledgment of the importance of this intuitively convincing, locally obtained evidence of the effectiveness of many social programs, how are program practitioners and evaluators to reconcile the claims of external evaluators who sometimes say that a program does not work, or that it works differently from the way that local staff members see it?

These conflicting views generally occur when data gathered by program staff at the local level are collected only on participants in the program. Those data often show that participants improve on relevant outcome measures, and hence program staff are convinced that their program is helping participants. They can see the improvements with their own eyes, and they have data to show that "gains" or "improvements" occur. And they are correct in these observations! Where they go wrong, and where Campbell would caution them, is that they attribute any improvements to the program they are running without considering alternative explanations. The function of impact evaluations—helping us understand whether a particular treatment caused changes in program participants—often is misunderstood by practitioners. To better see how these conflicts between evaluators and program staff occur, consider the following discussion. It is not a far-fetched dialogue.

Practitioner: [*resigned*] I have to appear at a school board meeting in a few months and defend my program for teenage mothers and their preschool children. I don't understand why they're giving me such a hard time. It's obvious that my program works.

Evaluator: How do you know?

Practitioner: I can see it with my own eyes. The mothers I work with are happier, and their children are learning new information. The children are better behaved in preschool.

Evaluator: Do you have any data to back up what you see?

Practitioner: [*thoughtful*] No, but I'll collect some data.

[*a few months pass*]

Practitioner: [*excited*] I'm all set! I collected data on families before they entered my program, and again at the end of the program. The families improved a lot! The mothers I work with report that they are happier after being in my program! Even better, their children's test scores went up! And to top it all off, family income went up!! (To be honest, I didn't expect THAT to happen!)

Evaluator: I see. The data DO show that your families improved. But do you know why they improved?

Practitioner: [*frustrated*] Because they were in my program!

Evaluator: Maybe so. But how do you know?

Practitioner: [*now angry*] Are you crazy? The families were in my program! The data I collected show that they improved over time! What other explanation is there?

Evaluator: THAT is a really tough question to answer. But it is exactly the kind of tough question that critics of social programs ask.

Practitioner: Well, how can I convince the critics that my program works?

Evaluator: One way is to do an experiment. Let's talk about why and how an experiment can provide convincing evidence about the impacts of your program.

Before offering advice on how to help bridge this gap between evaluators and practitioners, we need to acknowledge several additional factors that can contribute to the differences in perspective (Kennedy, 1997). First, evaluators may too often focus on *methods over message*. The evaluation literature is replete with methodological debates over research designs, statistical techniques, the choice of appropriate covariates in explanatory models, and other technical details. Although there is a large literature on the use of evaluation information (e.g., Patton, 1998; Weiss, 1998), an economist would say we have focused too much on the producer with little attention to the consumer, who, in fact, drives "the market" for research information.

Second, *scientific skepticism* has become more pervasive. As scientists and evaluators debate the technical details of research design and analysis, decision makers and the public are left confused about who to believe or whether to put any trust in science. The recent debate over the usefulness of mammograms is an excellent example of this problem, wherein a political recommendation superseded a scientific consensus and the public was left confused about what to believe (Plotkin, 1996). Similar examples are ubiquitous, and daily newspapers are filled with confusing and conflicting scientific results (e.g., claims about the nutritional benefits of particular foods change daily as new studies are done and researchers debate the validity of competing results). These debates among scientists lead the public to worry about the scientific process, and maybe worse, to worry that the problems we face are so intractable that solving them may be beyond our grasp.

Third, the problem is not that we are using the wrong method or research design, but that although our work is important to some research consumers (e.g., research funders at the federal, state, or local levels), it sometimes is *irrelevant* to other research consumers (e.g., program practitioners) because we are focused on questions that are unimportant to those who must deal with the daily realities of social programs (e.g., teachers in the classroom, case workers in the welfare office, health care professionals in hospitals and clinics, and social workers and criminologists dealing with modern-day dysfunctional behavior). It is not surprising that ever more precise answers to the wrong questions, or to questions that don't reflect practical situations, are ignored.

Finally, evaluation results are frequently *inaccessible* to those who need or are expected to use research findings. This is partly a problem of poor dissemination (physical inaccessibility) but is more often a consequence of conceptual inaccessibility (i.e., information not being presented in ways that people can understand). This inaccessibility is particularly problematic when research results suggest a need for a fundamental change in current practice or what Kuhn (1962) has referred to as a paradigm shift. For example, asking reading teachers to switch to whole language instruction from phonics is bound to meet with resistance that will not be helped by unclear and conflicting research findings. Research and evaluation findings need to be validated in the context of real-world programs for them to be accepted by practitioners.

IMPROVING THE DIALOGUE BETWEEN EVALUATORS AND PRACTITIONERS

Evaluators interested in furthering the experimenting society have an obligation to work more closely with and to establish an improved dialogue with program practitioners and program funders. Such an improved dialogue should both present the need for strong research and evaluation (using experimental methods where feasible to increase the validity of the attribution of observed effects) and seek to help us understand the different perspectives and contexts within which each of us work. These perspectives can be summarized as follows:

- *Evaluators* are focused on methodological issues that make our research more or less valid. We want to have confidence in the conclusions that we draw from our work, and as a result, we want policymakers and practitioners to listen to what we have to say.

- *Funders* are concerned about the use of their money to achieve the ends or social changes/improvements that they seek to attain, and about their public image and continuing ability to raise funds.
- *Practitioners* are concerned about the clients they are seeking to help and about doing their job in a way that minimizes the burden on themselves.

As Robinson (1998) has pointed out, "Narrowing the research-practice gap is not just a matter of disseminating research more effectively or using more powerful influence strategies. Such approaches assume that our research does speak to practice, if only the right people would listen" (p. 17). What is needed instead is both a better understanding of the needs of practitioners and a greater effort on the part of evaluators to place our work directly into the real-world experiences of program staff.

One part of our role as evaluators should be to teach program funders and practitioners the pitfalls of relying on simple evidence (e.g., one-group designs) and to help them understand the benefits that can be gained through implementation of randomized experiments. The experimenting society can make continuing progress if we help practitioners understand why they ought not to rely solely on their own observations to assess program effectiveness. In doing this, we have found it useful to present illustrations. Following are some examples that address the question raised by the program implementer at the end of the earlier discussion—"What other explanation is there?" Another way to put the question is "How do we know that the program caused the observed outcome?"

Example 1: Smashing a Telephone Pole

Imagine that a car is hurtling out of control down a steep hill. The car smashes into a telephone pole, which breaks into several pieces. An observer sees the whole thing and reports to the police that "the car broke the pole." An alternative way of saying this is that the program implementer (the car) administered the treatment (a smashing blow) to the program participant (the telephone pole), causing an effect (the pole breaking).

Hearing this story, we are pretty certain that the car caused the pole to break. We don't worry about alternative explanations or plausible rival hypotheses for the pole breaking. We don't wonder, for example, whether the pole was rotten and happened to fall over just when the car hit it, or whether the pole was hit and smashed by a meteorite at the same time as the car hit it. Both of these are rival hypotheses, but neither is very plausible. Nor do we feel the need to run an experiment to test the effect of a car hitting a pole. If we wanted, we could set up a hundred telephone poles, side by side, observe all the poles at pretest to be sure they all are standing, administer the treatment to half of the poles (smash them with a car) and not administer the treatment to the "control" poles, and then observe all the poles again at posttest, to see which ones had changed. Any difference between the poles in the two groups could be attributed to the treatment.

This experiment sounds pretty silly. Why? Because we expect telephone poles to stay standing unless something happens to them, like being smashed by a car. We don't expect telephone poles to spontaneously fall over. In other words, we don't expect to see any change in the outcome of interest unless a "treatment" is applied to it, and when the treatment is applied, we don't need to run an experiment to help us conclude that the treatment caused the outcome. We are willing to make that attribution on the basis of what we already know about the stability of telephone poles.

Example 2: Growing a Corn Stalk

Suppose we are interested in whether the use of fertilizer "X" leads to taller corn stalks. We plant a corn seed, apply fertilizer "X" weekly for 10 weeks, and measure the height of the corn stalk after the 10-week period. We see that the corn stalk has grown to a height of 3.0 feet. Would we attribute all of the 3-foot growth to the use of fertilizer "X"? No, because there is a plausible rival hypothesis—that the corn stalk would have grown and achieved a certain measure of height even without the fertilizer.

Determined to find the real effect of using fertilizer "X," we do an experiment. We plant two corn seeds (treatment and control) and apply fertilizer to the treatment seed weekly for 10 weeks. We do not use any fertilizer on the control seed. At the end of the 10-week growing period, we measure the height of each corn stalk and find the treatment corn stalk to be 3.0 feet tall and the control corn stalk to be 2.5 feet tall. We then calculate the "effect" of fertilizer "X" to be the difference between the gain in height of the treatment corn stalk and the gain in height of the control stalk:

effect of fertilizer "X" = 3.0 feet – 2.5 feet = 0.5 feet.

We would be wrong to attribute the entire 3.0-foot height gain to the use of fertilizer "X," because we expect to corn stalks to grow, even in the absence of the special fertilizer. If we didn't have the control corn stalk, we would have overestimated the amount of growth resulting from use of the fertilizer by 2.5 feet.

Example 3: Enhancing Child Development

Children are more complicated than telephone poles and corn stalks. They grow and develop in many ways and on different schedules. Furthermore, we are not as good at measuring children's growth and development as we are at measuring telephone poles and corn stalks.

Assume that we are interested in finding out whether a family literacy program enhances the cognitive development of preschool-age children. We find a family with a 4-year-old child, enroll the family in a high-quality family literacy program, administer a pretest to the child (let's say that the child scores 50 on the pretest), have the family participate in the program for a year, and then administer a posttest to the child (let's say that the child scores 70 on the posttest). We then calculate that the child gained 20 points (70 points at posttest –50 points at pretest) and draw the conclusion that the family literacy program caused the gain.

This is a very common analysis, one being conducted as part of hundreds of evaluations nationwide right now, and there is nothing wrong with the analysis—with saying that the child gained 20 points. There is something wrong with the conclusion—with the attribution of the 20-point gain to the family literacy program. Why? Because there is an alternative hypothesis for the gain—children develop and learn even in the absence of a special family literacy program. Even without a program, they would learn from their mothers, from their fathers, from their siblings, from their friends, from television, and so on. So the 20-point gain cannot be attributed entirely to the family literacy program.

How would we design a study to help us understand how much children really gain from participating in a family literacy program? We could find many families, each with a 4-year-old child, administer a pretest to each child, and find that the average child scores 50. Then we assign half of the families to participate in the family literacy program (treatment) and the other half to a control group. After a year, we posttest all the children and find that the children in the family literacy program had an average score of 70, while children in the control group had an average score of 68. Now, we calculate the effect of the family literacy program as the difference between the gain for the treatment children and the gain for the control children:

$$\begin{aligned}\text{effect of family literacy program} &= (70 - 50) - (68 - 50)\\ &= 20 - 18 = 2 \text{ points.}\end{aligned}$$

Our experiment controlled for normal maturation and showed that the effect of the family literacy program is not 20 points, as was assumed from the simple pre-post analysis, but is only 2 points. The experimentally derived findings lead us to very different conclusions than the non-experimental findings.

A Real-World Example

We have described some contrived examples demonstrating that if we do not think about plausible rival hypotheses for the outcomes that we measure, if we ignore normal maturation and development (of corn stalks or of human beings), then we can draw misleading conclusions about the effectiveness of a program. But is there "real" evidence that maturation/development of subjects can affect research results? Are there any data to support our assertions? Such evidence is very helpful in convincing program practitioners that what they often call a program effect is at least partly a developmental effect.

One example comes from a large-scale evaluation of the Comprehensive Child Development Program (St.Pierre et al., 1997). CCDP projects provided a wide range of educational, social, and health services to low-income families with the intent of improving the economic self-sufficiency of families and the cognitive, socioemotional, and behavioral development of children. A 5-year evaluation of this program was conducted in 21 projects; 4,000 randomly assigned treatment- and control-group families were included in the study.

The data from this evaluation show that the families participating in the evaluation improved over time on almost every outcome measure. For example, children's test scores increased, more mothers were employed, average household income increased, and the percentage of families receiving AFDC benefits and food stamps decreased.

Any program would be proud to have evaluation results like these, but the sobering fact is that these statements represent the progress, over time, of families in the randomly assigned control group. Hence, these are the normal levels of improvement that we can expect from low-income families in the absence of the CCDP program. An evaluation that ignored these normal levels of maturation and development would reach terribly wrong conclusions about the effectiveness of a program. In fact, the study found that the growth of the treatment group was identical to that of the control group and concluded that the program had no effect. This was despite the accurate observations of many local practitioners that something positive was happening to the children and families they were serving.

The CCDP evaluation provides just one example. Any large-scale random-assignment evaluation of a social science program that presents information about control group families reaches the same conclusion—control group families mature and develop over time, and the only way we can attribute effects to intervention programs is if the treatment group changes at a rate that is statistically different from that pattern.

ACCEPTANCE OF THE EXPERIMENTING SOCIETY BY EVALUATORS

Campbell was careful to warn that we should be open to the possibility that the experimenting society is unworkable. The clearest way in which the experimenting society will be shown unworkable is if we evaluators, those to whom Campbell handed much of the responsibility for building the experimenting society, are unable to convince ourselves that it is feasible or worthwhile to conduct strong research studies that allow us to control rival hypotheses when we are interested in assessing the effectiveness of a program. If we evaluators do not believe that research methods allowing the strongest possible causal attributions are possible, are important, and are necessary for understanding the effects of social programs, and if we are willing to draw conclusions about the effectiveness of social programs in the absence of experimental evidence (i.e., based on single-case studies, anecdotes, or flawed quasi-experiments), then we cannot hope to convince others of the value of the experimenting society.

In the past three decades, we have seen a great amount of research attesting to the feasibility of implementing field experiments in the social sciences (Boruch, 1974; Boruch & Wothke, 1985; Conrad, 1994). There is no question that randomized experiments are feasible and are being implemented on a daily basis in a variety of settings throughout the United States; however, there are practical problems to overcome.

It often is claimed that randomized studies are expensive, and it is true that randomized studies are more costly than weak alternatives. For example, it costs more to conduct a randomized evaluation or a strong quasi-experimental study than to conduct a one-group, pre-post study in which there is no control group. Furthermore, randomized studies can be difficult to implement, requiring time and effort on the part of the evaluation team to recruit field sites, to explain the nature and necessity of the randomization, and to conduct the randomization. The counterargument to these worries mirrors a popular bumper sticker: "If you think education is expensive just try ignorance." The fact is that any study that provides strong information about the effectiveness of a program is likely to be expensive and difficult to implement, and although a weak study can be cheap and relatively easy, why should we want to spend money on a study that does not provide the information we need?

Another charge sometimes leveled against randomized designs is that they are capable only of providing answers to simple questions. Some of the research questions currently being asked by funding agencies are quite complicated, taking the following form: "What components of the program work best with what type of clients under what circumstances?" Providing a serious answer to such a question would require a multifactorial randomized design, a study that is rarely practical in the real world. The problem here is that the question is complicated, not that the appropriate research design is defective.

Finally, it sometimes is argued that randomized studies take a long time. This can be true. It often takes on the order of 3 to 5 years to design, conduct, and report on the results from a randomized study of educational or social programs. Certainly, there are low-cost, short-turnaround studies that offer certain pieces of information more quickly than randomized experiments; however, there is no getting around the fact that those studies are limited to answering fundamentally different questions about program processes rather than questions about program effectiveness.

Given the pervasive push on the part of Congress, state and local agencies, and private foundations to obtain information about program effectiveness, there is a great temptation for funders and program practitioners to push evaluators to try to draw conclusions about a program's effectiveness through analyses of data from case studies or weak quasi-experiments. Given the need to make a livelihood by obtaining continued funding, there is good reason to worry about the willingness of evaluators to be hard-headed when assessing program effects.

Declaring programs effective on the basis of one-group, pre-post studies or one-group, post-only studies is common practice in evaluations conducted at the local level. For example, a recent study of the research designs used in more than 100 local evaluations of Even Start Family Literacy programs showed that 76% of the evaluations used a one-group, pre-post design, 31% of the outcome studies used a one-group, post-only design, 10% of the local outcome evaluations used a nonequivalent control group design, and no local evaluation used a randomized experimental design (St.Pierre, Ricciuti, & Creps, 1998).[3] Without exception, these studies variously declared the program being evaluated to be effective, to improve the lives of families, to raise the achievement of children, and so on, all in the absence of any evidence about what would have happened had families not been in the program.

If there is a group of social scientists who ought to be attuned to the issues involved in drawing causal conclusions about the effects of programs and who we might expect to practice the principles of the experimenting society, it is members of the American Evaluation Association (AEA). Evidence collected through attending sessions at recent AEA annual meetings, however, leads to worrisome conclusions. In particular, many studies presented by AEA researchers are based on one-group, pre-post designs or other quasi-experimental designs that are weak for making causal attributions.[4] Some of these are studies of highly visible programs, evaluated by well-known AEA members. There is nothing wrong with an evaluation that is not designed to make causal attributions; evaluations serve many other purposes. Even when the main purpose of an evaluation is formative, however, there is pressure to say something about whether or not a program "works." Faced with this pressure, it is amazing how often trained evaluators are willing to conclude that a program is "effective" on the basis of simple pre-post data.

As an example, one session at a recent AEA meeting focused on the evaluation of a well-known state-level program for early childhood family education. A nationally known evaluator has conducted research on this program for more than a decade. The evaluation was designed to collect a wide range of information, which has been used to help understand and improve the program. A sample of families was drawn and analyses were conducted to be sure that the sample did not differ from the population of families on selected characteristics. Data were collected on parents and children through interviews and videotaping in the fall and spring of a school year. The data were carefully and thoroughly analyzed. The findings were reported in attractive, well-formatted, well-written reports. The reports describe many economic, educational, and social characteristics of the families involved in the study. The reports also describe the extent to which families participated in the program. So far, a good study.

Then the reports discuss "outcomes." Parents reported many improvements in the ways in which they deal with their children, and independent raters confirmed those improvements through observations. In the body of the report, the evaluators are careful about presenting findings without making causal attributions; for example, "x percent of the parents showed improvement" on a given measure, and "y percent of parents demonstrated positive score change" on some other measure. These are straightforward, accurate statements about pretest/posttest changes in parents. When the report summarizes the findings, however, the perfectly reasonable descriptive statements about how parents changed over time were transformed into statements concluding that the program *caused* those changes! For example "Data from parents show that (the program) made a positive difference in . . ." and "[The program's] approach is effective with many different low-income families."

There was no randomly assigned group in this study, nor was there a quasi-experimental nonequivalent comparison group. In fact, there was no way to estimate the extent to which parents would have improved had they not participated in the program. Still, the evaluators were willing to make the leap from "parents improved" to "the program caused parents to improve."

This is not an isolated example. Indeed, declaring a program effective based on evidence gathered through simple pre-post studies is a common practice, and the fact that it occurs in evaluations presented at AEA meetings shows that even highly visible evaluators in the main evaluation association are sloppy about attribution and causality. These are important topics for evaluation and for the experimenting society. If AEA-based evaluators err in attributing normal maturation of families and family members to programs (or are pressured into doing so), imagine how difficult it is for program staff to avoid those mistakes. Causal attributions easily and seductively creep into our conclusions, even though we know better.

CONCLUSIONS

Directions for the Experimenting Society

We have provided evidence showing that, at least with respect to programs for low-income families, our society has advanced toward the active, experimenting approach advocated by Campbell, and that we have invested in high-quality impact evaluations of major programmatic initiatives. Where our society has not lived up to Campbell's hopes is in the way that evaluations are used. Maybe we have to accept that it is not reasonable to expect that evaluations will be used to quickly and easily sway program advocates and political figures. Programs, and programmatic approaches, have a substantial amount of momentum. They take years to develop, years to change, and years to abandon. But programs do change, programs do grow, and programs do get abandoned. Witness the recent large-scale change in the welfare system after 40 years of tinkering with the same basic policy. Certainly, some of the negative public and political opinion of the welfare system was due to the string of evaluation studies showing that the program had little or no impact.

Perhaps the process of changing programs and policy is just messier, longer-term, and more cumulative than Campbell envisioned. The growth of systematic research reviews (meta-analyses) has been made possible by the accumulation of several decades of evaluation and research studies, and we currently are in a time when syntheses of research studies are greeted by policy analysts and program practitioners with at least as much interest as studies of individual programs.

On one hand, it probably is easier for program advocates funders to accept criticism and suggestions for improvement when they are aimed at a more general level (e.g., early childhood programs or parenting programs) rather than at a particular example of that approach (e.g., the High/Scope curriculum or the Parents as Teachers program). On the other hand, acceptance of evaluation results is easier when the findings support the effectiveness of a particular intervention than when the evaluator tries to tell the more difficult story of "no program effect." As a consequence, instead of individual studies being used to promote, change, or de-fund individual programs, the way in which our experimenting society appears to work is to rely more on the accumulation of studies and research results, and it is these aggregations of research that are used to affect not just a single program but the direction of an entire stream of programmatic efforts.

Engaging in the Public Debate

In an effort to retain our objectivity, we evaluators have tended to stand outside the public dialogue about the strengths and weaknesses of various intervention programs in favor of a more limited dialogue that is shared only with our colleagues and with staff from the funding agencies that finance our work. If we seek to carry on the message and hopes of Don Campbell, we need to be active in bringing about the experimenting society, both by participating more fully in public dialogues about the utility of different programmatic approaches and by speaking more directly to the consumers of the information we produce.

Evaluators/researchers can make substantial contributions to program development and improvement. When we think about the major knowledge contributions that have been made with respect to programs for low-income families, we see that they come not only from program practitioners but also from "objective" researchers who decided to become involved with the systematic development, evaluation, and improvement of programmatic approaches. For example, Craig Ramey and his colleagues have conducted two decades of high-quality development and experimental research on programs for preschoolers and have produced some of the most influential and important findings in the child development literature (Infant Health and Development Program, 1990; Ramey & Ramey, 1992). David Weikart and his colleagues at the High/Scope Educational Research Foundation developed and ran the Perry Preschool Program, evaluated it using a randomized experiment, and conducted 30 years of follow-up studies (Schweinhart et al., 1993). The result has been the single most influential piece of research for children from low-income families. Judy Gueron and her colleagues at the Manpower Demonstration Research Corporation have conducted a long string of development and evaluation studies of welfare initiatives (Gueron & Pauly, 1991). This program of research provided some of the most important information in the recent policy debate about the federal welfare program. David Olds and his colleagues at the University of Denver are in the middle of systematically investigating the effectiveness of different programs for teenage mothers, and this research provides the most hopeful results in the field for improving the lives of young mothers (Olds, 1992; Olds, Henderson, Phelps, Kitzman, & Hanks, 1993). After years of building reputations as educational methodologists, Tony Bryk from the University of Chicago and William Cooley from the University of Pittsburgh became closely allied with large school systems (Chicago and Pittsburgh, respectively) in order to have more of an impact on policy. Few people would question the objectivity of these and other dedicated evaluator/program developers in spite of the fact that each of them decided to engage in program development and in policy debate, to the benefit of all of us.

Where to Go Next

We can do a better job of training evaluators to recognize and understand the issues involved in making causal attributions, and to understand the importance of high-quality research designs in dealing with plausible rival hypotheses and hence helping us to assess program impacts. At the same time, evaluators should engage in informal educational efforts with their clients and with program staff, so that these individuals have a better understanding of the normal maturational processes that operate when studying human development and other areas where outcomes change in the absence of any special treatment.

Although evaluators have to be able to resist the perfectly reasonable desires of program staff and funders to draw conclusions about programs based on research designs that do not allow those conclusions (most typically, the one-group, pre-post design), we need to spend more time learning from practitioners about the day-to-day experiences of their clients to provide greater insight into the levels and patterns of "normally" expected growth and change.

We believe in the value of the experimenting society. Twenty-five years after Campbell presented *Methods for the Experimenting Society* to the American Psychological Association, there is substantial evidence that many aspects of the experimenting society are indeed workable. Although there is much work left to do (see the viewpoint offered by Carol Weiss, Chapter 11, this volume), there is every reason to feel optimistic about the experimenting society. We hope that in another 25 years, other researchers will be able to document several additional steps taken toward the achievement of Campbell's dream.

NOTES

1. We advocate strong quasi-experiments as vehicles offering the possibility of drawing strong causal inferences; however, such approaches have been underutilized. There is a large literature attesting to the feasibility and frequent use of randomized experimental studies (e.g., Boruch, 1974; Boruch & Wothke, 1985), but to our knowledge there is no comparable literature pointing out the extent to which strong quasi-experimental approaches such as the regression-discontinuity design or the interrupted time series design have been used in field research. Rather, we most often see only the weakest quasi-experimental designs being implemented, for example, the one-group post-only design or the one-group pre-post study.

2. A problem arises when qualitative approaches are seen as a substitute for, or superior to, experimental approaches *for the purposes of drawing causal inferences*. There has been legitimate and substantial disagreement about this topic in the research community (Cronbach et al., 1980). We side with Cook and Campbell (1979) when they conclude that critics of experimental methods must be able to offer viable alternatives for drawing causal inferences, and that although many qualitative researchers have attempted to avoid causal explanations, "they are rarely successful. Their understandings, insights, meanings, analysis of intentions and the like are strongly colored by causal conclusions even when the terms 'effects,' 'gains,' 'benefits' and 'results' are carefully avoided" (pp. 93-94).

3. Percentages add to more than 100 because some local evaluations used more than one type of design.

4. Campbell viewed certain quasi-experiments as good fallback options when randomized studies could not be implemented; however, there is great variation in the inferential power of different quasi-experiments, and Campbell often referred to weak quasi-experiments, those that offer little ability to draw causal inferences, as "queasy experiments." Campbell and Stanley (1966) and Cook and Campbell (1979) inventoried and catalogued quasi-experiments as a way of discussing their strengths (few) and weak-

nesses (many) relative to randomized studies, not as a means of encouraging researchers to rely on quasi-experiments in place of randomized experiments.

REFERENCES

Barnett, W. S. (1995). Long-term effects of early childhood programs on cognitive and school outcomes. *The Future of Children, 5*(3), 25-50.

Barnett, W. S., & Boocock, S. S. (1998). *Early care and education for children in poverty.* Albany: State University of New York Press.

Boruch, R. F. (1974). Bibliography: Illustrative randomized field experiments for program planning and evaluation. *Evaluation, 2*, 83-87.

Boruch, R. F., & Wothke, W. (1985). Randomization and field experimentation. *New Directions for Program Evaluation* (Whole No. 28).

Campbell, D. T. (1969). Reforms as experiments. *American Psychologist, 24*, 409-429.

Campbell, D. T. (1971, April). *Methods for the experimenting society.* Paper presented at the annual meeting of the American Psychological Association, Washington, DC.

Campbell, D. T. (1979). "Degrees of freedom" and the case study. In T. D. Cook & C. S. Reichardt (Eds.), *Qualitative and quantitative methods in evaluation research* (pp. 49-67). Beverly Hills, CA: Sage.

Campbell, D. T. (1984). Can we be scientific in applied social science? In R. F. Connor, D. G. Attman, & C. Jackson (Eds.), *Evaluation studies review annual* (Vol. 9, pp. 26-48). Beverly Hills, CA: Sage.

Campbell, D. T. (1988). *Methodology and epistemology for social science: Selected papers* (E. S. Overman, Ed.). Chicago: University of Chicago Press.

Campbell, D. T., & Stanley, J. C. (1966). *Experimental and quasi-experimental designs for research.* Chicago: Rand McNally.

Clarke-Stewart, A. (1983). Exploring the assumptions of parent education. In R. Haskins & D. Adams (Eds.), *Parent education and public policy* (pp. 257-276). Norwood, NJ: Ablex.

Clarke-Stewart, A. (1988). Parents' effects on children's development: A decade of progress? *Journal of Applied Developmental Psychology, 9*, 41-84.

Conrad, K. J. (1994). Critically evaluating the role of experiments. *New Directions for Program Evaluation* (Whole No. 63).

Cook, T. D., & Campbell, D. T. (1979). *Quasi-experimentation: Design and analysis issues for field settings.* Boston: Houghton Mifflin.

Cronbach, L. J., Ambron, S. R., Dornbusch, S. M., Hess, R. D., Hornick, R. C., Phillips, D. C., Walker, D. F., & Weiner, S. S. (1980). *Toward reform of program evaluation.* San Francisco: Jossey-Bass.

Duncan, G. J. (1991). The economic environment of childhood. In A. C. Huston (Ed.), *Children in poverty: Child development and public policy* (pp. 23-50). New York: Cambridge University Press.

Fischer, R. L., & Cordray, D. S. (1995). *Job training and welfare reform: A policy-driven synthesis.* Unpublished manuscript, Vanderbilt Institute for Public Policy Studies, Vanderbilt University.

Greenberg, D., & Shroder, M. (1997). *The digest of social experiments* (2nd ed.). Washington, DC: Urban Institute Press.

Gueron, J., & Pauly, E. (1991). *From welfare to work.* New York: Russell Sage Foundation.

Infant Health and Development Program. (1990). Enhancing the outcomes of low-birthweight, premature infants. *Journal of the American Medical Association, 263*(22), 3035-3042.

Karweit, N. L. (1994). Can preschool alone prevent early learning failure? In R. E. Slavin, N. L. Karweit, & B. A. Wasik (Eds.), *Preventing early school failure: Research, policy, and practice* (pp. 58-77). Boston: Allyn & Bacon.

Kennedy, M. M. (1997). The connection between research and practice. *Educational Researcher, 26*(7), 4-12.

Kuhn, T. J. (1962). *The structure of scientific revolution.* Chicago: University of Chicago Press.

Love, J. (1998). *Overview of the Early Head Start research and evaluation project.* Princeton, NJ: Mathematica Policy Research.

Olds, D. (1992). Home visitation for pregnant women and parents of young children. *American Journal of Public Health, 146*, 704-708.

Olds, D., Henderson, C., Phelps, C., Kitzman, H., & Hanks, C. (1993). Effect of prenatal and infancy nurse home visitation on government spending. *Medical Care, 31*(2), 155-174.

Patton, M. Q. (1998). *Utilization-focused evaluation* (3rd ed.). Thousand Oaks, CA: Sage.

Plotkin, D. (1996, June). Good news and bad news about breast cancer. *The Atlantic Monthly*, pp. 53-82.

Quint, J. C., Bos, J. M., & Polit, D. F. (1997). *New Chance: Final report on a comprehensive program for disadvantaged young mothers and their children.* New York: Manpower Demonstration Research Corporation.

Ramey, C. T., & Ramey, S. L. (1992). Early educational intervention with disadvantaged children—To what end? *Applied and Preventive Psychology, 1*, 130-140.

Ramey, C. T., Ramey, S. L., Gaines, R., & Blair, C. (1994). Two-generation early interventions: A child development perspective. In S. Smith (Ed.), *Two-generation programs for families in poverty: A new intervention strategy* (pp. 199-228). Norwood, NJ: Ablex.

Robinson, V. M. (1998). Methodology and the research-practice gap. *Educational Researcher, 27*(11), 17-26.

Schweinhart, L. J., Barnes, H. V., & Weikart, D. P. (1993). *Significant benefits: The High/Scope Perry Preschool Study through age 27* (Monograph 10). Ypsilanti, MI: High/Scope Educational Research Foundation.

Smith, S. (Ed.). (1995). *Two-generation programs for families in poverty: A new intervention strategy* (Advances in Applied Developmental Psychology, no. 9). Norwood, NJ: Ablex.

St.Pierre, R. G. (1998, March). Testimony at *Head Start: Is it making a difference?* hearings of the Senate Subcommittee on Children and Families. Washington, DC: Government Printing Office.

St.Pierre, R. G., Layzer, J. I., & Barnes, H. V. (1998). Regenerating two-generation programs. In S. Boocock & W. S. Barnett (Eds.), *Early care and education for children in poverty* (pp. 99-121). Albany: State University of New York Press.

St.Pierre, R. G., Layzer, J. I., Goodson, B. D., & Bernstein, L. S. (1997). *National impact evaluation of the Comprehensive Child Development Program: Final report.* Cambridge, MA: Abt Associates.

St.Pierre, R. G., Ricciuti, A., & Creps, C. (1998). *Summary of state and local Even Start evaluations.* Cambridge, MA: Abt Associates.

U.S. Department of Health and Human Services (1998). *Trends in the well-being of America's children and youth.* Washington, DC: U.S. Government Printing Office.

U.S. General Accounting Office. (1997). *Head Start: Research provides little information on impact of current program* (GAO/HEHS-97-59). Washington, DC: Author.

Weiss, C. H. (1998). Have we learned anything new about the use of evaluation? *American Journal of Evaluation, 19*(1), 21-33.

Yoshikawa, H. (1995). Long-term effects of early childhood programs on school outcomes and delinquency. *The Future of Children, 5*(3), 51-75.

CHAPTER 8

The Honestly Experimental Society

Sites and Other Entities as the Units of Allocation and Analysis in Randomized Trials

Robert F. Boruch

Ellen Foley

1. INTRODUCTION

This chapter exploits one of Donald T. Campbell's alarmingly prescient observations. In "Reforms as Experiments," he said:

> Where policies are administered through individual client contacts, randomization at the person level may often be inconspicuously achieved . . .

AUTHORS' NOTE: Excerpts from this chapter were presented at the annual meeting of the American Society for Criminology (1996), the American Evaluation Institute (1997), the Max Planck Institute (Berlin, 1998), the Swedish Center for Social Work Research (Stockholm, 1998), and the Harvard Provost/University-Wide Seminar (1998). Work on this topic has been supported by the U.S. Department of Education, the National Science Foundation, and the National Institute of Mental Health and has been presented at the American Criminology Society's annual meeting, the Swedish Center for Evaluation of Social Services, and the Max Planck Institute on Human Development in Berlin. Criticism and ideas were graciously given by colleagues, notably David Huizenga, Haluk Soydan, and Karin Tengvald, and we are grateful. We are also indebted to Len Bickman, Lee Sechrest, and an anonymous reviewer for their comments on earlier versions of this chapter.

> But for most social reforms, larger administrative units will be involved, such as classrooms, schools, cities, counties or states. We need to develop the political postures and ideologies that make randomization at this level possible. (Campbell, 1969, p. 425)

Campbell's treatise recognized the use of entire countries, states, and smaller jurisdictions as the units of analysis in estimating the effects of new programs. The impact of divorce laws in Germany, breathalyzer tests in Britain, traffic crackdowns in Connecticut, and larceny law enforcement in Chicago fell within his ambit, at least with respect to time series-based estimates of each program's effect.

Our colleague, however, did not consider deeply the use of jurisdictions as units in *randomized* experiments (trials in the vernacular used here), perhaps because there were few experiments that involved units larger than the person or family when he wrote his "Reforms as Experiments" paper. Boruch (1997) merely identified examples of such studies and failed to pursue the topic.

In what follows, we depend on his insight, build on others' recent work, and educe the implications of these for the future. The topic is germane to contemporary investments in complex social programs that are designed to enhance health and well-being, welfare, and education level including adult literacy, as well as those intended to reduce crime. The topic is at times controversial, and debates are not always well informed. Consequently, this chapter is designed to be a prophylactic for uninformed claims as well as to advance our understanding about field trials that involve entities as the units of allocation.

1.1 Definitions

A *randomized experiment*, or *randomized field trial* (RFT) here, is a study that attempts to produce a fair comparison of the relative effectiveness of two or more approaches so as to understand what works, for whom, and for how long. In particular, such trials are designed to produce estimates of relative differences that are unbiased and allow a statistical statement of one's confidence in the results. In this chapter, the emphasis is on randomized trials that produce two or more equivalent groups of sites, organizations, or other entities that are then treated differently.

The *unit of allocation and analysis* refers to who or what is randomly assigned to treatments in a trial. Conventional textbooks in psychology, for instance, typically handle experiments in which *individuals* are the units of allocation. Books on the design of randomized clinical trials that assess surgical innovations and pharmaceutical drugs also focus mainly on individuals as the units. Here, we focus on sites and entities, broadly defined, rather than on individuals. Sites include geopolitical jurisdictions such as police precincts, neighborhoods, or communities. Entities may comprise institutions such as schools, colleges, or factories, or organized groups such as scout troops, local labor unions, school boards, or stores.

For instance, education organizations may be eligible to receive federal funds for implementing a new model of an adult literacy program. Twenty such sites might be willing and able to participate in a controlled experiment on the program's effectiveness. Ten of them would then be randomly allocated to receive special funding to employ a program model; 10 randomly allocated sites would not receive funds and would serve as a control group. This and better designs are discussed in what follows.

1.2 The Contents of This Chapter

We direct attention to the following topics:

- Rationale: Why should a federal agency or private foundation sponsor studies that allocate sites to different programs in randomized field trials on the programs?
- Precedents: How and by whom have sites been allocated in randomized trials?
- Difficulties: What are the complications and resolutions in doing randomized trials that involve sites as the units of allocation and analysis?
- Summary: To what effect have such trials been run?

The illustrations in this chapter are taken from education, health, employment and training, civil and criminal justice, housing, and other arenas. They are diverse, partly to demonstrate that useful policy experiments involving entities, rather than individuals, as the primary units of allocation and statistical analysis have been and can be carried out in a variety of settings. We also use a hypothetical example on the design of a randomized field trial on adult literacy programs. The effectiveness of such programs is poorly understood, and the design of such an experiment is complicated.

1.3 Presumptions

The first presumption in this chapter is that government agencies and private foundations are interested in estimating the relative effect of new programs that they sponsor. Put another way, we assume that one of their objectives is to discern the effectiveness of different services. Their objective is to answer the question "What works better, for whom, and for how long?"

A second presumption is that a defensible estimate of an innovation's effect depends on determining how sites or other entities would behave in the absence of an innovation. As a practical matter, one might, for example, develop such an estimate from a time series. One makes a forecast about literacy rates of communities in the future based on the data from the past on the literacy rate in the same communities. Dependable forecasts are possible in some countries and contexts, but not often. Here, we assume that time series data are unavailable. We assume further that the theory, resources, and data that would sustain alternative analyses of program impacts are unavailable. Some of the alternatives to randomized experiments were covered by Campbell and Stanley (1963) in a classic treatise. These were extended and explored more deeply in a book edited by Cook and Campbell (1979).

In the absence of good theory and time series data, the estimate of what would happen absent the new intervention must then be based on a control or comparison group. A simple, sturdy, and scientifically defensible method of composing a comparison group, one that permits fair estimates of the relative differences among programs, is considered here. It is the method of random assignment. For instance, a sample of eligible literacy centers might be randomly selected from the pool of eligible centers and engaged in the new program. The outcomes at these centers would then be compared to the eligible centers that were randomly selected from the pool to continue operating under the existing programs. The random assignment ensures that the two groups of centers are comparable, apart from the influence of the different literacy programs under which they operate. Such randomized field trials have been covered by Rossi and Freeman (1993) and Boruch (1997), among others.

A third presumption is more speculative, much as Campbell's (1969) remarks were. It is that part of the future of evaluation in the United States and other countries lies with controlled experiments, this being in the interest of producing evidence that will help in choosing programs (Boruch, 1995). This third forecast is empirically based in the sense that such experiments have been undertaken. Moreover, their frequency has increased. Table 8.1 for example, lists more than 50 different studies involving communities or sectors, schools or classrooms, or organizations as the units of allocation in a randomized field trial.

2. RATIONALE: WHY USE SITES AS THE UNITS?

Why should we consider sites, organizations, or other entities as the primary units of assignment to programs in evaluating the effect of a program? That is, why use entities in an RFT? The reasons outlined here include program theory, law and ethics, policy, statistical theory and rules of evidence, and the counsel of advisory groups.

TABLE 8.1 Studies Using Controlled Randomized Field Trials

The studies listed in this table use sites, organizations, institutions, or other entities, such as classrooms or neighborhoods, as the unit of allocation and analysis.

Aplasca et al. (1995)
Bain and Achilles (1986)
Bain, Achilles, Zaharias, and McKenna (1992)
Barcikowski (1981)
Basen-Engquist et al. (1997)
Bauman, LaPrelle, Brown, Koch, and Padgett (1991)
Bickman (1985)
Bloom, Bos, and Lee (1998)
Boruch (1993a)
Boruch (1993b)
Boruch (1995)
Botvin, Baker, Dusenbury, Botvin, and Diaz (1995)
Botvin, Baker, Dusenbury, Tortu, and Botvin (1990)
Botvin, Dusenbury, Baker, James-Ortiz, and Botvin (1992)
Bracey (1995)
Campbell (1969)
Clayton, Cattarello, and Johnstone (1996)
Cook et al. (1999)
Cook, Hunt, and Murphy (1999)
Coulson (1978)
Coyle et al. (1996)
Coyle, Boruch, and Turner (1991)
Dean, Searle, Galda, and Heyneman (1981)
Dennis and Boruch (1989)
Dent, Sussman, and Flay (1993)
Ellickson (1994)
Ellickson and Bell (1990)
Ellickson and Bell (1992)
Ennett, Tobler, Ringwalt, and Flewelling (1994)
Fairweather, Sanders, and Tornatzky (1974)
Fairweather and Tornatzky (1977)
Farrington (1997)
Finn and Achilles (1990)
Flay et al. (1985)
Freedman and Takashita (1969)
Fuchs and Fuchs (1999)
Garfinkel, Manski, and Michalopoulos (1990)
Gay (1996)
Glaser and Coffey (1967)
Goldberg et al. (1996)
Gosnell (1927)
Gottfredson (1986)
Graham, Flay, Johnson, Hansen, and Collins (1984)
Grossman et al. (1997)
Hansen and Graham (1991)
Hauck, Gilliss, Donner, and Gortner (1991)
Hill, Stycos, and Back (1959)
Hopkins (1982)
Hornik, Ingle, Mayo, McAnany, and Schramm (1972)
Jamison, Searle, and Suppes (1980)
Jason, Johnson, et al. (1993)
Jason et al. (1992)
Kaftarian and Hansen (1994)
Kellaghan, Madaus, and Airasian (1982)
Kelling, Pate, Dieckman, and Brown (1974)
Kelly, Lawrence, and Diaz (1991)
Killen et al. (1988)
Kirby, Korpi, Adivi, and Weismann (1997)
Kirby, Korpi, Barth, and Cagampang (1997)
LaPrelle, Bauman, and Koch (1992)
Leviton et al. (1990)
Leviton et al. (1999)
Lorian (1994)
McKay, McKay, Sinnestera, Gomez, and Lloreda (1978)
Moberg, Piper, Wu, and Serlin (1993)
Moskowitz et al. (1984)
Moskowitz, Malvin, Schaeffer, and Schaps (1984)
Mosteller (1995)
Murray and Hannan (1990)
Murray et al. (1994)
Murray and Wolfinger (1994)

(Continued)

TABLE 8.1 Continued

Murray, Moskowitz, and Dent (1996)
Nye, Achilles, Zaharias, DeWayne Fulton, and Wallenhorst (1993)
Olds (1988)
Pentz (1994)
Perry et al. (1992)
Peterson, Hawkins, and Catalano (1991)
Peterson, Hawkins, and Catalano (1992)
Raab (1999)
Randolph, Basinsky, Leginski, Parker, and Goldman (1997)
Raudenbush (1997)
Reichardt and Rindskopf (1978)
Rhett (1998)
Riccio (1998)
Riecken et al. (1974)
Ritter (1999)
Rosenbaum et al. (1991)
Rosenbaum, Flewelling, Bailey, Ringwalt, and Wilkinson (1994)
Rosenbaum and Hanson (1998)
Roth, Sholz, and Witte (1988)
Schaps, Moskowitz, Condon, and Malvin (1982)
Schonfeld (1998)
Sherman and Weisburd (1995)
Simpson, Klar, and Donner (1995)
Soumerai et al. (1998)
Stoker, King, and Foster (1981)
St.Pierre et al. (1995)
Turner, Miller, and Moses (1989)
University of Maryland, Department of Criminology and Criminal Justice (1997)
Wagenaar et al. (1997)
Wagenaar, Murray, Wolfson, Forster, and Finnegan (1994)
Wagenaar and Wolfson (1993)
Wasserheit, Aral, Holmes, and Hitchcock (1991)
Weisberg (1978)
Wolins (1982)
Woodruff (1997)
Word et al. (1990)
Wyatt et al. (1998)

2.1 Program Theory

By "theory," we mean how one supposes that programs and processes have the effects that we believe they will have. In other words, the theoretician proposes a "logic model" to explain tentatively what happens when a program is implemented, or lays out a formal path model or a causal chain. Some of these theories or models depend heavily on the idea that entities play a role in affecting the behavior of individuals. In what follows, brief examples are given from education, health risk prevention, employment and training, diffusion of innovation in mental health, and criminology.

Numerous theories of societal change posit that a program will work if it is delivered by organizational elements acting in concert. For example, Schorr and Schorr's (1988) book on the future of social services argues that the most successful social programs are comprehensive and "offer a broad spectrum of services" (p. 256). Similarly, the model of "full-service schools" advocated by

Dryfoos (1994) incorporates a broad range of community services including health, mental health, and adult instruction in the school environment so as to create school community centers that offer a wide array of services.

In the adult literacy field, for example, Development Associates (1992) found that most adult education providers funded through the Adult Education Act provided a multitude of services to their clients. These services included transportation, child care, mental health and career counseling, and other services. Because many community elements may be engaged sequentially or simultaneously to reinforce, augment, or enhance a course-based effort, organizations or clusters of them become a more appropriate unit of analysis than individuals who are served.

Contemporary research on preventing sexually transmitted diseases also depends, implicitly or explicitly, on theories about what institutional and group factors influence behavior. State-of-the-art monographs have tried to make plain how one might think about structural influences, including national policy on access to condoms, community and other localized societal norms, and individuals' cognitive processes based on theory of reasoned action. See the work by Wasserheit, Aral, Holmes, and Hitchcock (1991) generally and Hornik's (1991) chapter in particular. Randomized field trials undertaken in California and Texas have employed 20 schools as the unit of allocation and analysis so as to test programs based on several such theories (Basen-Engquist et al., 1997; Coyle et al., 1996).

A variety of RFTs have used schools as units to assess theory-driven programs that were designed to prevent or reduce substance abuse. The Midwestern Prevention Project (Pentz, 1994), for example, was based on a theory that adolescents' drug use depends on their characteristics, such as prior drug use, and on the adolescents' ability to handle peer pressure toward using drugs. More important, the theory also recognized that environmental and situation factors beyond the individual are important. School and community norms, for instance, can influence adolescent behavior. Efforts to influence such norms can, in principle, be made in a program deployed in multiple sites so as to reduce the prevalence of drug use by adolescents.

Consider, as another example, the RFTs conducted by Wagenaar and colleagues (Wagenaar et al., 1997; Wagenaar, Murray, Wolfson, Forster, & Finnegan, 1994; Wagenaar & Wolfson, 1993) on a community-based program directed toward reducing underage drinking. The theory underlying the program's design puts less emphasis on individual psychological factors than has been the case in earlier works. Rather, the theory posited first that influences on informal community policies, such as media and mobilization efforts, are important and manipulatable. These events then theoretically would lead to formal community policies, including enhanced law enforcement, altered service practices, and changed norms among adolescents. These, in turn, theoretically affect the underage youths' access to alcohol, measurable through surveys and in other ways. This is then expected to reduce injuries attributable to alcohol use.

In a tantalizing paper on experiments in welfare and employment, Garfinkel, Manski, and Michalopoulos (1990) propose that social interactions, which are important in some economic theory, cannot be captured by micro-experiments (i.e., by studies that use individuals as the primary units of allocation). One of their four illustrations of different theories that are relevant is based on Myrdal's idea that "a push on any of the components of a composite entity. . . such as an economy—will have effects on other components of the some entity" (p. 6). Further, one may suppose that the effects cascade. For instance, a program that is designed to elevate the employment status of a minority group and that does so will then lead to employment among still others in the group, and this further elevation will lead to other enhancements. Macro-experiments that involve entities as the units of allocation and analysis could capture some of this phenomenon. Garfinkel et al.'s (1990) other economic illustrations are kin to examples given above inasmuch as they bear on how social norms influence decisions.

Theory has also driven multistage research on how to engage and encourage mental hospital practices shown in earlier research to be more effective for treating certain forms of mental illness. Such theory involved ideas about the level at which the hospital might first be engaged (top down or professional staff up) and the best mode of engagement. The latter included involvement in workshops or demonstration projects as opposed to merely sending brochures, in the expectation that people react differently to these various engagement strategies (Fairweather, Sanders, & Tornatzky, 1974; Fairweather & Tornatzky, 1977).

A final illustration emerged in Sherman and Weisburd's (1995) experiment on intensive police patrols to reduce crime. Opposition to the implementation of intensive police patrols might be based on this theory: that the whereabouts of police officers is important and that they must spread their efforts so as to inhibit or detect crime uniformly. Focusing on crime "hot spots" will merely lead to a displacement of criminal activity. An alternate theory might be construed from the perspective of the police officer and the perpetrator. If officers are present or likely to be present in an area, the risk of detection of crime is greater in this area than in others; that is, the presence of police in hot spots would reduce one's interest in illegal acts and reduce the crime rate in the hot spots and more generally.

2.2 Law and Ethics

One reason why sites might be used as the units of assignment in RFTs is that the random assignment of individuals to alternative programs *within* a site may not be legal or ethical. Random allocation of entire sites to alternative programs might be regarded as both legal and ethically responsible.

For instance, in a randomized trial testing the Drug Abuse Resistance Education model (D.A.R.E.), researchers randomly assigned schools to treatment and control groups partly because it would have been difficult to get cooperation from schools if some of their students received the program and some did not (personal communication, T. R. Curtin, April 3, 1996). A kind of institutional ethic prevailed. Using schools as the unit of assignment also helped ensure the cooperation of control schools in the RFT: These schools were promised access to the D.A.R.E. program the year after the completion of the study.

Similarly, a local juvenile facility—one of 80 or so such facilities in Sweden, for example—may object to random allocation of their clients to different programs so as to discern which program is most effective in reducing recidivism. Other ethical values in the local facility may take precedence, such as giving the "same" service to everyone in the facility. A randomized trial in which eligible and willing institutions try out different approaches may then be regarded as more just. This point was made by Karin Tengvald at Stockholm's meetings on evaluating social service programs (Soydan, 1998). Again, the emphasis is on comparing *alternative* interventions in different communities, not on giving one set of these groups a "treatment" and leaving the others high and dry. The focus, then, is not simply on whether a treatment works but on which treatment works better.

2.3 Policy

As a matter of policy, the government agency or foundation that sponsors site-based programs make rules that affect organizations rather than individuals directly. Such rules require sites or organizations to provide information, take particular actions, create transactions, and so on. The implication is that a study of the effects of a national program has to recognize sites as the first targets in an evaluation design. The individuals within sites are the secondary targets.

For example, federal policy on demonstration projects in the United States has been clear, at times, in emphasizing that communities are essential in ameliorating certain social problems. Preventing substance abuse is a case in point. The Center for Substance Abuse Prevention (CSAP) was created to reduce the incidence of alcohol, tobacco, and drug use in the United States. It has tried to do so through efforts such as the Community Partnership Demonstration Program. This program's focus is on learning, through research and development, how diverse community-based organizations can be engaged in effective intervention and how to do research on this topic well. An array of ways to do so were described by Kaftarian and Hansen (1994), and many of these anticipated this chapter's topic. The emphasis was on communities as the unit of allocation and analysis in controlled randomized field trials (Ellickson, 1994; Lorian, 1994; Murray & Wolfinger, 1994; Pentz, 1994; Wagenaar et al., 1994).

Other examples of programs in which the most direct connection is between entities and government or foundation assistance *rather* than between individuals and such assistance are easy to identify. They appear in education, notably Chapter I/Title I Compensatory Education and other programs sponsored by the U.S. Department of Education; in Head Start, sponsored by the U.S. Department of Health and Human Services; in employment and training programs sponsored by the U.S. Department of labor; and in vocational rehabilitation programs sustained by the U.S. Department of Health and Human Services. They also appear in private efforts such as the Robert Wood Johnson Foundation's "Fighting Back."

2.4 Statistical Theory and Analysis

A fundamental assumption of statistical theory underlies all statistical analysis procedures: the observation on any given individual or entity is independent of observations on all others. When the assumption does not hold and the analyst fails to recognize this, the analysis will be compromised. When the analyst recognizes the relationships among observations, analyses will be complex.

That observations will be independent is plausible in many settings. For example, a criminal's commission of a property crime in Jersey City (or Tokyo or Bremen) usually can be assumed to be independent of another criminal's property crime in the same city if the offenders are not co-offenders or are unacquainted with one another in ways that influence their committing this kind of crime. One cancer patient's death (or survival) as a result of being given a new drug is presumably independent of another's if the patients are not acquainted and are not treated by the same physician or in the same hospital.

Assuming that the units of observation are independent is not plausible in other settings. A given gang member's response to a crime reduction program may not be independent of gang members' responses if the program involves all other members. A child's grade on a test of ability to work in teams presumably will not be independent of grades given to other children on the same team. Witnesses to a miracle are not independent to the extent that they share a culture or religion that influences their propensity to see a miraculous event (Kruskal, 1988).

As suggested earlier, the theory underlying some social interventions is that individuals will respond to the intervention when it is delivered to a collection of people rather than to a single individual. Part of the effect of a program is thought to be exercised through a group. Their responses will be related to one another.

For the statistician, all of this implies that it is not individuals who ought to be randomly assigned to programs, and it is not individual-level data that ordinarily must be used to estimate the program's effect. Rather, allocation and analysis should focus first on entire groups or organizations, and second on individuals within each entity.

Ignoring the fact that individuals do not react independently has serious consequences. Failing to recognize that individuals are members of groups can lead to mistakes. Two kinds of mistakes have been covered thoughtfully over the last two decades by, among others, Wolins (1982) and Murray (1998). These mistakes involve false declarations about the statistical likelihood of finding program differences when there is no real difference (Type I error) and false declarations about the statistical power of a randomized experiment.

Consider a simple experiment involving two treatments to determine whether a new adult education curriculum enhances literacy relative to a more traditional program. Assume that the new program is assigned by the government to one literacy center and a second literacy center serves as the control. Assume further that the new program really does not work better than the old one and that the two centers are perfectly substitutable for one another.

Many people would be tempted to analyze differences by focusing on the individuals as the units of analysis. They might compare the mean score on a standardized test in one literacy center to the mean score in the second, for example. A simple statistical *t* test undertaken to assess differences in performance at the nominal alpha level of .05 would often yield "significance" even if the programs do not differ in effectiveness. That is, they would declare the programs different in their effect. The conclusion is wrong more than 5% of the time (i.e., more often than the nominal alpha level would imply). One reason for this error is subtle but has been well understood for years (Wolins, 1982).

The "units" (students) making up each group of literacy centers are not independent. Students are similar to one another to some degree within a center. The estimate of the variability among students that is made in conventional statistical tests, and used in the denominator of the *t* test, is based on the independence assumption. When the units are not independent, this conventional variance estimate is too low. The result is that the observed *t* in a conventional statistical test is too large; that is, "significance" will be declared more frequently than 5% of the time when in fact there is no real difference in effects of the programs.

How badly one errs in declaring that there are significant differences when in fact there are none depends on how similar the individuals within each group are to one another. One index of similarity and of nonindependence that is commonly used is the intraclass correlation coefficient (Haggard, 1958; McGraw & Wong, 1996). For example, suppose that the relatedness among individuals is moderate (intraclass correlation is .20). When there are 100 students in the two

TABLE 8.2 Actual Levels of Significance in a Two-Treatment ANOVA With Nominal Alpha = .05, Intraclass Correlation at Various Levels, for Various Group Sizes (nG)

	Intraclass Correlation				
nG	$\rho = 0$	$\rho = .01$	$\rho = .05$	$\rho = .20$	$\rho = .40$
10	.05	.06	.11	.28	.46
25	.05	.08	.19	.46	.63
50	.05	.11	.30	.59	.74
100	.05	.17	.43	.70	.81

SOURCE: Adapted from Barcikowski (1981).
NOTE: The scenario involves testing a null hypothesis about the equality of two population means based on two samples whose means (averages) differ. The test statistic is a conventional (or *F*) whose value is judged against the distribution of *t* (or *F*) under the assumption that the population means are the same. The intraclass correlation here is designated as ρ. For example, when $\rho = 0$, the nominal alpha level (.05) of the test is the same as the actual alpha level. When $\rho = .20$ and size of the group is 100 (nG = 100), the actual alpha level is .70.

centers and the nominal significance level in a conventional *t* test is .05, the actual significance level is .70. That is, there is a 70% chance of finding a significant difference between programs when no real difference exists. Table 8.2, adapted from Barcikowski (1981), illustrates more generally how the nominal Type I error size may be misleading when individuals within groups assigned to programs in a field test are not independent of one another.

The simplest analysis of data that are generated in an experiment using entities as units employs the mean level of performance of each entity as an observation. For instance, the average literacy levels in each of 10 literacy centers operating program A would be compared to the 10 levels in as many centers that operate under program B. Such an analysis is correct given that the centers were randomized, but it is also rudimentary in that other approaches to analysis are likely to be more sensitive to small program differences.

Newer approaches to analysis exploit data at the entity level and, moreover, permit the exploitation of cross-sectional data on individuals within each center as well as repeated measures on these individuals. They also can take into account the fact that the centers operate in different parts of the country (for example) and that individuals vary in their demographic character. These statistical approaches fall under the rubrics of mixed models and hierarchical models. They are very well described in the state-of-the-art monographs by Murray (1998) and Bryk and Raudenbush (1992). The relevant computer-based software described in these sources and in articles such as Basen-Engquist et al. (1997) cover specific applications.

In recent years, a fine series of papers has been published on analysis of randomized field trials that involve entities as the first level of random allocation and analysis and subordinate (secondary and tertiary) units of analysis. These are apart from Murray's (1998) fine account. The best of these articles showed how different analyses at different levels influence one's conclusions about what works, and for whom. See, for example, Woodruff (1997) on nutrition education; Rosenbaum and Hanson (1998) on Project D.A.R.E.'s drug-use resistance program in schools; Moberg, Piper, Wu, and Serlin (1993) on health promotion programs in schools; and Stoker, King, and Foster (1981) on a case that appears to be hypothetical but is instructive.

A reminder. Our distrust of comparing only two entities, such as literacy centers, is warranted on grounds other than simple statistical considerations. For example, comparing a literacy center in Newark to one in Jersey City is suspect because the people in each place differ demographically. The contexts also differ. Comparisons of programs to reduce delinquency rates in Yokahama against Kyoto, or Bremen against Berlin, are also suspect for the same reasons. There are only two units of observation, one center each nested in a different city. The observations on Type I error just made are important, but they are less important than the matter of ensuring that the entities being compared are comparable. Because it is probably impossible to select two specific sites that are perfectly comparable, the argument here is that comparing many sites to one another—one set of sites operating under regime A and the second set operating under regime B—is the right way to go about comparing the regimes.

2.5 The Counsel of Advisory Groups on Research and Evaluation Policy

Some 30 years after Donald Campbell suggested that administrative units ought to be considered as the units in randomized experiments, a variety of expert panels have offered similar counsel to governments. It is remarkable that this advice has emerged in disparate scientific and political arenas, including sexually transmitted diseases, crime and substance abuse, and health risk reduction, among others. Preventing dangerous diseases, including sexually transmitted ones, at times requires that programs be deployed through organizations or geopolitical jurisdictions. As a consequence, the National Academy of Sciences Panel on Evaluating AIDS Prevention Programs suggested that diagnostic testing and counseling sites be considered as the units in controlled experiments to improve the services (Coyle, Boruch, & Turner, 1991). In a similar spirit, Leviton et al. (1990) mounted randomized experiments using small groups of gay men as the units to understand the effect of different approaches to AIDS counseling. Gay (1996), among others, used school classrooms as the units in a randomized trial to assess the effectiveness of AIDS education programs designed by the Red Cross.

Multidisciplinary conferences on sexually transmitted diseases (STDs), sponsored by the National Institute on Allergy and Infectious Diseases (NIAID), have led to the observation that clinical practices, factories, churches, and other organizations, as well as communities, might properly serve as the units in randomized trials (Green & Washington, 1991). In fact, NIAID continues to sponsor such studies. Some are identified below.

In considering approaches to preventing abuse of controlled substances, the participants in the "Communities That Care" Evaluation Design Conference, sponsored by the federal government, said, "A rigorous evaluation of a comprehensive community intervention requires an experimental design whereby communities are randomly assigned to experimental and control conditions" (see Peterson, Hawkins, & Catalano, 1991, p. 582). Debates over the design of a related study supported by England's Joseph Rowntree Foundation has been influenced by similar concerns (Farrington, 1997).

The National Research Council's Panel on the Understanding and Control of Violent Behavior offered the following:

> Recommendation 4: The panel calls for a new *multi-community* program of developmental studies of aggressive, violent, and antisocial behaviors, intended to improve both causal understanding and preventive interventions . . . (Reiss & Roth, 1993, p. 25)

This panel's report argued that "randomized controlled field experiments usually have important advantages as an evaluation strategy" (Reiss & Roth, 1993, p. 320). The report, however, did not examine deeply the nexus of randomized experiments and multicommunity programs.

Finally, consider that "Design and Analysis Issues in Community Trials" was the primary topic on the agenda of a 1992 National Institutes of Health conference. The participants agreed that the use of communities as the units of allocation and analysis presented challenges but that there were a variety of techniques for overcoming these challenges (Murray et al., 1994).

Each set of these recommendations was supported by evidence that organizational and community factors influence individual behavior. Each was driven by an interest in producing better evidence about how programs and contextual factors beyond a particular program influence institutional and individual behavior.

3. PRECEDENTS

Not all scientists or policymakers would agree that it is possible to execute good RFTs involving organizations or other entities as the units of random allocation. This disagreement is important, but it ought to be informed. In what follows, we give evidence on the feasibility of such RFTs using examples. Inasmuch as these studies are driven partly by theories about how individuals behave in larger institutional or geopolitical contexts, the ideas identified in section 2 are extended a bit here.

The examples are take from field research that involves schools and other education entities as units, communities and other geopolitical entities, and private and public organizations. Identifying precedents of each sort can be very difficult, and the compilation given here was undertaken with considerable effort so as to allow the reader to absorb the available literature quickly. The problems of identifying such studies and possible solutions also are discussed here.

3.1 Schools, School Districts, and Classrooms as the Units

In the research literature on health risk prevention, it is relatively easy to find examples of large-scale RFTs that use schools, school districts, or classrooms as the units. In what follows, we consider illustrative studies that use schools and classrooms in this and other arenas.

Schools and classrooms, for instance, have been randomly assigned to different approaches in educating children about avoiding substance abuse (Botvin, Baker, Dusenbury, Botvin, & Diaz, 1995; Moskowitz, 1984; Murray, Moskowitz, & Dent, 1996; Schaps, Moskowitz, Condon, & Malvin, 1982). In tests of the D.A.R.E. program in Illinois, members of 12 pairs of schools were randomly assigned to different programs in the interest of fair comparison (Rosenbaum et al., 1991; see also Clayton, Cattarello, & Johnstone (1996) for a long-term study). Other entity-based experiments on this program were reviewed in Ennett, Tobler, Ringwalt, and Flewelling (1994). Ellickson and Bell (1990, 1992) compared approaches to preventing drug and tobacco use against a control condition based on random assignment of 30 schools to the treatments in two states.

In their effort to evaluate a theory-driven program to reduce alcohol use by underage youths, Wagenaar et al. (1994) mounted a randomized field trial involving 15 school districts. Each of these districts contained three or fewer communities from which students were drawn. Seven of these districts in Minnesota and Wisconsin employed a special community-based prevention program. Eight of them constituted the control group.

Schools have also been the units in at least two smoking prevention experiments. The Television, School and Family Smoking Prevention Project used multiattribute balancing to randomly assign 35 Los Angeles area schools to different media-based smoking prevention campaigns. Flay et al. (1985) randomly assigned 22 matched schools to experimental and control conditions in the Waterloo Study, a Canadian smoking prevention effort. Tests of schoolwide cardiovascular risk reduction programs for children also have been undertaken. For example, schools have been randomly assigned to such programs in four states (Hansen & Graham, 1991; Killen et al., 1988; Perry et al., 1992).

Children's difficulty in confronting the opportunity to use controlled substances is not the only problem they face, of course. To understand how to reduce the psychological and educational risk of children who are moved from one context to another, Jason, Johnson, Danner, Taylor, and Krasaki (1993b) focused on children who transferred into new schools and who were, as a consequence, vulnerable. The project involved randomly assigning members of 10 matched pairs of schools to an innovative treatment program or to a control condition in order to determine whether their special transition program worked. See also Jason et al. (1992).

Until the late 1990s, high-quality evaluations of violence-reduction programs in schools were rare and randomized field trials rarer still. Among the notable exceptions is the Grossman et al. (1997) study of the effectiveness of violence-prevention curricula for second and third graders. Six matched pairs of schools were randomly assigned to employ the curriculum or to serve as control. Despite the small sample size, differences in behavior were discernible and persisted for 6 months.

New experiments have been planned by Korpi and Kirby (1998) to assess the relative effects of schoolwide programs developed under the Communities of Caring (C.C.) rubric. The programs, complex amalgams tailored to each site but with common themes, are designed to change the school's culture so as to enhance character development and positive behaviors among children and adolescents in high schools. The program's object is to reduce high-risk behavior, including substance use and violence. The evaluation design involves more than 40 schools being screened and then randomly allocated to the program and to control conditions.

Learning how to prevent children's use of controlled substances and how to prevent violence in schools is important, of course, but there is a more pervasive feature of children's lives that has an effect and has received some attention: achievement testing. Until the 1970s, no controlled field experiments of any scale appear to have been run to understand the effects of standardized testing on students in any country. In 1975, an opportunity presented itself in that the Irish Republic decided to consider for the first time standardized testing for children in the republic's 4,000 elementary schools. Kellaghan, Madaus, and Airasian (1982) and their colleagues at St. Patrick's College (Dublin) mounted a study in which 175 eligible schools, matched and stratified, were allocated randomly to five treatments. The basic control condition involved no standardized testing. Treatments included the introduction of standardized testing with and without feedback to teachers on student performance.

During the 1970s, the U.S. Department of Education sponsored a large-scale study to understand whether funding could be employed effectively by schools to reduce racial isolation and enhance the achievement of students. Eligible schools that were willing to participate in the experiment were randomly allocated to a special funding opportunity and to a control group whose members received no special treatment. The methodological problems encountered in this massive test were substantial, there having been no earlier work of this kind and no substantial pilot (pre-experiment) research. They have been covered ably by Coulson (1978), Reichardt and Rindskopf (1978), and Weisberg (1978).

The Tennessee School Improvement Incentives Project that Bickman (1985) described also involved rewarding randomly selected samples of schools for their performance and further providing a random subset of both groupings with technical assistance programs. This effort gained the cooperation of about 100 schools for the random allocation plan. It was apparently superseded by other state policy experiments.

Certain kinds of programs entail interactions among classroom children that are expected to facilitate each child's learning. For instance, effectively learning how to "avoid" or "refuse" sex involves developing an avoidance and refusal repertoire and having an opportunity to rehearse tactics, as well as learning about conditions under which various tactics might be used effectively. The learning is supposed to be facilitated by group interaction, and this in turn is supposed to affect knowledge, attitudes, and behaviors. The interaction also means that outcomes of children within a classroom are not statistically independent of one another. How much they are related usually is unclear. This has led some experimenters to use classrooms as the unit of allocation and analysis.

A number of such RFTs have been mounted to understand what kinds of programs might be deployed in education settings so as to enhance children's understanding of high-risk sexual behavior and how to avoid it. They vary considerably in scale. Gay's (1996) dissertation research, for example, involved matching eight middle-school classrooms and allocating half to a new Red Cross program and half to a control condition in which no such education effort existed. Aplasca et al. (1995) reported on an RFT on programs with similar objectives, targeting classrooms within schools in the Philippines. In a large-scale trial in California, Kirby, Korpi, Adivi, and Weismann (1997) randomly assigned 102 classrooms in six middle schools to a theory-driven SNAPP program that relied heavily on young "peer education" to implement the program. Another California-based program, Postponing Sexual Involvement (PSI), was evaluated using a complex research design in which classrooms were randomized in one component (Kirby, Korpi, Barth, & Cagampang, 1997). More than 50 schools were involved, but the number of participating classrooms is unclear.

A different stream of health-related work has concerned nutrition education. Woodruff (1997), for instance, described a San Diego experiment that involved eight intervention classes and nine control classes being randomly assigned to a new nutrition program from three community colleges. The article is interesting in that it compared four different ways of analyzing such data.

Randomized experiments that involve many classrooms in many schools and are directed toward enhancing children's achievement have been rare. Tennessee's study of the effects of reducing class size, the largest effort to date on the topic, is remarkable. In this STAR experiment, three treatments were compared: reduced class size (13-17 students), conventional classroom sizes (maximum of 22-25), and conventional classrooms augmented with teacher aides. About 100 classes of each type were composed of students and teachers who were randomly assigned to classroom, classrooms being one level of analysis (Bain & Achilles, 1986; Bain, Achilles, Zaharias, & McKenna, 1992; Bracey, 1995; Finn & Achilles, 1990; Mosteller, 1995; Nye, Achilles, Zaharias, DeWayne Fulton, & Wallenhorst, 1993; Word et al., 1990).

The Tennessee RFT on class size is a splendid example in many respects but is not a precedent in at least one respect. Classrooms in Nicaragua, for instance, have been randomly assigned to radio-based mathematics education and to conventional education so as to learn whether the former would enhance mathematics achievement and reduce education costs relative to the latter (Dean, Searle, Galda, & Heyneman, 1981; Jamison, Searle, & Suppes, 1980). A similarly designed experiment in El Salvador disintegrated; Hornik, Ingle, Mayo, McAnany, and Schramm (1972) gave an admirably candid description.

3.2 Communities and Other Geopolitical Entities as the Units

In a study of how to encourage voter registration in Chicago, for example, Gosnell (1927) *appears* to have randomly assigned distinct neighborhoods in political precincts to treatment or control conditions. The "treatment" involved publicity, mail, and in-person contacts, provided at times in different languages to diverse ethnic neighborhoods, to provide information about voter registration and to encourage registration. This was a noteworthy study in being one of the first (apparently) large-scale randomized field tests in the applied social sciences. It was almost certainly the first using voting jurisdictions as the unit of randomization and analysis.

Communities have clearly been the units of allocation and analysis in more recent evaluations of health related programs. LaPrelle, Bauman, and Koch (1992), for instance, reported on a study of the relative effectiveness of three media campaigns to prevent cigarette smoking among adolescents. They screened, matched, and then randomly assigned communities from a sample of 10 communities to one of three treatments and to a control group. Another smoking prevention project, the Community Intervention Trial for Smoking Cessation (COMMIT), assigned 11 matched pairs of communities to its treatment and comparison groups (Freedman, Green, & Byar, 1990, as cited in Peterson et al., 1992). In the broader health arena, the Minnesota Heart Health Program matched six communities to study efforts to prevent coronary heart disease.

In RFTs on fertility control policy in the Far East, communities and villages have been randomly assigned to different approaches to understand how to decrease birth rates (Freedman & Takashita, 1969; Hill, Stycos, & Back, 1959; Riecken et al., 1974). Small numbers of communities have been used as units in randomized studies of HIV risk prevention tactics (Kelly, Lawrence, & Diaz, 1991).

In media-based smoking prevention campaigns, standard metropolitan statistical areas (SMSAs) have been allocated randomly to the campaigns or to control conditions (Bauman, LaPrelle, Brown, Koch, & Padgett, 1991). Education studies in Colombia involved assigning small geographic areas in the *barrios* randomly to a cultural enrichment program for preschoolers to determine its effect relative to results in randomly selected control areas (McKay, McKay, Sinnestera, Gomez, & Lloreda, 1978).

Some randomized trials, directed toward the mentally ill, are mounted because the integration of multiple services within the community is thought to produce better service. So, for example, an experiment on Access to Community Care and Effective Service Supports (ACCESS) involves eight cities. Each city contains two independent jurisdictions that are randomly assigned to the ACCESS or to the control condition (Randolph, Basinsky, Leginski, Parker, & Goldman, 1997). About 50 agencies within each jurisdiction are supported to cooperate on integrating services for seriously mentally ill people who are homeless.

Finally, consider research on crime prevention. In the Kansas City patrol experiment, 15 police beats were matched and randomly divided into three groups of 5 beats each. This precedent compared the relative effects of reactive, proactive, and control (normal) patrols on victimization (Kelling, Pate, Dieckman, & Brown, 1974). Twenty years later, Sherman and Weisburd (1995) managed to execute a better randomized trial in Minneapolis, building on the earlier work. The researchers identified more than 100 "hot spots," local areas of predictably high crime and randomly allocated half of these areas to more intensive police patrol or to a normal patrol activity. Roughly speaking, the high-dose treatment involved twice as much police patrol time. There was substantial variation across areas and over time.

3.3 Other Private and Public Organizations

In some countries, a sensible way to enhance the well-being of individuals is through private organizations. Programs designed to reduce the risk of sexually transmitted diseases, for example, might be more effective if based in work sites rather than, say, field offices or schools. It is partly for this reason, in the United States, that the National Institute of Allergy and Infectious Diseases (NIAID) has invested in tests of factory-based peer education. The project involves some 40 industrial organizations in Zimbabwe, half being randomly assigned to programs designed to reduce incident HIV infection and the remaining to a control condition. No one knows whether peer education among factory workers will reduce infection, and this uncertainty justified the study (NIAID, 1997). Other randomized experiments have used work sites as units in assessing nutrition programs and weight-control and smoking cessation programs (Simpson, Klar, & Donner, 1995).

Private nonprofit organizations have, at times, committed resources to randomized field trials. For instance, more than 30 years ago, Goodwill Industries agreed to participate in controlled experiments on how to improve the management of the organization's stores (Glaser & Coffey, 1967). In this instance, independent stores were the units of allocation. An abstract of the out-of-print final report on the study is given in Riecken et al. (1974). Similarly, specialized adult literacy centers agreed to participate in a program aimed at enhancing literacy. Funded by the U.S. Department of Education, this study involved eight sites in the state of Oregon. Sites were to have been assigned randomly to the new programs or a control condition (Rhett, 1998). The trial was not executed because of its low statistical power and scientific yield.

In the medical arena, hospitals in Minnesota have cooperated in a study to understand how to reduce a particular kind of health risk. Nearly 40 community hospitals agreed to participate in an experiment to discover whether local medical opinion leaders and a formal feedback system could influence the rate at which the hospitals adopted new beneficial therapies for acute myocardial infarction patients (Soumerai et al., 1998). The theory underlying the program is that the entire hospital staff's understanding, not just the physicians' education, together with the monitoring of therapy, is necessary to produce change. Hence, allocating hospital physicians randomly to a program was not sensible. The experiment's design then involved the random allocation of 20 hospitals to this approach to clinical education and 17 hospitals to a control condition. Hospitals were the units of analysis in assessing the program's impact.

Our final illustration involves a program designed to enhance employment of individuals at high risk of unemployment who live in low-income public housing developments in communities that need economic revitalization. In each of seven cities, the experiment involves the random allocation of one multiple housing development to the program and one or two sites to a control condition. There are then 7 complex entities in the treatment condition and a total of 12 in the control condition. The presumptions underlying the program's design are that local collaboration and collective decisions are essential in transforming local communities in ways that affect, among other things, education, training and employment, and wage rates (Bloom, Bos, & Lee, 1998; Riccio, 1998).

3.4 Identifying Precedents: Vernacular and Other Problems

Identifying studies in which sites, organizations, or other entities were the units of allocation and analysis in randomized field trials is not easy. One cannot rely entirely on contemporary bibliographies, the Internet, or World Wide Web–based search engines.

Among other difficulties, the vernacular that is used to describe RFTs has not been standardized, so keyword searches of the literature must be inventive and sophisticated. For instance, some of the experiments discussed here are denominated as "cluster and randomized" in epidemiological reviews, such as Simpson et al. (1995). Murray (1998) prefers the phrase "group randomized trials" to refer to studies in which intact groups of individuals are assigned to different treatment conditions. For economists, labels such as randomized "saturation experiment" (Riccio, 1998) or "macro-experiment" (Garfinkel et al., 1990) are informative at times. In other disciplines, labels such as "community trials" and "multisite experiments" are used. Finally, as Murray (1998) points out, using entities as units in randomized experiments has natural links to statistical models and data generation that fall under the rubrics of "nested designs" and "hierarchical models and designs." The latter do not necessarily involve random assignment.

Complicating the problem of uncovering precedents, the entities that might be randomized vary in character. Dennis's (1988) Ph.D. dissertation, for example, focused on criminal and civil justice experiments. He developed fine descriptions of 41 well-documented studies published between 1972 and 1987. Interesting variations emerged from his work. The temporal units randomly assigned have included "days" on which status offenders appeared in court and "time periods" in a felony arrest diversion experiment. The legal units included police "beats" in the Kansas City preventive police patrol experiment. Simpson et al.'s (1995) conscientious review of health-related cluster experiments includes "time periods" as well as "standard metropolitan statistical areas" and various kinds of organizations as units.

The Riecken et al. (1974) book gave abstracts of reports on more than 40 randomized field experiments that had reached the attention of the Social Science Research Council's Committee on Social Experimentation. It included illustrations of the random assignment of neighborhoods (voter registration), neighborhoods, villages, and communities (fertility control); "cases" (therapy, criminal justice, etc.); groups of cases (mental health); and local vocational rehabilitative agencies. All these entries are cited in the references at the end of this chapter. The Riecken et al. (1974) volume, however, overlooked some earlier RFTs that are relevant here.

Of course, families that are the target of some social service and education programs also can be construed as entities. In some controlled randomized experiments, families have been the units of allocation. Precedents include Olds's (1988) Prenatal/Early Infancy Project, the Nauta and Hewett (1988) Child and Family Resource Program, and the St.Pierre et al. (1995) experiments on family literacy.

Finally, consider that even excellent reviews of impact evaluations cannot be used easily to identify what the units of allocation and analyses were. For instance, the University of Maryland's (1997) report to Congress on effective crime prevention programs is conscientious in rating the quality of studies on program effectiveness. No tables or text made plain, however, when the entities, rather than individuals, were the units in an RFT. If the units are identified, the specification is embedded in descriptive text. Further, it is sometimes hard to determine whether units were randomized at all. See Gottfredson (1997), for instance, whose review is instructive but not always specific.

One implication of the foregoing is that contemporary vernacular ought to be clarified. Emulating Murray's (1998) attempt to make the language more precise, for instance, we might establish a convention in which the phrase "randomized experiment" or "randomized field trial" must always include an adjective or noun that makes plain what was randomized. Such a convention would encourage key phrases or titles of the following sort: randomized factories experiment, randomized stores experiment, randomized field trial using hospitals as units, and so on. As awkward as this seems, analogous conventions in genomic research are far more complicated.

4. DIFFICULTIES AND POSSIBLE RESOLUTIONS

Challenges to using sites or other entities as the units of analysis in an RFT are numerous. Strategies that have been invented to surmount obstacles are valuable. Both are considered in what follows.

4.1 Choice of the Units of Allocation and Analysis in a Controlled Experiment

As Mosteller (1978) suggests, a fundamental problem of statistical analysis is deciding "what unit shall be regarded as independent of what other units" (p. 217). This problem has been confronted often, and in a variety of areas, by social scientists whose interest lay in estimating the relative effects of new programs. Consider the following.

Sherman and Weisburd (1995), for example, took pains to explore whether, in their experiment on intensive police patrols designed to reduce crime, to use "hot spots" rather than conventional police "beats" (patrol areas) as the units. Their judgment was based partly on concerns about the independence of each type of unit. Lind (1985) recognized analogous difficulty in choosing between "cases" or "related cases" as the units in designing randomized trials to understand how to reduce the burden on courts. Collins and Elkin (1985) had to handle the problem of how many members of a small therapy group could or should be chosen when the groups (rather than individuals) were to be randomly assigned to alternative therapeutic regimens.

In the Midwestern Prevention Project (MPP), there were at least two choices of units in the evaluation of a drug abuse prevention program (Pentz, 1994). One might have chosen the 26 communities that constitute the Kansas City, Kansas/Kansas City, Missouri/Indianapolis metropolitan area as the units of allocation, or one might have chosen the schools within these communities as the primary units.

The MPP researchers chose schools as the primary unit of allocation rather than communities for two reasons. First, the 26 communities in the Kansas City area are contiguous and receive similar media communications. One might argue that the communities are influenced strongly by local economic conditions and affected by them. That is, the MPP research group judged that communities, as defined by MPP, would not be the appropriate unit of allocation and analysis because communities are insufficiently independent. A second reason that schools were chosen as the unit in the MPP, according to Pentz, was that schools were assumed independent of one another. Pentz and her colleagues selected schools as the unit in part because the sample size of *schools* in the MPP area is larger than the sample size of communities.

In tests of variations on literacy programs, there are also choices. Junior colleges might be the units in some experiments. Community organizations or neighborhoods or high school districts might be the proper unit in others. These choices are made more difficult when considering the array of venues for adult literacy programs. The programs can be run by local education agencies, community colleges, volunteer organizations, community service groups, and technical institutes, among others, with or without connections to other entities (Young, Fitzgerald, & Morgan, 1994).

TABLE 8.3 Number of Units Per Treatment for Power ≥ .80 in a Two-Treatment Design When All Units Are Independent

	Alpha Level					
	.01		*.05*		*.10*	
Effect size	.10	.25	.10	.25	.10	.25
Sample size	586	95	393	64	310	50

SOURCE: Adapted from Barcikowski (1981).
NOTE: For example, when the alpha level is .05 and all units are independent, nearly 400 cases are required in each treatment to ensure, with a probability of .80, that a real difference between treatments is discerned in a formal statistical test, using the *t* distribution as a reference.

4.2 Statistical Power

Consider a randomized field trial in which two literacy programs are compared to each other to establish which is less costly. Suppose that a sample of literacy centers is screened for eligibility, matched and stratified properly, and then allocated randomly to the two literacy curricula. Assume that these organizations are independent of one another. A sample of students is drawn within each organization. Assume that the students within each school are not necessarily independent of one another.

How many centers might be required in this experiment to ensure that its statistical power is about .80? Assume, as is likely, that the true difference between the programs is small (.10) and fix the statistical threshold (alpha) at .05. If all the students within schools were independent, about 400 students for each plan would have to be sampled to discern the effect of the treatments under these conditions (see Table 8.3).

When the similarity among students within a school is low, the intraclass correlation will be low, but a larger sample size will be necessary to ensure that real differences between the programs are detected. Assuming a low intraclass correlation of .05, one might then use 85 schools with a sample of 10 students each for each treatment (program) in a formal test, or one might use 44 schools with 40 students each (see Table 8.4).

It will be sensible, at times, to assume that the intraclass correlation will be larger. Table 8.3 illustrates that the larger the intraclass correlation, the greater the number of schools in each treatment, or number of students, or both, is re-

TABLE 8.4 Number of Units Per Group (nG) and Number of Groups (G) Necessary for Power ≥ .80, Alpha = .05 When Intraclass Correlation (ρ) Is .05 and .20

	G	
nG	ρ = *.05*	ρ = *.20*
10	85	107
20	58	92
30	49	87
40	44	85

SOURCE: Adapted from Barcikowski (1981).
NOTE: For example, when two treatments are compared, the comparison may involve allocating 85 groups, each group containing 10 individuals, to the different treatments. When the intraclass correlation is .05 (ρ = .05), this scenario yields an experiment whose statistical power is .80, when the alpha level is .05. When the intraclass correlation is .20 (ρ =.20), the number of groups required is 107.

quired. For instance, if the intraclass correlation was at about the same level as "family" membership, say .20, then 107 schools would be required in each program group, and at least 10 students within each school would have to be randomly sampled. Alternatively, 170 schools could be sampled, with half (85) randomly allocated to each program, and at least 44 students sampled in each school. All this is, again, in the interest of discerning a small difference (.10) at conventional alpha level (.05).

In the opinion of LaPrelle et al. (1992), their experiment on community-based substance use prevention in citywide programs was underpowered. Four treatments in an experiment were spread over 10 communities. Their thoughtful post-experiment analysis suggested that about 40 communities per group would have been required to detect an important difference in the effectiveness of smoking prevention programs. Similarly, Curtin (personal communication, April 3, 1996) noted that the problem of low power of the school-based analysis of the D.A.R.E. program "forced" the analysis to depend on statistical tests that were based on students as the unit of analysis rather than on schools as the study progressed.

Other experiments appear to have relied successfully on at least three strategies to ensure adequate statistical power. First, entities that are independent were screened for eligibility and a reasonable level of homogeneity. Second, the entities were matched and then randomized. A third strategy is implicit: Engage as many entities as possible. Consider some examples.

The Communities Mobilizing for Change on Alcohol (CMCA) project (Wagenaar et al., 1994) screened school districts eligible to receive funding for a program to identify those that were at least 25 miles from other eligible districts so as to ensure independence of the units. Fifteen such districts were willing to cooperate in the experiment. These were then formed into matched pairs (and one triplet) on the basis of a pretest survey on underage alcohol use, population size, and the presence of a college or university in the district. This matching was done in the interest of enhancing statistical power. The matching also prevented peculiar, albeit unlikely, combinations that could occur by chance with full randomization.

Sherman and Weisburd (1995) were also sensitive to the problem of ensuring statistical power in a multisite police patrol experiment in Minneapolis. Their objective was to understand whether intensive police patrols in certain sites, notably high-crime "hot spots," could reduce criminal activity. Hot spots are presumed independent of one another and are numerous. Moreover, hot spots could be screened for eligibility based on conventional police records on calls, arrests, and so on. The implication was that hot spots, more than 100 of them, could be (and were) randomly assigned to intensive patrol so as to estimate the effect of such patrols on ambient crime rates.

In designing randomized field trials to assess the effects of different programs, one needs to specify the "expected effect size" so as to ensure adequate power. The expected effect size, however, may be difficult to specify because we are less familiar with using organizations in field tests than we are with using individuals.

Of some help would be an archive that lists randomized trials that use entities as the units, of the sort prepared for this article, and the effect sizes that were found in each. That is, each trial permits an estimate of effect size. The distribution of these effect sizes is a basis for the statistical power calculations. Such information can be archived in a continuously updated website, for example. A more sophisticated approach to power might depend on the specific theory underlying the program and its services and how the units are likely to change. Such theory is usually insufficiently precise.

Aside from building an empirical archive on effect sizes from earlier trials, other approaches to empirical anticipation of statistical power in designing such experiments are possible. The work by Bloom et al. (1998), for instance, properly recognizes that statistical power can be enhanced by identifying and employing good covariates, including prior measures. More interesting, the authors used data from schools in Rochester (New York) to understand the value of different ways to estimate power.

Among other things, we learn that intraclass correlations in the Rochester setting are stable over time regardless of the school grade of the children tested or the academic subject. As important, we learn that controlling for earlier test scores reduces the intraclass correlation substantially, implying that the number of organizational units need not be as large as one might select based on more simplistic analyses. Such control is apparently better exercised if one has individual-level data (children's test scores) than if one has institutional-level data (mean achievement levels for schools).

4.3 Measurement Systems and Theory

By a theory of "what should happen," we mean laying out the way that the programs being compared are each expected to engage and affect the entities. That is, the logic of how the thing is supposed to work needs to be made plain. Such theory, of course, will be based on fragmentary evidence, as all science is. It will, for new societal problems, be based on self-consciously critical speculation of the sort that Campbell (1969, 1988a) encouraged. More to the point, the theory guides us in selecting what should be measured and, at its most sophisticated, whether and how well it might be measured.

No such theory has been proffered for literacy programs and many other community-based efforts in the United States, but noteworthy illustrations appear in other quarters. Consider the multisite experiment described by Wagenaar et al. (1997). It was designed to understand whether a community-based program could reduce the use of alcohol by underage youths. Inasmuch as mobilization of communities was regarded theoretically as important to creating alcohol use policy, observations were made of community power structures and the attitudes of students and youths. Analyses were undertaken of media coverage. Changes in community practice were also measured on the supposition that these would follow community mobilization. Among other efforts, this stage included surveys of retail alcohol outlets to determine if indeed they failed to ask proof of the age of customers whose appearances were youthful. This was done because, in theory, decreasing youths access to alcohol would result in fewer alcohol-related traffic accidents. Further, the latter were assessed using state and local record systems.

Consider a second example. Contemporary research on preventing sexually transmitted diseases (STDs) is often based on a theory or logic model. One such model characterizes the incidence of infection, R, as a multiplicative function of transmissibility of the agent (an infectivity rate), β; the duration of exposure to infection, D; and partner change rate, C (Padian et al., 1991). All this happens in a social context, of course. Each is manipulable and measurable, in principle, in longitudinal experiments that focus on individuals as the unit of allocation. Each is manipulable and may be somewhat easier to measure on account of existing administrative records when entities are the units.

At the community level, the transmissibility parameter, β, might be observable routinely through probability sample surveys of anonymous respondents who are asked to report on risk factors such as gender; the use of tobacco, drugs, and alcohol; and age of coital debut. The rate of selecting new partners might also be determined this way. The duration of exposure, D, depends on the infecting organism. Some of these are reported in existing administrative measurement systems (e.g., syphilis and gonorrhea), and some are not (e.g., chlamydia). Many of the social context variables that are posited to be important in preventing STDs are measured routinely and in limited senses at the community level. These include socioeconomic status, ethnicity, education level, and, of course, temporal context (Hornik, 1991; Padian et al., 1991).

Reliability and Validity of Observations

The reliability of the observations of the outcome variables in randomized field trials affects the statistical power of trials. When the units of allocation are sites, the primary observations then analyzed are sites' means. The reliability of these means then becomes important.

The reliability of the means of sites depends on a number of factors. Feldt and Brennan (1993) provide a formula showing that this reliability depends on the empirical variance of individual observations across all entities. The empirical variance for the entity means (which is not the same as the ratio of individual variance and sample size) depends on the total number of observations, the number of observations within each entity, and the number of entities, among other things.

The order of magnitude of sites' means depends most heavily on the number of within-entity observations for experiments of the sort considered here. Consider, for instance, a scenario in which the dependent variable of interest is quantitative thinking ability. Using Feldt and Brennan's (1993) data on nearly half a million students and 400 schools, assume that in an experiment the variance for individuals is 3 and the variance of site means is 3.

Suppose that in designing a study we have a choice of using 4 to 20 *classrooms* with 30 students each, or 4 to 20 *schools* with 100 students each. The standard for choice is the estimated reliability of means for each type of entity.

If the classrooms are chosen as the unit of allocation and analysis, the reliability of the means will be, under these conditions, about .50. If schools are chosen as the unit, there being 100 children, say, in each school, the reliability of group means will be considerably higher, about .86, if 4 to 20 schools are used. If communities are the unit and 500 or so individuals are sampled in each, then the reliability of the means increases beyond this.

Management Information

In one sense, measurement of outcomes in an RFT is easy when organizations are the units of observation. Organizations, such as junior colleges, usually generate records on some indicators of output, such as the number of students who complete a course. They may also collect data on measures of outcome as a matter of routine (e.g., the rate at which students acquire jobs). Measurement may be more difficult in another sense. Each college or school may use different definitions for a student dropping out or obtaining a job. Each may use different measurement methods unless guided by a uniform set of rules. It is not uncommon, for example, for adult literacy programs to use different measures of persistence in a program. Further, the different programs being tested may influence the quality of measurement differently. The quality of measurement may improve in one system and not in another. This makes comparison difficult (as is often the case in police jurisdictions; e.g., Skogan & Lurigio, 1991).

When the object of the randomized field trial is to inform policy, the variables that ought to be measured will, at times, differ from variables measured in the interest of local operational management and control. In the student loan arena, Haines (1993) puts this distinction as a contrast between management information systems (MIS) for control purposes versus information systems for policy purposes.

In MIS at the local institutional level, for example, persistence rates may take priority. Other variables valued by the policymaker, such as academic achievement level, may take a lower priority in management information systems.

The unique staffing circumstances of most adult literacy programs also create difficulties in measuring both individual and program outcomes. More than a third of the programs supported by the Adult Education Act have no full-time staff (Young et al., 1994). Additionally, staff turnover is high (Foster, 1990). These circumstances make data collection and management difficult.

Finally, there is, in the adult education arena as in other arenas, some resistance and limited capacity for data collection. Kutner, Webb, and Herman (1993), for instance, note several challenges to implementing management information systems for adult education programs, among them the limited computer capacity and expertise of program staff. Further, there are difficulties in convincing staff of the value of collecting data. In the National Evaluation of Adult Education Programs, because participation in the outcome study was voluntary, only about half of the programs chose to report client outcome data to the evaluators. Those that did provide test scores did not necessarily follow guidelines on ensuring data quality. Roughly speaking, about 18,000 adult students were targets of literacy programs in the study, but data sufficient for learning about program effects could be produced for only 1,300 of them (Development Associates, 1994).

4.4 Engaging Organizations in Randomized Field Trials

Engaging sites in a controlled randomized field trial that involves many sites requires considerable skill. High-quality descriptions of obstacles and how they are surmounted are not common in the research literature. More generally, there has been little published documentation on why and when an institution agrees to be a unit in a randomized trial in any arena, suggesting that more work on the topic is warranted. Some exceptional studies are summarized in what follows to illustrate the products of such research.

Consider first Ellickson's (1994) paper on the conduct of Project ALERT. This RFT involved 30 schools randomly assigned to ALERT or to a control condition. Its object was to determine how well the project worked to prevent substance abuse among children and how long the project's effects last. Recruiting entire schools into an RFT must recognize natural limits on their capacity to participate. Ellickson (1994) reported that 11 schools that were invited to participate declined to do. One school, for instance, could not participate because of a court order demanding considerable resource allocation on racial equity. Four of the 11 schools declined to participate because they already had prevention programs in place. The reasons for other declinations concerned their capacity, at least indirectly (e.g., inability to ensure community support for engaging in the experiment). According to Ellickson, only one school declined because of the random assignment feature of Project ALERT.

The Project ALERT experience with recruitment differs a bit from that in another multisite study, the Communities Mobilizing for Change on Alcohol Project (CMCA). In the latter, *school districts* were screened for eligibility in an evaluation of a program designed to reduce alcohol use by underage youths (Wagenaar et al., 1994). Twenty-four districts were deemed eligible. Nine districts decided not to participate. "The most common reason for refusal was recent participation in another alcohol-related survey" (Wagenaar et al., 1994, p. 83).

Gay's (1996) dissertation discusses how, in a school context, the author negotiated the design and execution of a multiclassroom study experiment on an AIDS education program. Those with whom good professional relations had to be developed were, as one might expect, a team of teachers, administrators, and grant agency representatives who themselves were deeply engaged in the program's deployment at the school level. Gay contributed to the program's evolution and, on this account, seems to have been well equipped to negotiate the design features of the experiment with the team. The random assignment of eight classrooms to treatment and control conditions had to be "sold." The process of treatment delivery had to be fitted to the school's schedule. These and other negotiation processes took about 3 months.

Until Ritter and Boruch's (1999) study, no one had probed the origins of Tennessee's multisite RFTs on class size. Their report suggests the following. The RFTs's early origins lie in university-based studies of class size undertaken in one school. The results of these were used by Dr. Helen Bain to argue in a legislative context for a major effort to reduce class size in the state. In the face of some lawmakers' resistance to this, and in the interest of furthering understanding of the effects of reduced class size, Bain and her colleagues promoted legislation that authorized a demonstration project and provided funding for it. The language of the statute did not include the phrases "randomized trial" or "randomized experiment." It did call for "demonstration classes" and control groups, and this allowed the influence of competent researchers on the design of a trial.

Finally, consider efforts to mount high-quality research on the use of increased police patrols to reduce crime. Sherman and Weisburd's (1995) efforts involved negotiating the experiment's design through the Minneapolis City Council, the police department, and other stakeholders. These researchers capitalized on their understanding of case law on police authority and how to develop working relationships with police officers. Few academic journal articles describe such engagement processes even tersely. Sherman and Weisburd's (1995) article does so and is admirable on this account.

4.5 Temporal and Structural Stability

We often expect the behavior of individuals within sites not to change much over a short period of time, but the stability of certain characteristics of sites may be low or trends may reverse direction. Bauman et al. (1991), for example, found high positive correlation over a 2-year period ($r = .77$) for adolescents' reported *rates of recent smoking* in a sample of 10 cities but a *negative* correlation ($r = -.31$) for adolescents' *rates of experimentation* with smoking in the same cities. Reasons for this ostensible contradiction are unclear.

One normally assumes also that the organizations or other entities that are targeted for a program will be structurally stable over the study's course. A school in Year 1, for instance, is expected to be a school in Year 2. To judge from precedent, it is nonetheless prudent to anticipate some change.

For example, the Midwestern Prevention Project involved randomly assigning schools to different conditions. Pentz (1994) reported that 8 of the initial 50 targeted middle schools and high schools "closed or consolidated with other schools over the first three years of the study" (p. 44). Further, feeder schools changed as a consequence of changes in busing patterns and the installation of magnet schools that drew students from areas outside the original catchment area schools. The researcher's resolution of the problem was based partly on being able to observe the fraction of children at any point in time who have been posed to the program in the current school or any earlier school. The analysis then was based on statistical models that took such changes into account, at least up to a point.

Similar problems have occurred elsewhere. In the Irish Standardized Testing experiment, after matching and randomly assigning schools based on census data, the researchers found that many important school characteristics had changed (Kellaghan et al., 1982).

Tennessee's experiment on school incentives encountered difficulties, inasmuch as schools closed or were consolidated with others (Bickman, 1985).

Structural changes, one might presume, are more likely to occur in fragile institutions. Community-based entities that provide adult literacy services, for instance, are vulnerable in their reliance on volunteer staff. This resource might decline when local employment rates increase. The financial resources for a community organization may also be unstable, leading to the disappearance of a service center or a merger with another center.

All of this engenders complex problems in designing randomized experiments and in their analysis. Part of the resolution may lie in adapting for randomized field trials and in multiple analyses that make different assumptions about the changes and their relevance to the outcome variable. The topic warrants research.

4.6 Regional Variation

To obtain a precise evaluation of smoking prevention programs, Bauman et al. (1991) focused attention on only one geographic region. Despite this attempt to work in a homogeneous context, the experiment was underpowered; that is, the sample of organizations within the region may have been too small to discern a real effect of programs because there was considerable variation within the region.

To complicate matters, even within an ostensibly "homogeneous" region, organizational units may vary considerably. For instance, the rates of recent smoking among adolescents across 10 cities in one region reported by Bauman et al. (1991) were in the range of 2% to 7% in 1985 and 13% to 20% in 1987. Rates of smoking in 1987 among 1985 nonsmokers were in the range of 3% to 14% across the cities.

Ellickson and Bell (1992), in another randomized field trial, focused on two states and 30 schools evaluating a school-based drug use prevention program. There was considerable regional variation. To accommodate this variation, Ellickson and Bell (1992) used the school district as a blocking factor in randomly assigning schools to treatments. Prior to randomly assigning schools in community-wide drug prevention trials, Dent, Sussman, and Flay (1993) suggested the use of commonly available archival data for selecting and matching.

When sites are spread out through many regions, managing the trial will be difficult. For instance, the randomized field trial on the Emergency School Assistance Act (ESAA) of the 1970s involved school districts across the country. At the time, there was no precedent for this research, so there were substantial administrative issues that then engendered methodological problems (Coulson, 1978). Reichardt and Rindskopf (1978) argued, in an ambitious review, that the experiment would have been better done with fewer sites given the resources then available for running it.

How to arrange our thinking about regions, when the object is to mount a macrolevel randomized field trial, is unclear. Stratification or blocking by region makes sense, but the definitions of region and the implications of a choice have not been investigated deeply. In any event, reconnaissance prior to mounting any randomized experiment—pilot tests and analyses of extant data—are warranted.

4.7 Unbalanced Groups and Restricted Randomization

Consider a randomized field trial in which a sample of communities that is provided with increased literacy resources is compared to a sample of communities that has been randomly allocated to a waiting list (i.e., have not yet been given the resources). The number of communities involved in such a study must often be relatively small, say 20 to 40, in each of the groups. For the analyst, this raises a concern that the two groups that are randomly composed will not be "equivalent" at the outset. That is, there is an imbalance between the groups that is attributable to chance. This "unhappy random configuration" will complicate comparisons. One approach used to reduce the problem in multisite RFTs is restricted randomization.

In restricted randomization, some configurations of the random allocation of sites to different treatments are defined as undesirable a priori; that is, all possible randomized configurations under a particular experiment's design are laid out beforehand and the "unhappy" ones are eliminated from eligibility. A random selection is then made from the remaining configurations. For the applied researcher, constraining the randomization options to sensible configurations prevents badly unbalanced groups of institutions from being assigned to different program variations. For instance, Ellickson and Bell (1992) linked "unlike schools from districts into pairs and randomly (assigned) the pairs to the experimental conditions" to achieve balance (p. 85). The strategy is summarized in Table 8.5.

The implication is that when a small number of sites, rather than individuals, are the unit of allocation in controlled randomized tests, we can enumerate all possible allocations of sites in advance of the experiment. Further, we can eliminate the possible allocations that are strange, out of line, and so on. Having

TABLE 8.5 Restricted Randomization

1. Suppose that we have two districts (blocking factor), and suppose that there are three schools per district. How does one assign schools to three treatments so as to achieve a balance?
2. Lay out all possible pairings of schools, with one from each district. That is, lay out "configurations."

A1	B1	A1	B3
A2	B2	A2	B2
A3	B3	A3	B1
A1	B2	A1	B2
A2	B3	A2	B1
A3	B1	A3	B3
A1	B3	A1	B1
A2	B1	A2	B3
A3	B2	A3	B2

3. Eliminate configurations that contain pairs of schools that differ remarkably. For instance, if A1 and B1 are low SES and A3 and B3 are high SES, the two pairs are unbalanced. The first configuration is then eliminated.
4. Regard as eligible configurations the "best" configurations, for example, as measured by a weighted sum of variances of means.
5. Randomly assign pairs of like schools in eligible configurations to treatment.

SOURCE: Adapted from Ellickson and Bell (1992).

eliminated the allocations that are out of line, we can randomly select a configuration, allocate institutions in accord with it, and develop a comparison of programs that is precise. The population of all eligible configurations becomes the basis for constructing a reference distribution for testing statistical hypotheses formally.

4.8 Implementation Fidelity and Measurement

It makes no sense to try to estimate the effect of new programs unless one can verify that the program activities occur and can be described. "Implementation fidelity" here refers to the degree to which a new program is actually delivered to target individuals. Its measurement refers to observing indicators of fidelity (e.g., determining whether administrative actions are taken, information systems are emplaced, and so on). Learning that actions are indeed taken is a prerequisite for any "impact evaluation."

Implementation fidelity is as important in RFTs that involve sites as units as it is in a single-site randomized experiment that uses individuals as the units. Multisite tests, moreover, give us a realistic view of the statistical distribution of implementation levels across sites. For policymakers, this empirical variation across sites is important. The implication is that we ought to have better empirical data than we have now on what planned activities occur and what unanticipated activities occur when programs are implemented.

Several aspects of measurement in this arena are important but usually are not dealt with in depth in the research literature. They concern time, eligibility for programs within site, definitions, and the task of measuring fidelity at reasonable cost. Each is considered briefly here, and examples are provided.

Consider first the variable called time, an obviously crucial one in science and policy. How much time it takes the organization or community to deploy a new program and how the time requirement varies across a sample of sites is important to understand the dispersal of literacy programs or any other institution-wide programs, for example. Evidence on deployment time is, however, often buried in research reports, rarely appears in published articles, and has not been the subject of meta-analyses. The research and policy committee should have ready access to empirical curves of the level of implementation plotted against time for many diverse kinds of programs. It deserves thoughtful treatment on policy grounds independent of its value for planning macrolevel experiments.

Consider next the eligibility criteria that are used in determining who ought to be provided the opportunity of a service (or be the target) within each site. In the earliest RFTs involving schools as the units, minority students within a randomly selected and treated school were identified as eligible for specially funded services or not. In principle at least, students in the randomly assigned control schools would also be identifiable as eligible or not. As one might expect, actually employing the eligibility criteria in a treated school to identify and provide service to students is not the same as being asked in control schools to merely identify the eligible students. This difference is subtle but important. Reichardt and Rindskopf (1978) argued that this may have led to average differences in pretest scores between treated and control schools. That is, despite the fact that schools were randomly assigned to treatment and control conditions, the average pretest scores on achievement level differed. The averages would have been the same if students who were identified as eligible in each group of schools were equivalent. They could not be equivalent if eligibility criteria differed in each group.

RFTs that involve "integration" or "coordination" of services across many agencies within an organization or community present special problems. Developing a coherent definition of integration and measurable indicators of integration is not easy. Moreover, different strategies may be involved. Consider studies of ACCESS's effect on the homeless and mentally ill, for instance. The various jurisdictional units may differ on whether and how they employ interagency coalitions, interagency teams for service delivery, and interagency management systems; interagency agreements and memorandums of understanding; funding arrangements; eligibility standards; and colocation of services (Randolph et al., 1997). Learning how to observe any of these reliably and to ensure fidelity in implementation and its measurement is demanding.

Finally, consider that learning how to inexpensively monitor the fidelity of the program at the site level is not easy. Ellickson and Bell (1992), for instance, monitored more than 900 of the 2,300 class sessions in evaluating a schoolwide health program—Project ALERT. They claim that most classes were held. All relevant activities were presented in 80 of the sessions, and no activities were rushed in 60% of cases. What the rate of monitoring activity should be, and why, is not clear. There has been little research on this and on the costs incurred in such measurement.

5. SUMMARY: TO WHAT EFFECT?

Designing randomized field trials that yield good evidence about which programs work better, for whom, and how long is often desirable. Randomized trials that focus only on individuals as the recipients of services and on the random allocation of individuals to alternative programs are helpful but limited. Focusing on sites—communities, organizations, neighborhoods, schools, or other entities—as the units of assignment in controlled field trials of national programs is sensible, at times, for theoretical, statistical, policy, and ethical reasons. Moreover, using individuals as the unit of random assignment and analysis is wrong at times.

The main theoretical rationale for this perspective is that programs work when organizational elements act in concert (e.g., community-wide literacy programs). A basic statistical rationale for focusing on institutions as units is that conventional statistical analyses of the effect of programs can be wrong when analyses are based on individuals who are not independent within an institution. Mistaking chance differences for real differences becomes a distinct possibility. Because sites are more likely to be independent of one another, using them as the units of analysis helps to avoid this problem.

The policy rationale for focusing on organizations and other sites as the units for study is that organizations are often the immediate target for action by federal agencies and foundations. Individuals are not. The ethical rationale is that, at times, controlled allocation of organizations rather than individuals to alternative regimens, in the interest of a fair comparison, is more acceptable under standards of a professional, political, or legal ethic.

The feasibility of using sites as units in controlled experiments is demonstrable at times. For example, entities have been allocated at random or in a rule-driven way in controlled tests of fertility control methods, law enforcement programs, and health-risk reduction programs. The units of random allocation have been neighborhoods, factories, classrooms and schools, and so on.

There are difficulties in executing RFTs, of course, implying that we need to do research to address the following questions.

- How can statistical power be better anticipated and, more important, enhanced in such studies?
- How can measurement systems be best exploited or augmented in the service of such studies?
- How can better decisions be made about choosing the kinds of units that are appropriate for randomization?
- How might we better anticipate instability in the structure of sites and the information they produce, and variation among them in performance?

Improving the state of the art in this arena will not be easy. Absent improvements in how we think about experimenting on organizations, we will, in Walter Lippmann's words, "imperil the future . . . by leaving great questions to be fought out between ignorant change on the one hand and ignorant opposition to change on the other" (Lippmann, 1936/1963, pp. 495-497). His plea was to statesmen, especially President Franklin D. Roosevelt, to be "honestly experimental" rather than dogmatic and ignorant in fostering societal change. It is encouragement with which Donald Campbell was in clear sympathy, and toward which Campbell's methodological inventiveness was directed. The approach to the design of RFTs considered in this chapter accords with the spirit of this legacy. The success of the ideas and the methodology depend heavily on national interests in being conscientious and honest about experimentation.

REFERENCES

Aplasca, M., Siegel, D., Mandel, J. S., Santana, R., Paul, J., Hudes, E. S., Monzon, O. T, & Hearst, N. (1995). Results of a model AIDS prevention program for high school students in the Philippines. *AIDS*, Supplement 1, 7-13.

Bain, H. P., & Achilles, C. M. (1986). Interesting developments in class size. *Phi Delta Kappan*, *67*, 662-665.

Bain, H. P., Achilles, C. M., Zaharias, J. B., & McKenna, B. (1992). Class size does make a difference. *Phi Delta Kappan, 74*(3), 253-256.

Barcikowski, R. S. (1981). Statistical power with group mean as the unit of analysis. *Journal of Educational Statistics, 6*(3), 267-285.

Basen-Engquist, K., Parcel, G. S., Harrist, R., Kirby, D., Coyle, K., Banspach, S., & Rugg, D. (1997). The Safer Choices Project: Methodological issues in school based health promotion intervention research. *Journal of School Health, 67*(9), 365-371.

Bauman, K. E., LaPrelle, J., Brown, J. D., Koch, G. C., & Padgett, C. A. (1991). The influence of three mass media campaigns on variables related to adolescent cigarette smoking: Results of a field experiment. *American Journal of Public Health, 81*, 597-604.

Bickman, L. (1985). Randomized field experiments in education. *New Directions for Program Evaluation, 28*, 39-54.

Bloom, H., Bos, J., & Lee, S. W. (1998). *Using cluster random assignment to measure program impacts: Statistical implications for the evaluation of education programs.* New York: New York University, Robert F. Wagner School of Public Service.

Boruch, R. F. (1993a). *Multi-site evaluation and the children's initiative.* Unpublished manuscript prepared for the Pew Charitable Trusts, Philadelphia, PA. (Available from R. F. Boruch, University of Pennsylvania, Philadelphia, PA 19104)

Boruch, R. F. (1993b, October). *Multi-site tests in the civil and criminal justice arena.* Invited presentation at the annual meeting of the American Society of Criminology, Phoenix, AZ.

Boruch, R. F. (1995). The future of controlled experiments: A briefing. *Evaluation Practice, 15*(3), 265-274.

Boruch, R. F. (1997). *Randomized experiments for planning and evaluation: A practical guide.* Thousand Oaks, CA: Sage.

Botvin, G. J., Baker, E., Dusenbury, L., Botvin, E. M., & Diaz, T. (1995). Long term follow-up results of a randomized drug-abuse prevention trial in a white middle class population. *Journal of the American Medical Association, 273*, 1106-1112.

Botvin, G. J., Baker, E., Dusenbury, L., Tortu, S., & Botvin, E. (1990). Preventing adolescent drug use through a multimodel cognitive behavioral approach: Results of a three year study. *Journal of Consulting and Clinical Psychology, 58*(4), 437-445.

Botvin, G., Dusenbury, L., Baker, E., James-Ortiz, S., & Botvin, E. M. (1992). Smoking prevention among urban minority youth: Assessing effects on outcome and mediating variables. *Health Psychology, 11*(5), 290-299.

Bracey, G. (1995). Research oozes into practice: The case for class size. *Phi Delta Kappan, 77*, 89.

Bryk, A., & Raudenbush, S. W. (1992). *Hierarchical linear models in social and behavioral research: Applications and data analysis methods.* Newbury Park, CA: Sage.

Campbell, D. T. (1969). Reforms as experiments. *American Psychologist, 24*(4), 408-429.

Campbell, D. T. (1988a). The experimenting society. In S. Overman (Ed.), *Methodology and epistemology for social science: Selected papers by Donald T. Campbell* (pp. 290-314). Chicago: University of Chicago Press.

Campbell, D. T. (1988b). *Methodology and epistemology for social science: Selected papers by Donald T. Campbell* (S. Overman, Ed.). Chicago: University of Chicago Press.

Campbell, D. T., & Stanley, J. C. (1963). Experimental and quasi-experimental designs for research teaching. In N. L. Gage (Ed.), *Handbook of research on teaching* (pp. 171-246). Chicago: Rand McNally.

Clayton, R. R., Cattarello, A. M., & Johnstone, B. M. (1996). The effectiveness of drug abuse resistance education (DARE): Five-year follow-up results. *Preventive Medicine, 25*, 307-318.

Collins, J. F., & Elkin, I. (1985). Randomization in the NIMH Treatment of Depression Collaboration Research Program. *New Directions for Program Evaluation, 28*, 27-38.

Cook, T. D.. & Campbell, D. T. (1979). *Quasi-experimentation: Design and analysis for field settings.* Chicago: Rand McNally.

Cook, T. D., Habib, F., Phillips, M., Settersten, R. A., Shagle, S. C., & Degirmencioglu, S. M. (1999). Comer's School Development Program in Prince George's County, Maryland: A theory-based evaluation. *American Educational Research Journal.*

Cook, T. D., Hunt, H. D., & Murphy, R. F. (1999). *Comer's School Development Program in Chicago: A theory-based evaluation report.* Evanston, IL: Northwestern University.

Coulson, J. E. (1978). National evaluation of the Emergency School Aid Act (ESAA): A review of methodological issues. *Journal of Educational Statistics, 3*(3), 1-60.

Coyle, K., Kirby, D., Purcel, G., Basen-Engquist, K., Banspach, S., Rugg, D., & Well, M. (1996). Safer Choices: A multicomponent school based HIV/STD and pregnancy prevention program for adolescents. *Journal of School Health, 66*(3), 89-84.

Coyle, S. L., Boruch, R. F., & Turner, C. F. (Eds.). (1991). *Evaluating AIDS Prevention Programs* (Expanded ed.). Washington, DC: National Academy of Sciences.

Dean, J., Searle, B., Galda, K., & Heyneman S. P. (1981). Improving elementary mathematics education in Nicaragua: An experimental study of the impact of textbooks and radio on achievement. *Journal of Education Psychology, 73*(4), 556-567.

Dennis, M. L. (1988). *Implementing randomized field experiments: An analysis of criminal and civil justice research.* Unpublished doctoral dissertation, Psychology Department, Northwestern University.

Dennis, M. L., & Boruch, R. F. (1989). Randomized experiments for planning and testing projects in developing countries: Threshold conditions. *Evaluation Review, 13*(3), 292-309.

Dent, C. W., Sussman, S., & Flay, B. R. (1993). The use of archival data to select and assign schools in a drug prevention trial. *Evaluation Review, 17*(2), 247-260.

Development Associates. (1992). *National evaluation of adult education programs: First interim report.* Arlington, VA: Author.

Development Associates. (1994). *National evaluation of adult education programs, fourth report: Learner outcomes and program results.* Washington, DC: Office of the Undersecretary, U.S. Department of Education.

Dryfoos, J. G. (1994). *Full-service schools.* San Francisco: Jossey-Bass.

Ellickson, P. L. (1994). Getting and keeping schools and kids for evaluation studies. *Journal of Community Psychology* (Monograph series/CSAP special issue), 102-116.

Ellickson, P. L., & Bell, R. M. (1990). Drug prevention in junior high: A multi-site longitudinal test. *Science, 247*, 1299-1305.

Ellickson, P. L., & Bell, R. M. (1992). Challenges to social experiments: A drug prevention example. *Journal of Research in Crime and Delinquency, 29*(1), 79-101.

Ennett, S. T., Tobler, N. S., Ringwalt, C. L., & Flewelling, R. L. (1994). How effective is Drug Abuse Resistance Education? A meta-analysis of Project DARE's outcome evaluations. *American Journal of Public Health, 84*(9), 1394-1401.

Fairweather, G. W., Sanders, D. H., & Tornatzky, L. G. (1974). *Creating change in mental health organizations.* New York: Pergamon.

Fairweather, G. W., & Tornatzky, L. G. (1977). *Experimental methods for social policy research.* New York: Pergamon.

Farrington, D. P. (1997). Evaluating a community crime prevention program. *Evaluation, 3*, 3-26.

Feldt, L. S., & Brennan, R. L. (1993). Reliability. In R. L. Linn (Ed.), *Education measurement* (3rd ed., pp. 105-146). Phoenix, AZ: Oryx.

Finn, J. D., & Achilles, C. M. (1990). Answers and questions about class size: A statewide experiment. *American Educational Research Journal, 27*, 557-577.

Flay, B. R., Ryan, K. B., Best, J. A., Brown, K. S., Kersell, M. W., d'Avernas, J. R., & Zanna, M. P. (1985). Are social-psychological smoking prevention programs effective? The Waterloo study. *Journal of Behavioral Medicine, 8*(1), 37-59.

Foster, S. E. (1990). Upgrading the skills of literacy professionals: The profession matures. In F. P. Chisman (Ed.), *Leadership for literacy: The agenda for the 1990s* (pp. 73-95). San Francisco: Jossey-Bass.

Freedman, R., & Takashita, J. T. (1969). *Family planning in Taiwan: An experiment in social change.* Princeton, NJ: Princeton University Press.

Fuchs, D., & Fuchs, L. (1999, July). *Peer-assisted learning strategies: Programmatic experimental research to make schools more accommodating of diversity.* Report presented at the University of Durham Conference on Evidence-Based Policy, Durham, England.

Garfinkel, I., Manski, C. F., & Michalopoulos, C. (1990). *Are micro-experiments always best? Randomization of individuals or sites?* (IRP Conference Paper). Madison: University of Wisconsin/U.S. Department of Health and Human Services.

Gay, K. E. M. (1996). *Collaborative school-based research: The creation and implementation of an HIV/AIDS prevention curriculum for middle school students.* Unpublished doctoral dissertation, University of Pennsylvania.

Glaser, E. M., & Coffey, H. S. (1967). *Utilization of applicable research and demonstration results.* Los Angeles: Human Interaction Research Institute.

Goldberg, L., Elliot, D., Clarke, G. N., MacKinnon, D. D., Moe, E., Zoref, L., Green, C., Wolf, S. L., Greffath, E., Miller, D. J., & Lapin, A. (1996). Effects of a multidimensional anabolic steroid prevention intervention. *Journal of the American Medical Association, 276*(19), 1555-1562.

Gosnell, H. F. (1927). *Getting out the vote: An experiment in the stimulation of voting.* Chicago: University of Chicago Press.

Gottfredson, D. C. (1986). An empirical test of school based environmental and individual interventions to reduce the risk of delinquent behavior. *Criminology, 24*(4), 705-731.

Gottfredson, D. (1997). School based crime prevention. In University of Maryland, Department of Criminology and Criminal Justice, *Preventing crime: Report to the Congress* (NCJ 165366, pp. 5-1–5-74). Washington, DC: U.S. Department of Justice, Office of Justice Programs.

Graham, J. W., Flay, B. R., Johnson, C. A., Hansen, W. B., & Collins, L. M. (1984). Group comparability: A multiattribute utility measurement approach to the use of random assignment with small numbers of aggregated units. *Evaluation Review, 8*(2), 247-260.

Green, S. B., & Washington, A. E. (1991). Evaluation of behavioral interventions for prevention and control of sexually transmitted diseases. In J. N. Wasserheit, S. O. Aral, K. K. Holmes, & P. J. Hitchcock (Eds.), *Research issues in human behavior and sexually transmitted diseases in the AIDS era* (pp. 345-352). Washington, DC: American Society for Microbiology.

Grossman, D. C., Neckerman, H. J., Koepsall, T. D., Liu, P., Asher, K. N., Beland, K., Frey, K., & Rivara, F. P. (1997). Effectiveness of a violence prevention curriculum among children in elementary school: A randomized controlled trial. *Journal of the American Medical Association, 277*(20), 1605-1611.

Haggard, E. A. (1958). *Intraclass correlation and the analysis of variance.* New York: Dryden.

Haines, R. (1993). *A policy driven management information system: Identifying information requirements for evaluating policy outcomes.* Washington, DC: U.S. Department of Education.

Hansen, W. B., & Graham, J. W. (1991). Preventing alcohol, marijuana, and cigarette use among adolescents: Peer pressure resistance training versus establishing conservative norms. *Preventive Medicine, 20*, 414-430.

Hauck, W., Gilliss, C. L., Donner, A., & Gortner, S. (1991). Randomization by cluster. *Nursing Research, 40*, 356-358.

Hill, R., Stycos, J. M., & Back, K. W. (1959). *The family and population control: A Puerto Rican experiment in social change.* Chapel Hill: University of North Carolina Press.

Hopkins, K. D. (1982). The unit of analysis: Group means versus individual observations. *American Educational Research Journal, 19*(1), 5-18.

Hornik, R. (1991). Alternative models of behavior change. In J. N. Wasserheit, S. O. Aral, K. K. Holmes, & P. J. Hitchcock (Eds.), *Research issues in human behavior and sexually transmitted diseases in the AIDS era* (pp. 201-218). Washington, DC: American Society for Microbiology.

Hornik, R. C., Ingle, H. T., Mayo, J. K., McAnany, E. G., & Schramm, W. (1972). *Television and education reform in El Salvador* (Report No. 14). Stanford, CA: Stanford University, Institute for Communication Research.

Jamison, D., Searle, B., & Suppes, P. (1980). *Radio mathematics in Nicaragua.* Stanford, CA: Stanford University Press.

Jason, L., Johnson, J. H., Danner, K. E., Taylor, S., & Krasaki, K. S. (1993). A comprehensive, preventive, parent-based intervention for high risk transfer students. *Prevention in Human Services, 10*(2), 27-37.

Jason, L. A., Weine, A. M., Johnson, J. H., Warren-Sohlberg, L., Filippelli, L. A., Turner, E., & Lardon, C. (1992). *Helping transfer students: Strategies for educational and social readjustment.* San Francisco: Jossey-Bass.

Kaftarian, S. J., & Hansen, W. B. (Eds.). (1994). *Community Partnership Program: Center for Substance Abuse Prevention* (Monograph series/CSAP special issue). *Journal of Community Psychology.*

Kellaghan, T., Madaus, G. F., & Airasian, P. W. (1982). *The effects of standardized testing.* Boston: Kluwer-Nijhoff.

Kelling, G. L., Pate, T., Dieckman, D., & Brown, C. E. (1974). *The Kansas City Preventive Patrol Experiment: A summary report.* Washington, DC: Police Foundation.

Kelly, J. A., Lawrence, J. S., & Diaz, Y. E. (1991). HIV risk behavior reduction following intervention with key opinion leaders: An experimental analysis. *American Journal of Public Health, 81*, 168-171.

Killen, J. D., Telch, M. J., Robinson, T. N., Maccoby, N., Taylor, C., & Farquar, J. W. (1988). Cardiovascular disease risk reduction for tenth graders: A multiple factor school-based approach. *Journal of the American Medical Association, 260*(12), 1728-1733.

Kirby, D., Korpi, M., Adivi, C., & Weismann, J. (1997). An impact evaluation of Project SNAPP: An AIDS prevention and pregnancy middle school program. *AIDS Education and Prevention, 9*(Supplement A), 44-61.

Kirby, D., Korpi, M., Barth, R. P., & Cagampang, H. H. (1997). The impact of postponing sexual involvement curriculum among youths in California. *Family Planning Perspectives, 29*, 100-108.

Korpi, M., & Kirby, D. (1998). *Community of caring evaluation* (Grant proposal). Santa Cruz, CA: ETR Associates.

Kruskal, W. (1988). Miracles and statistics: The casual assumption of independence. *Journal of the American Statistical Association, 83*(104), 929-940.

Kutner, M. A., Webb, L., & Herman, R. (1993). *Management information systems in adult education: Perspectives from the states and from local programs* (NCAL Technical Report TR93-4). Philadelphia: University of Pennsylvania, National Center on Adult Literacy.

LaPrelle, J., Bauman, K. E., & Koch, G. G. (1992). High intercommunity variation in adolescent cigarette smoking in a 10-community field experiment. *Evaluation Review, 16*(2), 115-130.

Leviton, L., Finnegan, J. R., & Zaplea, J. G. (1999). Formative research methods to understand patient and provider responses to heart attack symptoms. *Evaluation and Program Planning.*

Leviton, L. C., Goldenberg, R. L., Baker, C. S., Schwartz, R. M., Freda, M. C., Fish, L. J., Cliver, S. P., Rouse, D. J., Chazotte, C., Merkatz, I. R., and Raczynski, J. M. (1999). Methods to encourage the use of corticosteroid therapy for fetal maturation: A randomized controlled trial. *Journal of the American Medical Association, 281*(1), 45-52.

Leviton, L., Valdiserri, R., Lyter, D., Callahan, C., Kingsley, L., Huggins, J., & Rinalde, C. R. (1990). Preventing HIV infection in gay and bisexual men: Experimental evaluation of attitudes changes from two risk reduction experiments. *AIDS Education and Prevention, 2*(2), 95-108.

Lind, E. A. (1985). Randomized experiments in federal courts. *New Directions for Program Evaluation, 28*, 73-80.

Lippmann. W. (1963). The Savannah speech. In C. Rossiter & J. Lare (Eds.), *The essential Lippmann: A political philosophy for liberal democracy* (pp. 495-497). New York: Random House. (Original work published 1936)

Lorian, R. P. (1994). Epilogue: Evaluating the Community Partnership Program. Reflections on a name. *Journal of Community Psychology* (Monograph series/CSAP special issue), 201-205.

McGraw, K. O., & Wong, S. P. (1996). Forming inferences about intraclass correlation coefficients. *Psychological Methods, 1*(1), 30-46.

McKay, H., McKay, A., Sinnestera, L., Gomez, H., & Lloreda, P. (1978). Improving cognitive ability in chronically deprived children. *Science, 200*(4), 270-278.

Moberg, D. P., Piper, D. L., Wu, J., & Serlin, R. C. (1993). When total randomization is impossible: Nested random assignment. *Evaluation Review, 17*(3), 271-291.

Moskowitz, J. (1984). The effects of drug education and follow-up. *Journal of Alcohol and Drug Education, 3*, 45-49.

Moskowitz, J., Malvin, J., Schaeffer, G., & Schaps, E. (1984). An experimental evaluation of a drug education course. *Journal of Drug Education, 14*, 9-22.

Mosteller, F. (1978). Errors: Nonsampling errors. In W. H. Kruskal & J. M. Turner (Eds.), *International encyclopedia of statistics* (Vol. 1, pp. 208-229). New York: Free Press.

Mosteller, F. (1995). *The Tennessee study of class size in the early school grades.* Cambridge, MA: Harvard University Press.

Murray, D. M. (1998). *Design and analysis of group randomized trials.* New York: Oxford University Press.

Murray, D. M., & Hannan, P. J. (1990). Planning for the appropriate analysis in school-based drug use prevention studies. *Journal of Consulting and Clinical Psychology, 58*, 458-468.

Murray, D. M., McKinlay, S. M., Martin, D., Donner, A. P., Dwyer, J. H., Raudenbush, S. W., & Graubard, B. I. (1994). Design and analysis issues in community trials. *Evaluation Review, 18*(4), 493-514.

Murray, D. M., & Wolfinger, R. D. (1994). Analysis issues in the evaluation of community trials: Progress toward solutions in SAS/STAT mixed. *Journal of Community Psychology* (Monograph series/CSAP special issue), 140-154.

Murray, D., Moskowitz, J. M., & Dent, C. W. (1996). Design and analysis issues in community-based drug abuse prevention. *American Behavioral Scientist, 39*(7), 853-867.

National Institute of Allergy and Infectious Diseases. (1997). *Briefing book.* Washington, DC: U.S. Department of Health and Human Services.

Nauta, M. J., & Hewett, K. (1988). Studying complexity: The case of the Child and Family Resource Program. In H. B. Weiss & F. H. Jacobs (Eds.), *Evaluating family programs* (pp. 389-405). New York: Aldine de Gruyter.

Nye, B. A., Achilles, C. M., Zaharias, J. B., DeWayne Fulton, B., & Wallenhorst, M. P. (1993). Tennessee's bold experiment: Using research to inform policy and practice. *Tennessee Education, 22*(3), 10-17.

Olds, D. (1988). Common design and methodological problems encountered in evaluating family support service: Illustrations from the Prenatal/Early Infancy project. In H. B. Weiss & F. H. Jacobs (Eds.), *Evaluating family programs* (pp. 239-265). New York: Aldine de Gruyter.

Padian, N., Hitchcock, P. J., Fullilove, R. E., Kohlstadt, V., Brunham, R., & the NIAID Study Group. (1991). Issues in defining behavioral risk factors and their distribution. In J. N. Wasserheit, S. O. Aral, K. K. Holmes, & P. J. Hitchcock (Eds.), *Research issues in human behavior and sexually transmitted diseases in the AIDS era* (pp. 353-361). Washington, DC: American Society for Microbiology.

Pentz, M. A. (1994). Adaptive evaluation strategies for estimating effects of community based drug abuse prevention programs. *Journal of Community Psychology* (Monograph series/CSAP special issue), 26-51.

Perry, C., Parcel, G. S., Stone, E., Nader, P., McKinlay, S. M., Leupker, R. V., & Webber, L. S. (1992). The Child and Adolescent Trial for Cardiovascular Health (CATCH): An overview of intervention program and evaluation methods. *Cardiovascular Risk Factors*, *2*(1), 36-43.

Peterson, P. L., Hawkins, J. D., & Catalano, R. F. (1991). *Evaluating comprehensive risk reduction: A design conference* (report of the conference). (Available from J. D. Hawkins Social Development Research Group, University of Washington, Seattle, WA 98103)

Peterson, P. L., Hawkins, J. D., & Catalano, R. F. (1992). Evaluating comprehensive community drug risk reduction interventions. *Evaluation Review*, *16*(6), 579-602.

Raab, G. (1999, July). *Defining outcomes in a trial of school-based sex education.* Paper presented at the University of Durham Conference on Evidence-Based Policy. Durham, England.

Randolph, F., Basinsky, M., Leginski, W., Parker, L., & Goldman, H. H. (1997). Creating integrated service systems for homeless persons with mental illness: The ACCESS program. *Psychiatric Services*, *48*(3), 369-373.

Raudenbush, S. W. (1997). Statistical analysis and optimal design for cluster randomized trials. *Psychological Methods*, *2*(2), 173-185.

Reichardt, C. S., & Rindskopf, D. (1978). Randomization and educational evaluation: The ESAA evaluation. *Journal of Educational Statistics*, *3*(1), 61-68.

Reiss, A. J., & Roth, J. A. (Eds.). (1993). *Understanding and preventing violence.* Washington, DC: National Academy of Sciences Press.

Rhett, N. (1998, June 25). *Possible experiments using literacy centers as units* (e-mail memorandum). Washington, DC: U.S. Department of Education, Planning & Evaluation Service.

Riccio, J. A. (1998). *A research framework for evaluating Jobs-Plus, a saturation and place-based employment initiative for public housing residents.* Unpublished manuscript, Manpower Demonstration Research Corporation, New York.

Riecken, H. W., Boruch, R. F., Campbell, D. T., Caplan, N., Glennan, T. C., Pratt, J. W., Rees, A., & Williams, W. (1974). *Social experimentation: A method for planning and evaluating social programs.* New York: Academic Press.

Ritter, G. W., & Boruch, R. F. (1999). The political and institutional origins of a randomized controlled trial on elementary school class size: Tennessee's Project Star. *Educational Evaluation and Policy Analysis*, *21*(1), 111-125.

Rosenbaum, D. P., Flewelling, R., Bailey, S., Ringwalt, C. L., & Wilkinson, D. L. (1994). Cops in the classrooms: A longitudinal evaluation of drug abuse resistance education. *Journal of Research in Crime and Delinquency*, *31*(1), 3-31.

Rosenbaum, D. P., & Hanson, G. S. (1998). Assessing the effects of school based education: A six year multilevel analysis of Project D.A.R.E. *Journal of Research in Crime and Delinquency, 35*(4), 381-412.

Rosenbaum, D. P., Ringwalt, C., Curtin, T. R., Wilkinson, D., Davis, B., & Taranowski, C. (1991). *Second year evaluation of D.A.R.E. in Illinois.* (Available from D. P. Rosenbaum Center for Research in Law and Justice, University of Illinois at Chicago, Chicago, IL 60607)

Rossi, P., & Freeman, H. (1993). *Evaluation.* Thousand Oaks, CA: Sage.

Roth, J. A., Sholz, J. T., & Witte, A. D. (Eds.). (1988). *Taxpayer compliance: Vol. 1. An agenda for research.* Philadelphia: University of Pennsylvania Press.

Schaps, E., Moskowitz, J., Condon, J., & Malvin, J. (1982). A process and outcome evaluation of a drug education course. *Journal of Drug Education, 12*, 245-254.

Schonfeld, D. J. (1998). *Proposal to the William T. Grant Foundation: Comprehensive elementary school AIDS education.* New Haven, CT: Yale University School of Medicine and Child Study Center.

Schorr, L. B., with Schorr, D. (1988). *Within our reach: Breaking the cycle of disadvantage.* New York: Doubleday.

Sherman, L., & Weisburd, D. (1995). General deterrent effects of police patrol in crime "hot spots": A randomized controlled trial. *Justice Quarterly, 12*(40), 625-648.

Simpson, J. M., Klar, N., & Donner, A. (1995). Accounting for cluster randomization: A review of primary prevention trials, 1990 through 1993. *American Journal of Public Health, 85*(10), 1378-1383.

Skogan, W. G., & Lurigio, A. J. (1991). Multi-site evaluations in criminal justice settings. *New Directions for Program Evaluation, 50*, 83-96.

Soumerai, S. B., McLaughlin, T. J., Gurwitz, J. H., Guadgnoli, E., Hauptman, P. J., Borbas, C., Morris, N., McLaughlin, B., Gao, X., Willison, D. J., Asinger, R., & Gobel, F. (1998). Effect of local medical opinion leaders on quality of care for acute myocardial infarction. *Journal of the American Medical Association, 279*(17), 1358-1363.

Soydan, H. (Ed.). (1998). Evaluation as discourse in social work research [Special issue]. *Scandinavian Journal of Social Work Research.*

Stoker, H. W., King, F. J., & Foster, B. F. (1981). The class as an experimental unit. *Evaluation Review, 5*(6), 852-859.

St.Pierre, R., Swartz, J., Gamse, B., Murray, S., Deck, D, & Nickel, P. (1995). *National evaluation of the Even Start Family Literacy Program: Final report.* Cambridge, MA: Abt Associates.

Turner, C. F., Miller, H. G., & Moses, L. E. (Eds.). (1989). *AIDS: Sexual behavior and intravenous drug use.* Washington, DC: National Academy of Sciences Press.

University of Maryland, Department of Criminology and Criminal Justice. (1997). *Preventing crime—what works, what doesn't, what's promising? A report to the Congress* (NCJ 165366). Washington, DC: U.S. Department of Justice, Office of Justice Programs.

Wagenaar, A. C., Murray, D. M., Gehan, J. P., Wolfson, M., Forster, J. L., Toomey, T. L., Perry, C. L., & Jones-Webb, R. (1997). *Communities Mobilizing for Change on Alco-*

hol (CMCA): Outcomes from a randomized trial. Unpublished manuscript, University of Minnesota.

Wagenaar A. C., Murray, D. M., Wolfson, M., Forster, J. L., & Finnegan, J. R. (1994). Communities Mobilizing for Change on Alcohol: Design of a randomized community trial. *Journal of Community Psychology* (Monograph series/CSAP special issue), 79-101.

Wagenaar, A. C., & Wolfson, L. M. (1993). Tradeoffs between science and practice in the design of a randomized community trial: Community trials design issues. In T. K. Greenfield & R. Zimmerman (Eds.), *Experiences in community action projects: New research in the prevention of alcohol and other drug problems* (DHHS Pub. No. ADM 93-1976, pp. 119-129). Rockville, MD: Center for Substance Abuse Prevention.

Wasserheit, J. N., Aral, S. O., Holmes, K. K., & Hitchcock, P. J. (Eds.). (1991). *Research issues in human behavior and sexually transmitted diseases in the AIDS era.* Washington, DC: American Society for Microbiology.

Weisberg, H. (1978). How much does ESAA really accelerate academic growth? *Journal of Educational Statistics, 3*(1), 69-78.

Wolins, L. (1982). *Research mistakes in the social and behavioral sciences.* Ames: Iowa State University Press.

Woodruff, S. I. (1997). Random effects models for analyzing clustered data from a nutrition education intervention. *Evaluation Review, 21*(6), 688-697.

Word, E. R., Johnston, J., Bain, H. P., DeWayne Fulton, B., Zaharias, J. G., Achilles, C. M., Lintz, M. N., Folger, J., & Breda, C. (1990). *The State of Tennessee's Student/Teacher Achievement Ratio (STAR)*(Technical report). Nashville: Tennessee Department of Education.

Wyatt, J. C., Patterson-Brown, S., Johanson, R., Altman, D. G., Bradburn, M., & Fisk, N. M. (1998). A randomized trial of educational visits to enhance use of systematic reviews in 25 obstetrics units. *British Medical Journal, 317,* 1041-1046.

Young, M. B., Fitzgerald, N., & Morgan, M. A. (1994). *National evaluation of adult education programs: Executive summary.* Washington, DC: U.S. Department of Education, Office of the Undersecretary.

Young, N. K., Gardner, S. L., & Coley, S. M. (1994). Getting to outcomes in integrated service delivery models. In *Making a difference: Moving to outcome based accountability for comprehensive service reforms* (National Center on Service Integration, Resource Brief 7). Falls Church, VA: National Center for Service Integration.

CHAPTER 9

Rival Explanations as an Alternative to Reforms as "Experiments"

Robert K. Yin

I. THE POTENTIAL ROLE AND IMPORTANCE OF RIVAL EXPLANATIONS IN A NEW FRAMEWORK FOR DESIGNING EVALUATIONS

The Opportunity and Need: To Create a New Frame of Reference

For more than a generation of evaluation theory and practice, if not longer (Madaus, Stufflebeam, & Scriven, 1983), the experimental method and its quasi-experimental adaptations (Campbell & Stanley, 1963; Suchman, 1967) have been at the heart of evaluation design. The method pervasively dominates all evaluation thinking, as none of us can escape such concepts as "random assignment," "comparison groups," or the very notion of "groups," even if only trying to offer alternatives. For instance, most pleas for alternative qualitative strategies use the experimental method as a point of departure, by first observing its shortcomings (e.g., Denzin & Lincoln, 1994; Guba & Lincoln, 1989).

More than anyone else, Donald T. Campbell was associated with—some would say "led"—evaluation's charge in the experimental direction. He and his colleagues (Campbell & Stanley, 1963; Cook & Campbell, 1979) authoritatively documented the use of the experimental method and its quasi-experimental form. He alone, however, pointed to the additional relevance of quasi-experimental design to the evaluation of societal *reforms*, highlighted by his landmark work on "Reforms as Experiments" (1969). To summarize his claim in one sentence, Campbell (1969) presented

> an experimental approach to social reform, an approach in which we try out new programs designed to cure specific social problems, in which we learn whether or not these programs are effective, and in which we retain, imitate, modify, or discard them on the basis of apparent effectiveness on the multiple criteria available. (p. 409)

This chapter offers an alternative to experimental and quasi- experimental thinking under this very circumstance—but only when "reform," or complex social change (e.g., "comprehensive community initiatives" (see Connell, Kubisch, Schorr, & Weiss, 1995)—is the subject of evaluation.[1] The hope is to use "rival explanations"[2] to create a new frame of reference.

A Personal Note and the Presumed, Ongoing Transition in Donald Campbell's Thinking

Possibly different from any author in these commemorative volumes dedicated to Donald Campbell, I met him only once. We did not have real conversation, only exchanging pleasantries (he regarded me as the advocate of "quantitative" case studies). Our only other interchange, several years prior to this encounter, was one written correspondence. I began it by asking if he would write a foreword to my original manuscript on case study research. (I was advised against such a request by the staff of Sage Publications, who noted the frequency of authors' requests of Campbell and who feared the potential delays that could result.) To my surprise (and delight), Campbell agreed. He then asked me to draft a hypothetical foreword, so he could see what I thought he should say. In actuality, Campbell used the draft to learn more about my own thinking—but did not use it in his final work in any way. (Further, to the delight of the Sage editors, Campbell also finished the foreword in a timely manner.)

What Campbell (1994) wrote as a foreword to my book, only he could have conceived. I have read the three-page foreword innumerable times. Each time, something new pops out that provokes deep thought—about case studies, about evaluation, and about the scientific method itself. Whether favoring case studies or not, I heartily recommend that others consider such multiple readings.

Once again, a single sentence in Campbell's foreword suffices: "More and more I have come to the conclusion that the core of the scientific method is not experimentation per se, but rather the strategy connoted by the phrase, 'plausible rival hypotheses' " (p. ix).

The entire foreword was about this topic. Connected to it were ongoing transitions in Campbell's thinking, away from an exclusive—some would say "exclusionary"—focus on experimental or quasi-experimental methods. For instance, from this perspective Campbell had published a little-known article (Campbell, 1975). The article revisited his earlier condemnation of the "one-shot case study" (Campbell & Stanley, 1963), now suggesting that explanations in case studies are difficult to come by and noting that this represented an "extreme oscillation away from my earlier dogmatic disparagement of case studies" (Campbell, 1975, p. 191). Campbell's realization was that in classic, single-case anthropologic studies—like those produced by the eminent investigators with him earlier at Northwestern—the investigator was likely to raise and reject "innumerable alternative solutions." This practice did not logically emanate from his previous caricature of the one-shot case study, whereby there should have been "a surfeit of subjectively compelling explanations" (Campbell, 1975, p. 182). As a result, Campbell concluded that some process not covered by experimental design was at work.

As another example, Cook and Campbell's (1979) well-regarded text on quasi-experimental methods in field settings included one powerful design (compared to many of the other designs) that in retrospect can be considered a case study design: the nonequivalent, dependent variables design (p. 118). In this design, even if a study consists of only a single case (with no comparison cases), but the hypothesized outcomes include an array of different outcomes—not just one single dependent variable—the empirically observed pattern of outcomes can be compared to the hypothesized array as a way of interpreting the results. (Not mentioned by the text was that rival arrays also can be hypothesized, and that the empirical comparison can be made with alternative hypothesized arrays, thereby testing different hypotheses, even with a single case.)

The Use of Rival Explanations in Designing Case Studies

My own work on case studies (Yin, 1994a), recognized mainly as a contribution to the logic of design as distinguished from the logic of data collection (e.g., Platt, 1992), depends heavily on the use of rival explanations. *Within* a single case, the main explanation and its rivals need to be articulated prior to data collection—both to offset the loss of rigor resulting from the inability to use quasi-experimental designs and to provide a basis for later interpreting the findings. *Across* multiple cases, the design principle calls for replication (analytic generalization), not sampling (statistical generalization), and the alternative directions for replication again reflect rival explanations. Rival explanations therefore help to drive the design of single—as well as multiple—case studies.

The function of rivals in designing case study research parallels the function of experimental methods in designing evaluations of social reforms. First, both social reforms and case studies are complex entities. A case study deliberately deals with the context and not just the phenomenon or intervention being studied.[3] Social reforms, though cast by Campbell as "new programs," certainly connote a similarly complex concept of a changing, multifaceted intervention and not just change in a single variable.

Second, the same objective is involved in use of rivals or of experimental designs, though producing different levels of certainty in the results. Returning to the foreword to my book, Campbell distinguished between the function of experimental designs—and particularly of the "randomized assignment to treatments" model—as an effort to control "an infinite number of 'rival hypotheses,' [but] *without specifying what any of them are*" (Campbell, 1994, p. x). He then likened case study designs to an older paradigm—still coming from physical science laboratories—calling for the explicit identification of specific rivals and then rendering each one implausible, through the collection of data targeted specifically at supporting or rejecting each stated rival.

One difference between these two strategies is therefore whether the rivals must be explicitly identified or not. Another difference is that the experimental design's ability to rule out an infinite (large) number of rivals does produce a high level of certainty in the results; however, the loss of certainty by the case study's ability to rule out only one or two rivals is not necessarily intolerable. The possibility exists that, for many case study situations, testing and being able to reject even one or two plausible rivals may produce a satisfactory conclusion. In other words, a case study's real-world policy and practical situation may be sufficiently served if the most compelling and plausible rivals are addressed and rejected—even though the degree of certainty may not be as high as when using experimental designs. Put another way, social interventions are so context-bound that the search for useful evaluation lessons need not strive for the universal solutions implied by using experimental designs.

The Use of Rivals as the Basis for Design, Outside of Case Studies

The potentially broader importance of rivals, beyond their use in case studies, also surfaced in my review of exemplary evaluations (Yin, 1994b). Studies having wide-ranging differences in methods, concepts, and academic fields—including the use of Bayesian statistics, single case studies, and laboratory experiments—nevertheless shared four commonalities. Of the four, only one dealt with research design, but this was the existence of a constant awareness and testing of rival hypotheses.

Using Rival Explanations in Four Related Crafts

In addition, at least four other empirical crafts outside of evaluation achieve their objectives by focusing on rival explanations: journalism (e.g., Bernstein and Woodward's 1974 inquiry into the Watergate crisis), detective work, forensic science, and astronomy. All the crafts are empirical, attempting to amass evidence to explain real-life events, yet the application of experimental designs is not at the core of any of the crafts (although all may have occasions when experiments can be part of the inquiry process).

In each craft, the investigator defines the most compelling explanations, tests them by fairly collecting data that can support or refute them, and—given sufficiently consistent and clear evidence—concludes that one explanation but not the others is the most acceptable. New explanations may emerge years later and are then investigated—as in society's infatuation with the assassination of presidents or famous people, much less historians' continued interpretation and reinterpretation of the causes of the American Revolution. Thus, the original conclusion is never airtight.

Note, however, that new theories can emerge years later to challenge experimental findings, too. Even Campbell (1982) decried the image of social experimentation in which a single big national program evaluation was expected to settle a policy decision "once and for all." Our claim is simply that the conclusions from the practice of journalism, detective work, forensics, and astronomy may not be airtight but may be of sufficiently high certainty to lead to the policy and practical action needed at the moment. Moreover, in the four crafts as actually practiced, reversals produced by the later emergence of new theories or reinterpretations appear minimal rather than commonplace. In other words, one hears about some occasions when journalists, detectives, coroners, or astronomers may have their original findings reversed (again, not dissimilar from the frequency of hearing about the reversal of experimental findings), but this does not occur most of the time.

Ability to Address More Complicated Social Change

There is more to our claim: The reduced level of certainty—surely a loss—is substantially offset by other benefits when an investigation is freed from having to impose an experimental design. The main benefit accrues to something akin to "external validity." Now, rather than going begging, a whole host of societal changes may be amenable to empirical investigation—compared to the limited situations where an experimental or quasi-experimental design can be implemented. Further, these broader situations represent those very topics where social policies and stakes are currently the highest—for example, the comprehensive community initiatives (Connell et al., 1995) mentioned earlier; changes in state or federal policy that might produce systemic and institutional change in education and welfare, not just micro change; and the variety of economic development and reform initiatives that are in desperate need of some evaluation but that do not fit the experimental mold.

For instance, Lisbeth Schorr (1997) observes that a new focus on "results" marks social programming. She then points to earlier pitfalls that need to be avoided to meet this challenge: making the wrong outcomes (those that evaluators can count rather than those worth knowing) the subject of study; using designs implicitly assuming that "one circumscribed problem is [to be] addressed by one circumscribed remedy"; and standardizing treatments across sites, resulting in variations in neighborhood environments being treated as "contaminants" (pp. 141-143). Observing the typical, negative findings from evaluations based on the experimental method, and then drawing her own connection between program and evaluation design, Schorr poses the provocative question, "What if interventions that change only one thing at a time fail . . . *because* they change only one thing at a time?" (p. 144).

Four illustrative trends in American society provide added examples of significant social trends not the subject of widespread evaluation, despite their importance: the nationwide decline in crime rates and especially in violent crime starting in 1992-1993, the creeping rise in math SAT scores (but not verbal SAT scores) starting in 1990-1991, the reduction in welfare rolls that *preceded and exceeded* the amount predicted as a result of welfare reform starting in 1994-1995, and the decline (after years of increase) in adolescent pregnancy starting in 1995-1996.

In each case, authorities had been predicting the reverse of what actually happened. There was strong denial, as a result, that the initial change in trend was in fact real, and attention initially was devoted to searching for artifacts. Only after the trends continued to persist for 2 to 3 years did analysts start to look seriously for possible substantive explanations, but as yet no convincing arguments have emerged. Certainly, these four trends must be among the very conditions on which we would want evaluation to place its highest priority and resources.[4] Why, then, do we find such an absence of evaluation where such meaningful social change, if not reform, may have taken place?

The answer forms the premise of this entire chapter—that (a) these higher and more complex social conditions are not amenable to the experimental method, (b) evaluators have been locked into a mind-set that is governed by the experimental method, and (c) without an alternative mind-set, evaluation will be directed to trivial pursuits, not these significant changes—some would say "reforms"—in American society. As similarly observed by Schorr,

> When it comes to broad, complex, and interactive interventions (early childhood supports, school reform, and better links to employment, for example) aimed at changing multiple outcomes (such as school success, employment rates, and a reduction in the formation of single-parent families), traditional evaluation has been of scant help. (p. 141)

One interpretation of these words is that the traditional methods—including triangulation and multi-operationalism as espoused by many, including Campbell—need to be practiced better. An alternative interpretation is that the traditional methods may fall short of practice *and methodological theory*, thereby presenting the need for alternative approaches. While not rejecting the first interpretation, the purpose of the remainder of this chapter is to add to all evaluators' options by focusing on the second.

II. BUILDING A SYSTEMATIC UNDERSTANDING OF RIVAL EXPLANATIONS, FOR USE IN DESIGNING EVALUATIONS

The challenge now posed is, having realized the importance of being freed of the experimental method and its mind-set, what guidance can be given on the proposed alternative mind-set of dealing with rival explanations? How does an evaluator know when the "most compelling" rivals are surfacing and being addressed? Can any and all investigators be expected to practice the new mind-set, or is some training and preparation needed? Many of the broadly used texts (e.g., Kidder & Judd, 1986; Nachmias & Nachmias, 1992; Rossi & Freeman, 1993) present hardly a word about rival thinking, much less any exposition on the topic. Investigators trained only by such texts are not likely even to be sensitive to the need for defining rival explanations and certainly will have gained few lessons on how to do so. The purpose of the remainder of this chapter is, therefore, to begin such an exposition.

Four types of rivals are discussed.[5] As a preview, three of them may be considered "craft rivals," and only one reflects "real-world rivals" but is the one in greatest need of further articulation—and is therefore the subject of section III of this chapter.

Craft Rivals: 1. The Null Hypothesis

The classic methods rival in all of social science is the null hypothesis, with statistical "chance" being the rival explanation for a presumed "effect." How investigators are to understand and use the null hypothesis is a basic tenet of statistical work and has been discussed in virtually every social statistics book of note in the past 40 years (e.g., Blalock, 1960). For instance, Type I errors are those situations in which we reject the null hypothesis when it is likely to be true, and Type II errors occur when the null hypothesis is falsely accepted when it is likely to be untrue.

The application of statistics is likely to cover only a narrow range of the most important social changes in need of evaluation.[6] Further, building on the use of the null hypothesis does not advance our quest for understanding real-life rivals. The null hypothesis can be valuable when initially establishing whether an outcome (or "effect") exists. Unfortunately, the logic underlying the use of the null hypothesis does not extend to actual explanation building—for example, how potentially causal explanations and their rival explanations are to be developed (e.g., Rubin & Babbie, 1993, p. 486). For example, rare is the situation in which an evaluation explains how it might have avoided a Type II error, yet inquiry here could help to promote a fuller understanding of an intervention. Along these lines, Boruch (1997, p. 211) usefully points to four possible reasons for making Type II errors: The measurement of the response to the treatment is insufficiently valid, the measurement is insufficiently relevant to the treatment, the statistical power is too low, or the wrong population was targeted for treatment.

Craft Rivals: 2. Threats to Validity

Another commonly understood group of rivals are those enumerated when experimental or quasi-experimental designs are used (e.g., Campbell, 1969; Campbell & Stanley, 1963). The "threats to validity" also have been called rival hypotheses or alternative explanations (Reichardt & Mark, 1998). All are concerned with methodological artifacts that need to be addressed; however, the experimental and quasi-experimental designs have implied that addressing them is sufficient, when in fact they are only part of the story.

Craft Rivals: 3. Investigator Bias

A third type of rival sometimes appears on lengthier lists of the threats to validity (but was not listed in Campbell and Stanley's original 1963 work). Whether included as part of such a lengthier list or not, the rival deserves and has received much attention on its own right. The rival covers the situation in which the investigator's theories, values, or preconceptions—in some unknowing and unacceptable way—produce in laboratory studies an "experimenter effect." In field research, such potential bias has an additional counterpart: the effect of the researcher on the setting or individuals studied, generally known as "reactivity" (Maxwell, 1998, p. 91).

This third type of craft rival applies to many more situations than the first two and therefore needs to be addressed by nearly every evaluation. The need is possibly greatest when investigators follow a certain style of qualitative research, where "the intent of qualitative researchers to promote a *subjective* research paradigm is a given. Subjectivity is not seen as a failing needing to be eliminated but as an essential element of understanding" (Stake, 1995, p. 45).

Granting the important insights, understandings, and lessons to be derived by using this style of qualitative research, the rival still in need of testing and rejection is that the subjectivity has not led to totally egocentric lines of thinking, independent of what is being observed or studied. The fear is that—across different studies of different situations—subjective researchers are only telling us something about themselves, under the misleading guise of informing us about the external world. Thus, qualitative researchers who agree that the examination of rival interpretations is an important commitment for all kinds of research (e.g., Stake, 1994, p. 245, footnote) presumably include inquiry into investigator bias as among the collection of rivals.

A preliminary but not conclusive way of minimizing this bias is for investigators to state, prior to their inquiry, their own values and ideology and, more important, something about the theoretical underpinnings of what they are studying—for example, theory about how the programs being studied are supposed to work. Such theoretical propositions drive an investigator's important choices—that need to be made explicit—regarding the entire array of intervention, outcome, and contextual conditions to be studied (e.g., Boruch, 1998, p. 172). Unfortunately, this application of the use of theory—to permit explicit inquiry into rivals regarding investigators' biases—can but should not be confused with the broader use of theory (discussed later in this chapter), in which substantive insights are being developed. Those opposing the use of theory on the latter grounds, if leading to the absence of any theoretical propositions to cover investigator bias, must then show how investigator bias has been addressed.

Real-Life Rivals

The first three types of rivals all may be considered methodological or craft rivals. They derive from the designs or methods that were chosen for study, and they mainly deal with ruling out methodological artifacts.

The dominance of the experimental method in all of our thinking can lead and has led to the undesirable situation where these craft rivals have been equated with all possible types of rivals.[7] Cook (1997, p. 33), for instance, considers (experimental) design and statistical controls as being needed for "ruling out alternatives to the notion that the program under study is responsible for any observed relationships between the program and outcome changes," but he does not point out that design and statistical controls may not be sufficient or always available.

This narrower definition of rival explanations—limiting them to craft rivals—is possibly the reason why the previously cited texts in evaluation (e.g., Kidder & Judd, 1986; Nachmias & Nachmias, 1992; Rossi & Freeman, 1993) do not discuss rival thinking or hypotheses, believing that their discussions of research design incorporate this topic. Overall, the constrained definition is therefore a significant barrier to be overcome in escaping the mind-set of the experimental method.

The broader and in fact more common use of rival explanations covers real-life, not craft, rivals. Data that do not fit the original explanation, contrary examples or cases, or compelling alternative arguments all lead to the surfacing of real-life rivals. The investigator's commitment is not to disproving these rivals but to deliberately seeking data in support of them; failure to find strong supporting evidence for alternative or contrary explanations then helps to increase the confidence in the original, principal explanation (Patton, 1990, p. 462). Further, it is likely that comparing alternative real-life rivals will not lead to clear-cut "support" or "no-support" kinds of conclusions, but "a matter of considering the weight of the evidence and looking for the best fit between data and analysis" (Patton, 1990, p. 462).

Currently, the evaluation literature offers virtually no guidance on how to identify and define real-life rivals. One general approach starts with theory-based evaluation and the usefulness of logic models (Wholey, 1979) and program theories to establish the presumed causal flow of events and hence explanation (e.g., Bickman, 1987; Chen & Rossi, 1992; Weiss, 1997). The idea of rival explanations, however, has not been an integral part of this general approach. Most logic models, for instance, do not stipulate rival paths, and most program theories have their hands full defining the logical relationship between an intervention and the outcomes of interest, much less stipulating possible rivals. What is needed is a direct enumeration and description of real-life rivals, a task that has been left to the remainder of this chapter.

III. REAL-LIFE RIVALS: A PRELIMINARY TYPOLOGY AND EXAMPLES

Real-life rivals compete with the main substantive hypothesis of an evaluation. They do not address methodological or artifactual conditions. The real-life rivals are truly alternative explanations for the observed phenomenon or result. The ideal evaluation design confronts *both* craft and real-life rivals, but the purpose of this chapter is to expand our understanding of real-life rivals.

For simplicity's sake, the discussion assumes that some planned intervention—whether a direct service, a new policy, or some other complex initiative, and no matter how extended over time—is the subject of evaluation. The discussion also assumes that, given today's "results orientation" (e.g., the federal government's implementation of the Government Performance and Results Act, as well as parallel developments in other sectors such as private philanthropy), the evaluation includes a focus on outcomes. Thus, the evaluation is to establish the validity of any outcome claims and whether the planned intervention had a hand in producing the outcomes. This hypothetical planned intervention, for discussion purposes only, shows that rival explanations can deal with the most difficult evaluation condition but should not be taken as an indication that they are limited to this condition.

Under this hypothetical situation, the most important substantive rivals deal with claims that, although the observed results may have been legitimate, the targeted intervention did not produce them. Rather, other conditions are alleged to have led to the observed results. As a guide for considering the possible rivals, Exhibit 9.1 lists six types. The list is preliminary, the goal being to produce a fuller typology as experience cumulates.

EXHIBIT 9.1

Real-Life Rivals: A Preliminary Typology

Rivals Related to the Targeted Intervention

1. ***Direct (Practice or Policy) Rival:*** Claims that a different type of intervention, not the targeted intervention, accounts for the observed result.
2. ***Commingled (Practice or Policy) Rival:*** Claims that another intervention of the same type, co-occurring with the targeted intervention, accounts for the observed result.

Rivals Related to Implementation

3. ***Implementation Rival:*** Claims that the pattern of implementation accounts for the observed result.

Rivals Related to Theory

4. ***Rival Theory:*** Claims that an alternative theoretical perspective can explain the observed result better.

Rivals Related to External Conditions

5. ***Super Rival:*** Claims that the targeted intervention and the observed results are but part of a larger and more potent process that accounts for the observed result.
6. ***Societal Rival:*** Claims that some salient social, political, or economic condition accounts for the result.

On this preliminary list, the first two rivals are directly related to the nature of the intervention under evaluation (*direct* rival and *commingled* rival). A third rival (*implementation* rival) points to the possibility that the observed results occurred because of the implementation process, not the substance of the intervention. A fourth rival (rival *theory*) is related to the identification of differing theoretical perspectives, to guide the search for evidence and to interpret the results. Finally, a fifth and sixth type (*super* rival and *societal* rival) are related to external conditions within which the intervention operated.

The remaining text gives real-life examples of these six rivals. After discussing each type separately, the text gives an illustration where (and more desirably) a study simultaneously covered several rivals, including methodological rivals. A final illustration also covers several rivals and draws from ongoing work at COSMOS Corporation.

1. Direct (Practice or Policy) Rival ("The Butler Did It")

"The butler did it" has been a classic way of saying that some other person, not the originally accused person, was in fact the perpetrator of a crime. Analogously, this first type of real-life rival occurs when one or more alternative interventions (practice or policy) are claimed to have produced the same result as the target intervention (practice or policy).

A good example of a practice-oriented rival was the debate that occurred when crime rates started to decline across the country in 1993. At first, most experts denied the reality of the downturn. Crime rates had been rising steadily until then and had been expected to continue rising. When the downturn repeated and even accelerated the following year—gradually encompassing all types of crime but especially violent crime—various explanations started to be promoted to explain the new trend. Box 9.1 presents the rival explanations to account for the downturn in New York City and the arguments for favoring one of them, but not to the exclusion of the others. In the example, each of the claimed causes for the downturn in crime can be considered a practice rival. Each could have produced the observed result (an impact on crime rates), all are likely to have had some role, and a good evaluation is still needed to collect a more precise array of data in support of one or the other alternative.

A second example comes from case studies of small- and medium-sized manufacturers. The case studies investigated the importance of technical assistance (see box 3 in Exhibit 9.2) in improving a firm's performance (Yin, 1998, p. 256). Searching for rival practices and conditions (see boxes 12 and 13 in Exhibit 9.2) was an integral part of the case design and data collection. Note that the more such rivals are addressed, the less need is felt for a "comparison," much less "control," case.

BOX 9.1

Rival Explanations for Declining Crime Rates

In New York City, the initial theory pointed to a new practice, publicized by Mayor Rudolph Giuliani. He had ordered the police department to crack down on minor street infractions, hoping to create a tangible atmosphere of intolerance to criminal behavior. Such intolerance, it was hoped, would produce the reverse of the "broken windows" effect described by James Q. Wilson and George L. Kelling in 1982, whereby the authors claimed that broken windows in an environment portray a disorderly environment—signaling a tolerance for more serious crime. Giuliani's practice met the main conditions for being a plausible hypothesis: The new practice was widely implemented, it could be logically linked to the observed outcome (reduced crime), and it had started before the initial downturn in crime.

A rival hypothesis, which has now become accepted as the more compelling explanation for the downturn, was that New York (and other jurisdictions) began a new method of fighting crime at about the same time. The new method (Kelling & Coles, 1996) was based in part on a new computer-based capability providing local police with information—overnight and on a block-by-block basis—on the pattern of crime and related activities during the previous day. As another part of the new method, such rapid and detailed reporting was then combined, in New York City, with reorganization at the precinct level, whereby precinct commanders were held strictly accountable for reviewing the daily data and developing and putting into place immediate countermeasures. Further, precinct resources also were reorganized, giving the precinct commander more control over previously specialized and decentralized units such as homicide, drug-busting, and other police units that had tended to operate independently of each other.

Although something major such as a decline in crime inevitably results from many conditions, including the addition of new police (which occurred just at the end of the tenure of Giuliani's predecessor, Mayor David Dinkins) and strongly rising incarceration rates that have taken offenders "off" the streets, the new police method has received slightly more credit than the other initiatives in accounting for the downturn in crime. First and most important, the new method had been practiced in the New York Transit System, prior to its implementation in the police department, with the same result (and the result occurred prior to any downturn in crime more generally). Second, the timing of the new method and the occurrence of the strongest initial downturns were much more aligned. Third, jurisdictions not using the new method are believed not to have experienced nearly the same volume of decline as those that did. To date, however, there has been no formal evaluation to confirm more definitively the role of the new method or any of the rival explanations.

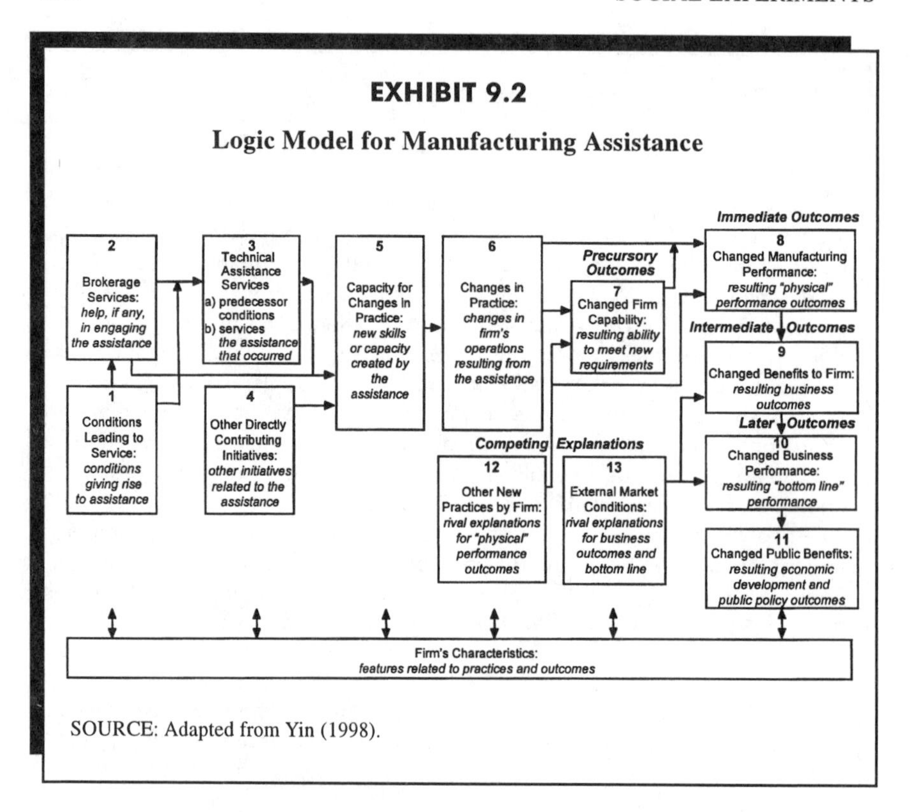

EXHIBIT 9.2

Logic Model for Manufacturing Assistance

SOURCE: Adapted from Yin (1998).

A third example of a direct rival works the same way, but the focus is on some public policy rather than a circumscribed practice. Policies can be much more diffuse in their impact than specific practices (the policy environment is usually complex and diffuse, so that any single policy is not likely to have a clear effect). In addition, most policies are not implemented evenly, making the tracking of a causal path highly variable. As a result, supporting or rejecting a rival policy is likely to be much more difficult than supporting or rejecting a rival practice.

Box 9.2 discusses two rival policies, investigated by Gregory Hooks (1990), regarding the persistent claim, in the 1980s, that Japan's national government assumed a proactive and highly supportive economic development posture, resulting in the (then) resurgent and dynamic economic growth of Japanese industry. The claim includes the notion that other countries' economic advancement did not benefit from such advantages, thereby "explaining" the Japanese success. Hooks's evidence, using a key example from the U.S. economy, suggests otherwise. His work is an apt illustration of how one can investigate targeted and rival policy actions.

BOX 9.2

Government Support of Key Industrial Development: Japan Only?

During the 1980s, countries (including the United States) knew that Japan, through its powerful trade ministry, both supported favorable international trade policies and supported and subsidized specific industries and firms. The common folklore was that this degree of intervention by the national government rarely existed elsewhere, especially in the United States, and that the Japanese system therefore created an unfair international market. The main hypothesis therefore became the claim that, to the extent that U.S. industries had continued to make themselves internationally competitive, this was a product of traditional capitalism, unfettered by government intervention.

This hypothesis was the subject of empirical study by Gregory Hooks (1990), who examined the U.S. government's role in two critical industries in the 1950s: aerospace and the early development of the microelectronics industries, especially the semiconductor industry. Whereas one industry (aerospace) was acknowledged by most to have been heavily supported by government policies and purchases, the other (semiconductors) was not.

In semiconductors, Hooks's investigations showed that a major breakthrough was the invention of the integrated circuit in 1959 (by Texas Industries) and then the ability to mass produce it (by Fairchild Semiconductor). Although the government did not directly finance these inventions or development, once they had occurred, the Pentagon "energetically financed production refinements" (Hooks, 1990, p. 358). Further, "the Pentagon voraciously consumed the expensive first generations of integrated circuits. . . . Not only did this guarantee a market for costly integrated circuits, but it proved their utility" (the Minuteman intercontinental missile program of the early 1960s was the government's application). Hooks used these revelations to challenge the traditional juxtaposing of the role of government in promoting U.S. and Japanese international competitiveness, having produced support for a rival hypothesis—that the U.S. government did indeed support industrial development in a manner parallel to the Japanese government.

2. Commingled (Practice or Policy) Rival ("It Wasn't Only Me")

Saying "it wasn't only me" implies that other parties also had a hand in the results, without denying that oneself also was involved. Analogously, this second type of rival stipulates that, in the same community and during the same time period as the targeted intervention, a similar kind of intervention also was occurring.

This situation arises frequently, for instance, with regard to drug prevention initiatives. The targeted intervention (practice or policy) may have been part of a particular federal program (e.g., Yin, Kaftarian, Yu, & Jansen, 1997); however, the complete inquiry needs to investigate whether similar prevention initiatives also were operating at the same time and with the same or similar target populations. If so, this rival could equally explain the observed outcomes. Unfortunately for evaluators, most communities do have a variety of initiatives going on at the same time—when support and priority are sufficiently high. In fact, a notable feature of community life is the desired redundancy of such efforts. To this extent, attributing the observed results to only one of them in a definitive manner may be extremely difficult.

This last point raises a more general observation with regard to how rival explanations may operate in real life: There are times when no single causal agent can be readily identified. Here again is an example when uncertainty is higher, but relevance to real local conditions is greater. A real-life community, recalling Schorr's (1997) work, should indeed be expected to be more complex than the experimental model, and answers that are more equivocating may nevertheless be more accurate and relevant.

3. Implementation Rival ("Did We Do It Right?")

The implementation process is now respected as being able to influence, or even produce, significant outcomes on its own (e.g., Yin, 1982). A logical rival to any substantive claim is therefore that implementation processes interceded and accounted for the results. Such a rival hypothesis has been especially relevant when the expected outcomes or results have not occurred.

Implementation *failure*—for example, insufficient time or resources, insufficient potency or "dosage" of an intervention, and managerial or human resource shortcomings—is clearly a plausible rival when interventions do not produce the expected results. Rather than assuming that a lesson has been learned about the substance of an intervention and its workings, findings in support of implementation failure provide a more mundane explanation of the outcomes. In real life, proper implementation continues to be a challenge, and testing for implementation failure therefore can be an important part of an evaluation. Similarly, lack of implementation *fidelity*—for example, evidence that the intervention was implemented in reality as it should have been on paper—also can be a plausible rival that needs to be addressed whenever the expected results do not occur.

4. Rival Theory ("It's Elementary, My Dear Watson")

Using rival theories calls for bringing different conceptual perspectives to the same set of facts. Sometimes such theories have been well developed and have heavily influenced the way everyone things about a topic. A study or evaluation can then be the occasion for critically testing or comparing such theories. The best example of this situation, now a classic in its field, is Graham Allison's (1971) study of the Cuban missile crisis—possibly the single encounter after World War II that might have produced worldwide nuclear holocaust. The evidence exhaustively examined by Allison supports two of three competing theories. The evidence offered little support for a "rational (single) actor" theory of decision making that had dominated academic and policy thinking for a long time (see Box 9.3).

Two other examples come from an earlier period of research when theory about the importance of the implementation process was in its early stages. One study showed how implementation problems, rather than any "barriers to innovation," accounted for the lack of change in a school (Gross, Giacquinta, & Bernstein, 1971). The study was significant because the "barriers" explanation had dominated thinking to that time. Similarly, the other study showed that the absence of implementation assistance was more likely to be the reason for the failure to put new research ideas into practice, in urban systems, rather than the more classic "two cultures" theory (the world of research versus the world of practice) that had prevailed to that time (Szanton, 1981).

5. Super Rival ("It's Bigger Than Both of Us")

This and the next type of rival deal with another perspective, when the plausible rivals are drawn not from competing interventions, from implementation, or from alternative theoretical perspectives (as in the first four types of rivals), but when, instead, some broader (bigger, stronger, or more pervasive) processes might be at work ("it's bigger than both of us"). Under these conditions, although the targeted intervention might appear to have been the main cause of an observed outcome, in fact the broader (bigger, stronger, or more pervasive) process was simultaneously in place and could be (at least equally) credited as the causal agent.

Investigation of this type of rival requires an understanding of the broader processes. One example is a study of the growth of microcomputer systems in school districts, which occurred during the 1980s when microcomputers were first being used by schools on a large scale (Yin & White, 1985). The study sought to explain the apparent differences in growth and spread within a school

BOX 9.3

Significant Social Science Theories as Rivals in Explaining the Cuban Missile Crisis

Graham Allison (1971) matched the facts and outcomes of the Cuban missile crisis to the events that would have been predicted by three rival theories: the rational (single) actor theory, the organizational process theory, and the political process theory. (His book is organized in exactly this sequence, with three consecutive chapters examining the match between the actual events and those predicted by each rival theory.) At the time, all three theories were salient in Allison's field (political science), although the rational actor theory was the longest standing. The opportunity to compare the predictions made by these three theories was therefore a genuine contribution to the entire field, and as a result, Allison's book has been become a classic in political science as well as public policy more broadly.

Allison collected an enormous amount of data about the events surrounding the Cuban missile crisis, which involved a confrontation between the United States and Soviet Russia in the fall of 1962 and which had worldwide implications. Allison came to be regarded as an expert about this event, also being given the opportunity to review Soviet documents and records much later, when access to these materials occurred because of changed (and more congenial) U.S.-Soviet relations. He showed how theory could focus on key questions:

- Why did the Soviets place *offensive* missiles in Cuba in the first place?
- Why did the United States respond with a *naval embargo* when the missiles already had been shipped to Cuba?
- Why did the Soviets withdraw the missiles, even though the naval embargo offered no obvious threat to them?

In examining the events, which unraveled rapidly over a period of only 2 weeks, Allison showed how the organizational and political process theories fit better with the observed events than the rational actor theory—even though the rational actor theory had tended to dominate thinking about the motives for international action until that time.

district. A variety of explanations related to microcomputer technology (e.g., software-hardware compatibility) were entertained; however, as shown in Box 9.4, the most plausible explanation appears related to a condition of school

BOX 9.4

The Growth of Microcomputer Systems in Schools: Bureaucratic Forces at Work Again?

Schools experienced different rates of expansion in their microcomputer systems in the early 1980s, when microcomputers were first becoming popular. Why these rates differed was the subject of a study of 12 school districts by Yin and White (1985). The authors first defined the outcome of interest operationally and precisely: the ratio of students to microcomputers in a school district, over time. They then collected data about this outcome in each of the 12 school districts, observed strong variations across the 12 districts, and asked why these rates might have differed.

A popular hypothesis, held by many educators, was that growth was inhibited to the extent that different (and hence incompatible) hardware and software had been purchased by the same school district. This hypothesis proved false. So did a developmental hypothesis that claimed that microcomputer systems simply grew as a function of the passage of time, with the study finding that the older systems were not necessarily the larger ones. A third rival hypothesis was that microcomputer systems grew at different rates as a function of the school subject for which the microcomputers were used (the early expectation, not supported at all, was that use of microcomputers in mathematics would lead to more rapid growth than use in other school subjects).

Instead of these explanations, the one most supported by the data derived from a fuller understanding of the bureaucratic settings of schools: Those microcomputer systems that grew fastest were those that had both instructional and administrative applications. With both types of applications, a broader base of familiarity and support for microcomputers was developed among the school *and administrative* staff, and more resources subsequently were committed to the purchase of new microcomputers and the expansion of the system—even in an era when most school districts' budgets were level or slightly declining. The development of a broader base, in turn, was not considered specific to the original issue of microcomputer implementation: Amassing a broader base can help in adoption and implementation of any number of innovations. To this extent, the most credible rival reflected a condition *exogenous* to the situation, not dealing with competing interventions or theoretical perspectives related to microcomputers.

districts more generally and not limited to microcomputers: Garnering a broad base of educational and administrative support, which is likely to increase any new educational innovation's growth.

6. Societal Rival ("The Times They Are A-Changin' ")

A societal rival goes even further than the processes covered by the super rival. Almost independent of any intervention, societal rivals point to the much broader social, political, or economic conditions that mark the "times." For example, rather than the working of specific interventions and processes, cultural differences could be the real explanation for any number of individual or group behaviors. Similarly, contextual conditions can produce significant social change and therefore cannot be overlooked in any rendition of possible rival explanations.

In history, Crane Brinton's description and analysis of major revolutions (the English, American, French, and Russian revolutions) is a classic. Brinton is explicit in his description and awareness of the use of the scientific method, even though his subject is a historical one (1960/1938, pp. 11-13). He assumes that his own methodology may be likened to that of clinical rather than experimental sciences. A key aspect of his explanation of why revolutions occur deals with the economic conditions of the time, and Box 9.5 describes how *rising expectations*, not immediate economic distress, formed an important contextual condition surrounding the actions of the revolutionaries.

Examples of Studies Examining Multiple Rivals

The preceding examples have highlighted different types of rivals, singly; however, the more powerful investigations entertain several plausible rivals within the same study, not just a single rival. Below are two such examples, one drawn from the literature and the other from current investigations at COSMOS Corporation. The two are intended to encourage future evaluations to investigate as many plausible rivals as possible.

Who Authored the Federalist *Papers?*

The *Federalist* papers were written to convince the citizens of New York State to ratify the U.S. Constitution and are part of this country's cherished heritage. There were 77 papers, and 43 were known to have been authored by Alexander Hamilton, 14 by James Madison, 5 by John Jay, and 3 by both Hamilton and Madison. Sole authorship of the remaining 12 papers was subsequently claimed by both Hamilton and Madison, and these competing claims (examples

BOX 9.5

What Economic Conditions Might Help to Produce Major Social Revolutions?

Crane Brinton (1960/1938) studied the four major revolutions in Western history: the English revolution of the 17th century, the American revolution of the 18th century, the French revolution of the 19th century, and the Russian revolution of the 20th century. All four revolutions occurred within "Western" states, and whether a more general "theory" of revolution could be assembled was Brinton's major task. His book, a classic in history, tracks the stages of a revolution as if it were a fever and covers cross-case examples and trends without presenting any of the cases singly.

Part of Brinton's inquiry focused on the economic conditions surrounding the onset of a revolution. A common explanation had been that underprivileged groups are under economic distress prior to a revolution, resulting in part in the impetus for the revolution. Brinton's investigations showed otherwise. Economic distress was not great prior to any of the four revolutions. More important than such distress was an element of a rival explanation—that significant segments of the population felt that prevailing conditions limited or hindered their economic activity and aspirations. This explanation, focusing on *rising expectations* rather than severe poverty conditions, not only was important in all four revolutions but also has become a precursor to be watched when expecting other acts of civil disobedience (e.g., the urban riots of the 1960s in the United States).

of *direct practice rivals*) were the subject of continued historical study and debate for the next two centuries. In the early 1980s, two statisticians, Frederick Mosteller and David L. Wallace, used Bayesian statistics to analyze the frequency of certain words (those that appeared more frequently in the papers known to have been authored by Madison compared to those that appeared more frequently in papers known to have been authored by Hamilton) to provide compelling evidence that the author of all of the disputed papers was Madison (Mosteller & Wallace, 1984). The study has since become a classic in its field.

In addition to testing for the main competing claims, the investigators also considered and rejected other, more subtle rivals (examples of *direct rivals*): that Madison edited what was originally a Hamilton paper, or that a third person, John Jay, was the actual author. The main study and investigation of these rivals formed only about two-thirds of the inquiry. The remaining third provided additional and extensive analyses addressing and rejecting a series of craft rivals—that the results were not dependent on parametric assumptions and heavy automated calculations, or the effects of outliers, or use of Bayes's theorem, or a variety of other possible artifacts. The thoroughness of the inquiry allowed the investigators to offer a general method for analyzing other classics whose authorship has been in dispute, including the Russian novel *And Quiet Flows the Don* and even the Bible.

Have U.S. Math SAT Scores Been Rising, and If So, Why?

From 1991 to 1998, mathematics scores on the Scholastic Aptitude Test I: Reasoning Test (SAT I) appear to have risen steadily if slowly, taking into account the fact that scores were recentered in 1996 to reflect the current population of test takers (see Exhibit 9.3). Most intriguing is the fact that the scores appear to have improved more so than the parallel verbal scores.

Because of the high priority devoted to improving U.S. students' performance in mathematics and science, COSMOS's staff has been conducting a preliminary study of these trends. Early results show that experts again initially downplayed the possible rise, attributing it to some characteristic of the SAT, not any true change in math performance. Only when comparable improvements showed up in other tests—for example, the National Assessment of Educational Progress and the ACT College Admissions Test—were people willing to pay closer attention to the possibility of a real rise.

Second, the inquiry focuses on important real-life explanations for the rise—that students are taking more math and science courses, courses are more rigorous, graduation requirements have become tougher, and the spread of computers and other educational technology also has helped. If true, these explanations would support the math and science reform efforts promoted by many local, state, and federal agencies during this period. Many of these efforts have been part of systemic initiatives (multiple planned interventions occurring at the same time). An important public policy question is whether the systemic initiatives, overall, have had any effect.

In examining this claim, the inquiry collected data showing that, from 1982 to 1992, the proportion of high school students taking algebra II grew from 37% to 56%, while the proportion taking geometry rose from 48% to 70%. The proportion taking biology rose from 79% to 93%, and for those taking chemistry, the proportion rose from 32% to 56%. Similar increases were found in other math and science courses. These changes in the earlier period could then account for the rise in SAT scores from 1991 on. Along the same vein, 1995 SAT takers reported having enrolled in record numbers of honors-level courses. Further, a Council of Chief State School Officers report found no evidence that courses were being "dumbed down" to attract more students.

EXHIBIT 9.3

Mean and Total SAT Scores for Males and Females, 1989-1997

	Mean Score					
	Math			*Verbal*		
Year	*M*	*F*	*Total*	*M*	*F*	*Total*
1989	523	482	502	510	498	504
1990	521	483	501	505	496	500
1991	520	482	500	503	495	499
1992	521	484	501	504	496	500
1993	524	484	503	504	497	500
1994	523	487	504	501	497	499
1995	525	490	506	505	502	504
1996	527	492	508	507	503	505
1997	530	494	511	507	503	505
1998	531	496	512	509	502	505

SOURCE: College Board Online (http://www.collegeboard.org).
NOTE: Scores are recentered.

The preliminary inquiry, however, also investigates a set of rival claims. The first and most common rival is that a smaller proportion of students is taking the test, which tends to produce a higher average score. The data show, however, that student participation, reflected by the number of students as well as the percentage of high school students taking the SAT, has been increasing. A related rival is that there is a higher proportion of males (who tend to do better than females) taking the test. The data also do not support this hypothesis, with a record-high number of females taking the test in 1993. Further, as shown in Exhibit 9.3, although females still score lower than males, they have made their own noteworthy gains in math scores, narrowing the gap with the males.

Another rival is that there is a higher proportion of Asian American students (who tend to do better in math than other ethnic groups) taking the SAT. Although Asian Americans did post the highest average math score on the 1996 SAT, their gain in math scores since 1987 is virtually the same as their gain in verbal scores during the same period of time. Thus, their participation alone cannot account for the somewhat higher increases in math scores than verbal scores previously noted in Exhibit 9.3.

Addressing these and other rivals and investigating in greater detail the pattern of improved scores remain subjects of study. For instance, states and school districts varied in their success in implementing systemic initiatives. To the extent that this variation has been documented, further inquiry could try to determine whether the more successful states and districts would show greater improvements than the less successful ones.

IV. CONCLUSION

This chapter has argued in favor of rival explanations as an alternative way of defining research designs. Not used in the design of any of the just-cited studies were techniques commonly associated with the experimental method or its quasi-experimental variants, such as the randomized assignment to treatments procedure, the manipulation in any way of real-life events by experimenters, or comparison or control groups (or even invoking the concept of "groups"). The chapter also presented a preliminary taxonomy of rival explanations to facilitate investigators' identification of important rivals. A final note is that the strength of the alternative research designs is related to the number of plausible rivals, both real-life and craft, that are addressed in an evaluation.

The demonstrated use of rival explanations has shown the feasibility of an alternative mind-set for designing and doing evaluations. The alternative does not displace the experimental method; rather, it permits evaluators to address more readily the larger societal interventions connoted by the term *reform*. The cited examples cover police reorganization in combating crime, federal support of industrial development, community drug prevention initiatives, decision making in an international crisis, change in urban systems, the growth of microcomputer systems in schools, the major revolutions in modern Western history, authorship of key papers in the nation's archive, and the possible effects of systemic reform in math and science education. Other examples not cited but that fall well within the alternative mind-set include ongoing changes in health care (in particular, managed care), the effects of "internationalization" on American businesses, comprehensive community initiatives, and the Internet's influence on all levels of social development and exchange. These types of social actions ought to be in the mainstream of evaluation's efforts but until now have been at the periphery or ignored altogether.

Beyond demonstrating feasibility, much work still remains. More precise criteria need to be developed to cover, for instance, what makes a rival most plausible, or the type or extent of evidence needed to reject (or support) a rival. The criteria should focus on real-life, and not just craft, rivals. Also possible is the need to expand cognitive perspectives, so that rivals going beyond existing theoretical constructions can be entertained (Guba & Lincoln, 1989).

The ultimate goal is to make rival explanations a useful alternative for designing evaluations of complex social actions, giving the evaluation field a tool of potentially great significance. The tool should now be recognized as the *rival explanation method*, which could be a most fitting centerpiece of Donald Campbell's legacy.

NOTES

1. The possible relevance of our alternative when other, less complex situations are the subject of evaluation is beyond the scope of this chapter.

2. This use of rival explanations to create an entirely different evaluation framework differs substantially from other contributions in these volumes, by Chip Reichardt, David Cordray, and David Rindskopf. Their discussions cover similar territory regarding the importance of, and how to think about, rival explanations, befitting the importance that Campbell devoted to the topic. I believe their contributions all exist within the original experimental and quasi-experimental framework, however, and are not attempts to develop a methodology of using rival explanations to serve as a general approach distinct from experimental methods—which is the main theme of the present chapter.

3. From a technical standpoint, case studies produce a condition I have recognized as a suitable technical definition of case studies, where "the number of variables of interest far outstrips the number of datapoints" (Yin, 1994a, p. 13). Accepting this definition renders irrelevant most parametric statistical techniques, which are therefore not avoided in case study research because of a lack of rigor but because of a lack of relevance.

4. Note that the four trends are not simply national trends but have strong variations among locales—providing an even greater opportunity to create potent evaluation designs. The Urban Institute is conducting a multi-million-dollar, multiyear evaluation of welfare reform, but whether the evaluation is addressing the issue of the excessive decline in welfare rolls—beyond that expected by any experts—is unclear. For the three other topics, the author is unaware of any comprehensive evaluations under way.

5. A possible fifth type, "counterfactuals," also is discussed in the evaluation of community initiatives (e.g., Hollister & Hill, 1995); however, the counterfactual is not defined as an alternative explanation of any sort but as the need to establish "what would have happened in the *absence* of the program initiative" (Hollister & Hill, 1995, p. 128, emphasis added) and is therefore not really another type of rival.

6. Lee Cronbach quotes Walter Deming as having observed that "the most important lesson we can learn about statistical methods in evaluation is that circumstances where one may depend wholly on statistical inference are rare" (Cronbach, 1983, p. 39).

7. Cook and Campbell (1979, pp. 20-23 and 59-64) do discuss substantive, or what we have called real-world, rivals; however, they readily admit taking Popper's falsification logic—originally aimed at true rivals—and instead "stress attempts to gain knowledge by pitting causal hypotheses *not against other explanatory or descriptive theories* but against mundane nuisance factors" (p. 31, emphasis added). This redirection toward

craft, not real-life, rivals then underlies the whole ensuing work. As one result, the only discussion of real-life rivals is brief and falls under a methodological topic—that of the "construct validity of putative causes and effects" (p. 59)—leaving investigators little idea of how to formulate any needed real-life rivals.

REFERENCES

Allison, G. T. (1971). *Essence of decision: Explaining the Cuban missile crisis.* Boston: Little, Brown.

Bernstein, C., & Woodward, B. (1974). *All the president's men.* New York: Simon & Schuster.

Bickman, L. (Ed.). (1987). *Using program theory in evaluation.* San Francisco: Jossey-Bass.

Blalock, H. M., Jr. (1960). *Social statistics.* New York: McGraw-Hill.

Boruch, R. F. (1997). *Randomized experiments for planning and evaluation: A practical guide.* Thousand Oaks, CA: Sage.

Boruch, R. F. (1998). Randomized controlled experiments for evaluation and planning. In L. Bickman & D. J. Rog (Eds.), *Handbook of applied social research methods* (pp. 161-191). Thousand Oaks, CA: Sage.

Brinton, C. (1960). *The anatomy of a revolution.* New York: Anchor Books/Doubleday. (Original work published 1938)

Campbell, D. T. (1969). Reforms as experiments. *American Psychologist, 24,* 409-429.

Campbell, D. T. (1975). "Degrees of freedom" and the case study. *Comparative Political Studies, 8,* 178-193.

Campbell, D. T. (1982). Experiments as arguments. *Knowledge: Creation, Diffusion, Utilization, 3,* 327-337.

Campbell, D. T. (1994). Foreword. In R. K. Yin, *Case study research: Design and methods* (2nd ed., pp. ix-xi). Thousand Oaks, CA: Sage.

Campbell, D. T., & Stanley, J. (1963). *Experimental and quasi-experimental designs for research.* Chicago: Rand McNally.

Chen, H. T., & Rossi, P. H. (Eds.). (1992). *Using theory to improve program and policy evaluations.* New York: Greenwood.

Connell, J. P., Kubisch, A. C., Schorr, L. B., & Weiss, C. H. (Eds.). (1995). *New approaches to evaluating community initiatives: Concepts, methods, and contexts.* Washington, DC: Aspen Institute.

Cook, T. D. (1997). Lessons learned in evaluation over the past 25 years. In E. Chelimsky & W. R. Shadish (Eds.), *Evaluation for the 21st century: A handbook* (pp. 30-52). Thousand Oaks, CA: Sage.

Cook, T. D., & Campbell, D. T. (1979). *Quasi-experimentation: Design and analysis issues for field settings.* Chicago: Rand McNally.

Cronbach, L. J. (1983). *Designing evaluations of educational and social programs.* San Francisco: Jossey-Bass.

Denzin, N. K., & Lincoln, Y. S. (Eds.). (1994). *Handbook of qualitative research.* Thousand Oaks, CA: Sage.

Gross, N., Giacquinta, J., & Bernstein, M. (1971). *Implementing organizational innovations.* New York: Basic Books.

Guba, E. G., & Lincoln, Y. S. (1989). *Fourth generation evaluation.* Newbury Park, CA: Sage.

Hollister, R. G., & Hill, J. (1995). Problems in the evaluation of community-wide initiatives. In J. P. Connell, A. C. Kubisch, L. B. Schorr, & C. H. Weiss (Eds.), *New approaches to evaluating community initiatives: Concepts, methods, and contexts* (pp. 127-172). Washington, DC: Aspen Institute.

Hooks, G. (1990). The rise of the Pentagon and U.S. state building: The defense program as industrial policy. *American Journal of Sociology, 96*, 358-404.

Kelling, G. L., & Coles, C. M. (1996). *Fixing broken windows: Restoring order and reducing crime in our communities.* New York: Free Press.

Kidder, L. H., & Judd, C. M. (1986). *Research methods in social relations.* New York: Holt, Rinehart & Winston.

Madaus, G. F., Stufflebeam, D., & Scriven, M. S. (1983). *Evaluation models: Viewpoints in educational and human services evaluation.* Norwell, MA: Kluwer-Nijhoff.

Maxwell, J. A. (1998). Designing a qualitative study. In L. Bickman & D. J. Rog (Eds.), *Handbook of applied social research methods* (pp. 69-100). Thousand Oaks, CA: Sage.

Mosteller, F., & Wallace, D. (1984). *Applied Bayesian and classical inference: The case of the* Federalist *papers.* New York: Springer-Verlag.

Nachmias, D., & Nachmias, C. (1992). *Research methods in the social sciences.* New York: St. Martin's.

Patton, M. Q. (1990). *Qualitative evaluation and research methods* (2nd ed.). Thousand Oaks, CA: Sage.

Platt, J. (1992). "Case study" in American methodological thought. *Current Sociology, 40*, 17-48.

Reichardt, C. S., & Mark, M. M. (1998). Quasi-experimentation. In L. Bickman & D. J. Rog (Eds.), *Handbook of applied social research methods* (pp. 193-228). Thousand Oaks, CA: Sage.

Rossi, P. H., & Freeman, H. E. (1993). *Evaluation: A systemic approach* (5th ed.). Thousand Oaks, CA: Sage.

Rubin, A., & Babbie, E. (1993). *Research methods for social work.* Pacific Grove, CA: Brooks/Cole.

Schorr, L. B. (1997). *Common purpose: Strengthening families and neighborhoods to rebuild America.* New York: Anchor/Doubleday.

Stake, R. E. (1994). Case studies. In N. K. Denzin & Y. S. Lincoln (Eds.), *Handbook of qualitative research* (pp. 236-247). Thousand Oaks, CA: Sage.

Stake, R. E. (1995). *The art of case study research.* Thousand Oaks, CA: Sage.

Suchman, E. A. (1967). *Evaluative research: Public service and social action programs.* Russell Sage Foundation.

Szanton, P. (1981). *Not well advised.* New York: Russell Sage and Ford Foundations.

Weiss, C. H. (1997). Theory-based evaluation: Past, present, and future. *New Directions for Evaluation, 76*, 41-55.

Wholey, J. S. (1979). *Evaluation: Promise and performance.* Washington, DC: Urban Institute.

Wilson, J. Q., & Kelling, G. L. (1982, March). The police and neighborhood safety. *The Atlantic*, pp. 29-38.

Yin, R. K. (1982). Studying the implementation of public programs. In W. Williams, R. F. Elmore, J. S. Hall, R. Jung, M. Kirst, S. A. MacManus, B. J. Narver, R. P. Nathan, & R. K. Yin (Eds.), *Studying implementation* (pp. 36-72). Chatham, NJ: Chatham House.

Yin, R. K. (1994a). *Case study research: Design and methods* (2nd ed.). Thousand Oaks, CA: Sage.

Yin, R. K. (1994b). Evaluation: A singular craft. *New Directions for Program Evaluation, 61*, 71-84.

Yin, R. K. (1998). The abridged version of case study research. In L. Bickman & D. J. Rog (Eds.), *Handbook of applied social research methods* (pp. 229-259). Thousand Oaks, CA: Sage.

Yin, R., Kaftarian, S., Yu, P., & Jansen, M. (1997). Outcomes from CSAP's community partnership program: Findings from the national cross-site evaluation. *Evaluation and Program Planning, 20*(3), 345-355.

Yin, R. K., & White, J. L. (1985). Microcomputer implementation in schools: Findings from twelve case studies. In M. Chen & W. Paisley (Eds.), *Children and microcomputers: Research on the newest medium* (pp. 109-128). Thousand Oaks, CA: Sage.

CHAPTER 10

Donald T. Campbell and the Art of Practical "In-the-Trenches" Program Evaluation

Burt Perrin

Knowing and describing Donald T. Campbell is akin to the proverbial blind men describing an elephant, with each touching a different part of the elephant and arriving at a very different portrayal. Campbell was not a single-issue person and defies easy categorization. As the *New York Times* indicated in its obituary (Thomas, 1996), when Campbell "took up his last academic post, at Lehigh University, university officials threw up their hands and simply designated him 'university professor.' "

This chapter focuses on Campbell's contributions to the *art* of evaluation, in particular to the art of practical, in-the-trenches evaluation that may be less "scientific" than other forms of evaluation for which Campbell may be better known. I present a very personal view of how I have been influenced by Campbell. In these respects, the perspective in this chapter may vary somewhat from that of other contributors to this volume. My hope is that it will be viewed as complementary and will help the reader see another side of Campbell.

To my mind, Campbell's greatest contributions may be to the *art* of research, and in particular to the art of social inquiry and evaluation. He has indicated that developing a research strategy, assessing the strengths and limitations of alternative methodological approaches, deciding which questions to ask and how to pose them, interpreting findings, drawing conclusions, and identifying next steps requires judgment, and soft judgment at that. There is no substitute for thinking, including the application of common sense.

It is not possible for me to do my work without drawing heavily on Campbell's contributions. I cannot conceive of any other program evaluator, knowingly or not, doing otherwise. It may appear incongruous for me to claim Campbell as the major inspiration and force behind my career and the approach to my own work. Campbell devoted the latter part of his career in particular to the exploration of seemingly esoteric philosophical issues and the sociology of science. He is best known as a methodologist, in particular for his advocacy of rigorous methodological research designs, usually quantitative in nature.

In contrast, my career has gone in a very different direction from that of Campbell. I work as a hands-on, practical, nitty-gritty consultant to governments and to other public and private organizations of various sizes. I carry out evaluations, needs assessments, and other forms of applied research, and also assist my clients with planning and organizational effectiveness. My clients are usually senior or line managers who have no interest in what they view as academic considerations and only wish information and advice to help them with the operation of their programs or the development of new policy approaches. Most of my contracts are low budget and short term, rarely exceeding more than a few months.

I use the simplest methods possible to increase the confidence of my clients in taking their next steps. During my more than 25 years as an evaluator, I have not had an occasion where direct use of an experimental design would be appropriate in addressing my clients' primary needs.

Campbell, however, has been subject to considerable misunderstanding, indeed even misrepresentation. For example, he has been cited, by supporters and critics alike, as a proponent of hardheaded, quantitative experimental designs for evaluation above all other methods. Campbell, however, also advocated the use of many *other* methods for evaluation, including qualitative approaches. As Shadish, Cook, and Leviton (1991) indicate,

> Campbell stresses the need for critical commentary on all measures, for measuring multiple constructs, for using multiple measures of any one construct, and for measuring process to get within the black box. . . . Campbell became cited within evaluation as simultaneously legitimating hardheaded experimental methods *and* their "softer" qualitative counterparts. (p. 132; italics in original)

Campbell has also been portrayed as a theoretician and methodologist, with little concern for how evaluation is used in practice. This, however, is not quite true. Many of his writings specifically devoted to program evaluation issues discuss constraints of "trapped administrators" and discuss ways in which evaluation can be used to influence social policy and to support program implementation. Although Campbell does not provide a "how to" guide, I find much practical guidance from Campbell's teachings for utilization-focused evaluation.

CAMPBELL'S CONCEPTS AND CONTRIBUTIONS MOST BASIC TO PRACTICE

In the balance of this chapter, I review some of the concepts and contributions of Campbell that strike me as most relevant to the art of practical program evaluation.

The Logic of (Scientific) Inquiry

How does one go about asking questions, choosing methods, designing methodological approaches, assessing "facts" and data, drawing inferences, determining causality, and reaching conclusions? These are basic considerations to every evaluator and issues I need to deal with on a daily basis.

Perhaps the most important thing I learned from Campbell is the recognition that these and related considerations, indeed the entire logic upon which scientific inquiry is based, are dependent on one's assumptions, values, culture, and expectations. Campbell, particularly in his later years, devoted more and more attention to exploration of philosophical considerations, in particular phenomenology, epistemology, and the philosophy and sociology of the social sciences. It may appear ironic for me, a very practically oriented practitioner, to claim as most helpful some of Campbell's most esoteric, theoretical, and philosophical work, but essential to the art of evaluation is making underlying values and assumptions as explicit as possible and taking them into account in the design, implementation, and analysis of one's work. I constantly attempt to identify my own assumptions as well as those of my clients and stakeholders. The process of doing this frequently results in changes to the evaluation approach. Philosophy and theory underlie practice—whether we are aware of it or not.

For example, Campbell helped me understand that something seemingly as basic as perception is not merely an "objective" process but is dependent on the interaction of prior knowledge, expectations, and context with sensory inputs. In his introductory social psychology lectures, Campbell cited evidence indicating that people who are hungry are more likely than those who are not to perceive ambiguous objects as representing food. Segall, Campbell, and Herskovits (1966) demonstrated the influence of culture on perception, finding, for example, that one's reactions to visual illusions such as the Müller-Lyer illusion are based on one's environment and previous experiences. Campbell's extensive work on pattern matching demonstrates that we can examine specific elements only if we assume, even temporarily, the validity of the overall context or background.

Campbell's work demonstrating how sensory inputs are mediated through the context and knowledge of the observer has major implications for the practice of evaluation. This is one of the major reasons why eyewitness accounts can be so variable and unreliable. Many evaluation designs, however, fail to take this into account, and reports of observers too often are taken at face value. I have learned the importance in my work of understanding context, through use of a range of qualitative techniques, in being able to determine which questions are appropriate to ask and in being able to interpret "data."

A major theme running throughout much of Campbell's work is that all measurement is theory laden. Objective knowledge or "proof" is impossible. As Kuhn (1962), for example, has demonstrated, the process of science is not as objective as one may be led to believe. Paradigms determine which methods are considered acceptable, indeed what constitutes "facts" or "evidence." Findings that are not consistent with the prevailing paradigm are not accepted as legitimate. This has very practical implications for the practice of evaluation. For example, it explains why survey-based customer satisfaction ratings or other performance indicator data, while quantifiable, often bear little resemblance to what people will say in interviews or what is apparent through observation. On another level, the concept of paradigms is helpful in considering how to present evaluation findings. I have learned that it is critical to be aware of the primary audience or audiences for evaluation reports, and to make sure that findings have face validity from their perspective.

Campbell's theory of evolutionary epistemology is consistent with the concept of falsification. He indicated that we can never "confirm" or "prove" anything, through experimentation or through any other means. Drawing on Popper's work on falsification, all we can do is put theories to the test and see if they escape being disconfirmed. To generate new knowledge, new ideas somehow need to emerge and be put to the test. Campbell's initial support for the development and evaluation of pilot projects follows from this, but he cautions that one should not evaluate too soon, until a project has had a chance to get up and running and to overcome initial glitches and bugs—a warning, alas, that seems to be observed more in the breach than in its observance. A frequent challenge I must face in the evaluation of newly created programs is helping both program advocates and funders realize that assessment of ultimate program impact will take some time. I make extensive use of program logic models to help focus evaluation activities on realistic outcomes, given the status of the program. Open-ended approaches to evaluation are consistent with the above concept, as being amenable to serendipity and to unanticipated findings.

As Shadish et al. (1991) discuss, Campbell initially proposed the use of experimental (and quasi-experimental) approaches for evaluation in assessing the impact of pilot/demonstration projects. As they point out, however, long-term, stationary demonstration projects in today's context are increasingly rare. Campbell (1994) himself recognized that most evaluation takes place in settings precluding good quasi-experimental or experimental designs. This is consistent with my own experience. Campbell has also indicated that there are alternative forms of causality besides manipulative causality through randomization.

Nevertheless, in my work I find absolutely critical an understanding of the *logic* of the scientific method, including the thought process by which one goes about assessing evidence, and familiarity with the many threats to both internal and external validity and to the drawing of causal inferences such as those discussed in Campbell and Stanley (1963) and Cook and Campbell (1979). Conversely, application of experimental or quasi-experimental designs without thinking through the meaning of causality and without an awareness of what can and what cannot be demonstrated through experimentation can result and has resulted in "hard" evaluations that nevertheless are inappropriate and/or of poor quality and consequently of no value.

As the above discussion illustrates, one cannot do program evaluation without an understanding of the values and assumptions on which our approaches are based. Data do not speak for themselves, and there is no such thing as "proof" or the "best" method. Campbell decried the misuse of many of the quasi-experimental designs provided in Campbell and Stanley (1963) and in Cook and Campbell (1979), without consideration of the many trade-offs required in selecting any design or method. As he said in one of his last publications (Campbell, 1994), "My methodological recommendations have been over-cited and under-followed. . . . I would rather that you would listen to my arguments" (p. 295). The art of evaluation requires thinking through many of the aspects of how knowledge is created.

Plausible Rival Hypotheses

Let me now move from philosophy to perhaps Campbell's most important and practical concept. As noted above, the heart of Campbell's philosophy is that we can never prove; all we can do is rule out potential hypotheses or explanations. This is expressed in Campbell's (e.g., 1979; see also Webb, Campbell, Schwartz, & Sechrest, 1966) concept of *plausible rival hypotheses.*

This concept is deceptively simple. It suggests that in evaluation and social research, we should strive to identify potential alternative explanations or threats to validity, then identify which of these are most plausible. There is no need to devote research effort or funds to attempt to prove the obvious. Campbell (e.g., 1979, also see Shadish et al., 1991) suggested that the best way of determining the plausibility of competing explanations is to use common sense.

Different stakeholders in an evaluation are likely to have varying ideas about what *they* consider "plausible" or not. I find that one of the major challenges in evaluation is to determine these various views, as well as what types and amount of evidence would be credible, in the design stage. It matters little what the *evaluator* considers as relevant or not: What counts is what the ultimate *users* of the evaluation think.

In my experience, the major way in which evaluations go wrong is by addressing the wrong questions. Program evaluation typically fails when researchers address questions of interest to themselves but not of primary interest to the evaluation users, or when they attempt to force onto a program a method that just does not apply. It is necessary to adapt the method to fit the needs of a particular evaluation, and not vice versa. In planning an evaluation, it is critical to seek the views of key stakeholders and users of the evaluation about what *they* view as the key questions about the program and what evidence *they* would consider convincing or not. I have argued that one should involve the harshest critics of a program in the evaluation design and process, to ensure that their rival views of the program are addressed in the evaluation. Participatory approaches to evaluation can help create support for the evaluation process and methods, enhance the relevance of the evaluation, and provide for buy-in to the results and to ultimate action.

Campbell has also advocated using the simplest methods possible to eliminate plausible rival hypotheses. In particular, he has argued that common sense frequently is sufficient to address the key questions (e.g., see Campbell, 1979, 1994; Shadish et al., 1991). In other situations, relatively simple methods, including use of descriptive data, may be all that is required. To employ methods that are more robust than needed or appropriate is to engage in methodological overkill. This increases the time and cost of the evaluation and can reduce its ultimate utility. This also can result in increased cynicism about and resistance to evaluation.

Campbell proposed the concept of quasi-experiments (Campbell & Stanley, 1963; Cook & Campbell, 1979) as a way in which plausible rival hypotheses sometimes can be addressed without resorting to true experiments. The concept recognizes that there is no such thing as the "perfect" method. As Reichardt (1996), for example, indicates, Campbell's work on quasi-experimentation quickly became widely accepted and has been recognized as having more influence on social science methodology than any other work since World War II. The concept, challenging randomized experimental designs as the only acceptable method for demonstrating causality, nevertheless was heresy when it was initially proposed.

The concept of plausible rival hypotheses has major implications for practical program evaluation. When evaluation is carried out for the purposes of contributing to decision making, program improvement, or policy development, there is never time to wait for the "definitive" answer. Decisions have to be—and are—made regularly on the basis of imperfect information. As Campbell has indicated, we should strive not for "truth" but for plausibility and for the reduction of uncertainty.

How much confidence is necessary? That depends on the situation, what is riding on the decision at hand, and what evidence the key users of evaluation would find credible. I have suggested that evaluations should strive to provide the *minimum* amount of information needed to provide the *minimum* amount of extra confidence required for decision making. The art of practical evaluation requires working with the client to assess the trade-offs of alternative approaches and to agree on what information would be most useful in a given situation.

Even though I rarely use quasi-experimental designs as such in my own evaluation work, I nevertheless find the concept critical to my thinking. Campbell (e.g., 1994) has indicated that more important than the application of any particular design is the rationale and thought process in the choice of methods. The analysis of sources of validity, the discussion of which makes up the bulk of Campbell and Stanley (1963) and Cook and Campbell (1979), is absolutely critical to any work in evaluation. As Campbell (1994) observed in his later years, "Do recognize that most of your evaluation tasks will be in settings precluding good quasi-experimental or experimental designs, and that even so, there is much of value that you can do to help improve program effectiveness" (p. 292).

Triangulation and the Importance of Using a Range of Diverse Methods

Campbell and Fiske's (1959) "Convergent and Discriminant Validation by the Multitrait-Multimethod Matrix" has been identified as the most widely cited paper in the social sciences. In essence, it made the case for triangulation, for the use of multiple methods to accommodate the limitations of any single approach.

As previously discussed, Campbell has indicated that *all* methods are flawed. All methods are theory laden and are based on questionable assumptions. All researchers bring their values and their biases to their work, and these influence their choice of methods and interpretation of data. There is no such thing as completely objective research or research methods. Quantitative methods are no more objective than qualitative methods; their biases and underlying assumptions are just less visible.

These biases and flaws are inherent in *every* method and can never be completely eliminated. They can be obviated to some extent by making biases and limitations as explicit as possible, but because values and assumptions are culture laden, this is not always as easy or possible as it may seem. Furthermore, users of evaluation studies are unlikely to take much notice of qualifications and disclaimers buried in dense evaluation reports.

As Campbell has indicated, the best way to control for the inherent biases and limitations of any single method is to use a range of complementary methods. Campbell was a methodological pluralist. We can have greater confidence in results when different methods come up with essentially the same findings. In particular, we can have the greatest confidence in the findings when different types of methods, such as a mix of quantitative and qualitative methods, are used.

Campbell was a strong supporter of innovative and varied methods. For example, in Webb et al. (1966), he advocated the use of unobtrusive or nonreactive methods. One commonly cited example of the "oddball" methods discussed in the book (Campbell's introduction to the book indicates that one of its working titles was *Oddball Methods*) is assessing the relative popularity of museum exhibits through identifying the rate at which floor tiles in front of various exhibits wear out. There are alternatives to questionnaires and focus groups!

Campbell supported the use of both quantitative and qualitative methods drawn from competing perspectives and views of the world. Campbell pointed to the futility of searching for the "perfect" research method or for the "correct" view of the world. He supported the use of meta-evaluation to draw on a wide range of different individual studies so as to be able to draw conclusions and generate knowledge not possible through any single study. He also recognized the importance of a systems approach in attempting to make sense of the complexities of social phenomena in the real world.

The concept of triangulation has important practical implications. It means that we usually are better off applying limited resources for evaluation to multiple imperfect methods rather than trying to develop the "perfect" design. Even in low-budget evaluations, such as those I typically conduct, I strive for the use of a range of methods, even if this means limiting to some extent what can be obtained from any particular method. For example, rather than trying to implement the most comprehensive survey possible, I might use a short, focused questionnaire with a small (but representative!) sample, complementing this with qualitative data such as obtained through a small number of in-depth interviews. Use of multiple methods can be more cost-effective than use of any single method, however perfected its execution, and can provide more accurate and useful information as well.

External Validity

Campbell made the critical distinction between internal and external validity (in particular, see Campbell and Stanley [1963] and Cook and Campbell [1979], although he made the distinction much earlier, e.g., Campbell [1957]). Internal validity refers to the ability to draw causal conclusions in a specific situation. External validity refers to the ability to generalize the findings of a particular study to other persons, settings, or times. Campbell indicated the importance of ensuring internal validity; without it, there is nothing to generalize.

Evaluation findings that cannot be generalized, however, if only to the same program with identical characteristics at a future time, are of little or no use. Without being able to identify what factors are responsible for impact, findings about impact have little or no practical value. Not only is it difficult or impossible to hold all variables constant except one in the real world, but evaluation based on this paradigm rarely is meaningful without the ability to consider the role of context and how multiple factors interact with one another. As Kurt Lewin has indicated, the whole is definitely greater than the sum of the parts. Without being able to establish external validity, it is not appropriate to generalize findings from one setting to another, irrespective of the degree of control and internal validity established, unless the exact same conditions are in place. In the real world, this makes little sense.

In the reality of policy and program delivery, especially in today's environment where change is happening at an accelerating rate, little remains stationary. Program activities change continuously, frequently in very substantial ways, in response to changes in the political and social environment, to needs and opportunities, and to events experienced to date. Indeed, as I have indicated elsewhere (e.g., Perrin, 1994, 1998), responsive programs *should* be changing and adapting. They *should* be reviewing whether their intended outcomes are still desirable, or if they need modification, supplementation, or replacement. They *should* be responding to feedback from clients, to new information, and to evaluations. Few programs either can or should stand still long enough to permit experimental evaluation.

What does this mean for practical program evaluation, where the generalizability of findings usually is critical? Rather than attempting to eliminate as many extraneous factors as possible, we should strive instead toward differentness rather than sameness in program elements and contexts. Going back to Campbell and Fiske (1959), we can have greater confidence in the veracity of our findings when they hold up across somewhat different contexts, as well as through different methods. One of the strengths of a cluster evaluation approach is that it can enable the drawing of conclusions across a frequently heterogeneous range of programs, giving us greater confidence in the findings and their generalizability than if we examined only a single program or a collection of copycat programs. It is also critical to recognize that programs are in a constant state of flux and that unless this is taken into account in the design and reporting of evaluation, findings will be dismissed as old news that no longer is relevant.

All Knowledge Is Qualitative

Campbell has been both praised and criticized for advocating the use of quantitative rather than qualitative methods, but this is a misrepresentation of his views. In fact, he referred to the "*mistaken belief that quantitative measures replace qualitative knowing*" (Campbell, 1988b, p. 323; italics in original). He indicated that "science depends upon qualitative, common-sense knowing even though at best it goes beyond it" (Campbell, 1979, p. 50). In my view, there has been insufficient attention to Campbell's rationale and support for qualitative approaches. Particularly (but not exclusively!) during his last two decades, he placed increasing emphasis on the importance of qualitative knowing. As Ginsberg (1998), for example, indicates, in the late 1970s he referred to parts of Cook and Campbell (1979) as "too statistical." Campbell was a supporter of qualitative methods, such as case studies (e.g., Campbell, 1989), observation (e.g., Webb et al., 1966), and other ethnographic methods (e.g., Ginsberg, 1998). He emphasized the importance of measuring process as a means of getting inside the "black box."

Campbell indicated that how constructs are conceived and how questions are asked limits the potential answers that can be obtained. This is equally true of quantitative and qualitative methodologies, but it is a particular danger with quantitative methods, which carry the aura of "objectivity," where assumptions are more frequently hidden, and where findings are frequently presented stripped of context. As Campbell (1979) said, for example,

> Quantitative results may be as mistaken as the qualitative. After all, in physical science laboratories, the meters often work improperly, and it is usually qualitative knowing, plus assumptions about what ought to be showing, that discovers the malfunction. (p. 53)

I find Campbell's work in this area of tremendous practical importance. I have learned not to take findings from either quantitative or qualitative sources at face value without exploring the context in which questions are asked. Whenever possible, I will combine my survey work with at least some intensive interviewing. I find that the qualitative information is invaluable, if not absolutely essential, in interpreting questionnaire responses, and more often than not it gives me more useful and actionable information than the more quantitative survey results.

It is important to recall that Campbell did not take a stand in favor of one type of methodology over another. He pointed out limitations of qualitative methods, just as he did of quantitative approaches. As indicated earlier, he consistently stressed the need for the use of multiple methods, both quantitative and qualitative.

Misuse of Performance Measures in Program Monitoring and Evaluation

Performance measurement, or the use of indicators to demonstrate the results of program performance, now is all the rage. It is widely touted as the new way of providing for a focus on the results of public programs. In recent publications about the use and misuse of performance measurement (Perrin, 1998, 1999), I have indicated that there has been little critical consideration of how performance measurement actually works in practice.

In fact, Campbell addressed this topic. Initially (for example, in his 1971 unpublished draft of "The Experimenting Society," later published in 1988), he tentatively supported the use of social indicators and administrative records for evaluation purposes, while recognizing their "squishy and biased nature" (Campbell, 1994, p. 293). Even then, he spoke of the dangers of "legislating the indicator" rather than the problem. In the 1970s, he warned that outcomes could get so politicized and so quickly corrupted that they would do more harm than good (Patton, 1998).

Campbell later (e.g., 1984, 1988a, 1994) became increasingly explicit in stating not only that indicators were subject to inevitable misuse if used for monitoring or decision-making purposes but also that they would inappropriately distort program activities. In 1984, he discussed how the impact of monitoring program activities results in changes in program activities to make the numbers look good. He cited, as one of the major lessons he learned from watching how evaluation works in practice, the error of specifying as program goals measures that are open to bureaucratic manipulation. He referred to "a discouraging law that seems to be emerging: the more any social indicator is used for social decision making, the greater the corruption pressures upon it" (Campbell, 1988b, p. 309). This is commonly referred to as goal displacement (e.g., Perrin, 1998).

In 1994, Campbell went further still:

> [The] importance for our field [of the corruption of indicators] now seems to me much greater. I have held off advocating an experimenting society until they can be solved. Any report form designed for a regularly used management information

system will be distorted from descriptive accuracy by the reporters' beliefs about the managerial use to be made of the reports. (p. 294)

Given the recent interest in and attention to performance indicators across the English-speaking world, coupled with the lack of critical consideration to how they can be misused, Campbell's observations and warnings are especially relevant and topical.

Campbell's Vision for Evaluation

Although Campbell may be known primarily as a theoretician and a methodologist, his papers "Reforms as Experiments" (1969b), "Methods for the Experimenting Society" (1971), and "The Experimenting Society" (1988a) have been cited as among the most influential papers in evaluation (e.g., Reichardt, 1996). As Campbell (1994) indicated, "Reforms as Experiments" was his first paper targeted to the program evaluation agenda. In these papers in particular, Campbell sets out his vision for evaluation. He calls for a society that would experiment with social reforms, learning from experience, with the help of evaluation, about which approaches are effective or not and which should be retained, imitated, modified, or discarded.

Campbell has provided us with a powerful vision, in essence suggesting the *raison d'être* of evaluation: to encourage innovation and to serve as a tool that would enable learning from experience and the application of this information in improving social policies and programs. This view has major implications for the art of evaluation practice. For example, it suggests a constructive approach, rather than an adversarial or audit approach, to evaluation. It is entirely consistent with the more recent concept of the learning organization (although Campbell's own focus was on societal rather than on organizational change and behavior). Evaluators are only starting to recognize the potential of evaluation to serve as a tool to generate learning (see, e.g., Preskill & Torres, 1999; Torres, Preskill, & Piontek, 1996).

Although Campbell himself did not explore the concept of evaluation use in detail, he recognized the realities of politics and the positions and constraints of "trapped administrators" (e.g., Campbell, 1969b). Although his own particular interest was in the identification of impact, he acknowledged the legitimacy of evaluation approaches that provide guidance for program implementation and management. Campbell set the stage for thinking about use, laying the groundwork for others to follow, in particular Michael Quinn Patton (e.g., 1997) and Carol Weiss (1998). These and other researchers have demonstrated convincingly that use is dependent more on the process of how evaluation is done than on its outcome, and that traditional, detached "scientific" approaches are least likely to result in use. Patton's work in particular (e.g., Patton, 1997) identified the importance of involving key participants in the evaluation process so as to build support and commitment for action on evaluation. This has major implications for methodology, in particular regarding the applicability of preordinate, detached, "objective" methods with limited face validity.

Understanding the "whys" and "hows" is more important to use than just knowing what did or did not happen. Campbell acknowledged the importance of getting inside the black box to understand causal relationships. As Mohr (1988) has indicated, program staff "deplore being told that their efforts are or are not having much effect. . . . What they want to know is *why*—how to make a weak program stronger or an effective program even more effective, or perhaps more efficient" (p. 26).

To some extent, what we know now about evaluation use may appear to be at odds with Campbell's desire for scientific demonstrations of impact, but in his later years in particular (e.g., see Campbell, 1994), he was explicit in recognizing that most settings preclude good experimental or quasi-experimental designs, and that evaluation nevertheless can do much of value to help improve program effectiveness. The art of evaluation requires making judgments about what types of information will be most useful in any given situation.

CONCLUSION: CAMPBELL THE PERSON

People who knew Campbell remember him as much for his humanity as for his intellectual and scientific contributions. In spite of his stature, he was known for his humility, for his incredible kindness and caring, and for his tolerance, which he demonstrated in all possible ways. He was an inspiration in how he lived his life and interacted with others. It was impossible not to be touched by him.

Campbell demonstrated the art of research by example as well as through his writings. He was not beholden to any particular theoretical approach; indeed, he warned against the dangers of biases promoted by those with vested interests in any particular theoretical approach. He was one of the few synthesizers of the 20th century who talked—and listened—to everyone. He was not dogmatic in his own thinking. He invited criticism of his own work, and he revised many of his own ideas over the course of time. He was open to new discoveries from all sorts of sources. I recall fondly how, during my tenure at Northwestern, he frequently welcomed me—a lowly and confused undergraduate—into his office for chats on every conceivable subject. I remember on one occasion, after I pontificated on some book or other, that he asked me to order it in his name at Great Expectations, his favorite bookstore, which was a very Campbellian place, a musty little space full of books on a range of esoteric topics flow-ing off the shelves onto tables, chairs, and the floor. Through both word and deed, Campbell encouraged me to keep an open mind and not to worry about messiness.

Even Campbell's "failings" were an inspiration to those who knew him. As Reichardt (1996) pointed out, Campbell (1988b) spoke not of his successes in his autobiographical introduction to an edited volume in his honor, but about his research mistakes and dead ends. He spoke, for example, of his early experience as an untenured professor at the University of Chicago, where he found the pressure to produce works of genius stultifying and how he only started to produce works that would promote granting of tenure at Northwestern, where he was originally hired to teach. Campbell occasionally encountered demons in his personal life, but he nevertheless persevered. He suffered from writer's block on occasion yet was a very prolific writer nevertheless, with more than 250 publications to his credit. Campbell's was the archetypical messy office, full of piles of papers and books on every conceivable surface; nevertheless, he seemed to have the unerring ability to place his hand in the middle of one of the many piles and locate just what he was looking for. He is an inspiration to those of us who sometimes have trouble getting going and who cannot keep any flat surface free of clutter for more than a few minutes at a time.

To overcome what he saw as somewhat arbitrary boundaries dividing social science disciplines, Campbell (1969a) proposed a "fish-scale model of omniscience," suggesting that professors should encourage their students to develop their own overlapping specialties, rather than aiming for "chips off the same block." He practiced what he preached. His students and followers are active in a variety of disciplines representing a range of theoretical approaches. Although my own career veered off in a somewhat different direction, he was always supportive. Indeed, at his 75th birthday party in 1992, held in Chicago in conjunction with an American Evaluation Association conference and bringing together dozens of his former students, he singled me out, with praise, as one of the few people present without a Ph.D.

Campbell originally was a farm boy. He never talked down to anyone, and he encouraged and welcomed questions and disagreements. He frequently was self-deprecating, making light of his own considerable accomplishments. As Reichardt stated, quite possibly Campbell never met a person he did not encourage. As Tom Cook indicated, he was a real mensch.

Campbell taught me that effective research and social inquiry is as much an art as a science. I, a practical, hands-on evaluator, cannot do my work, nor can I envision anyone else doing likewise, without drawing heavily on the contributions of Don Campbell—Campbell the philosopher, the theoretician, the methodologist, the visionary, and the human being.

REFERENCES

Campbell, D. T. (1957). Factors relevant to the validity of experiments in social settings. *Psychological Bulletin, 54*, 297-312.

Campbell, D. T. (1969a). Ethnocentrism of disciplines and the fish-scale model of omniscience. In M. Sherif & C. W. Sherif (Eds.), *Interdisciplinary relationships in the social sciences* (pp. 328-348). Hawthorne, NY: Aldine.

Campbell, D. T. (1969b). Reforms as experiments. *American Psychologist, 24*, 409-429.

Campbell, D. T. (1971). *Methods for the experimenting society.* Paper presented at the meeting of the Eastern Psychological Association, New York, and at the meeting of the American Psychological Association, Washington, DC.

Campbell, D. T. (1979). "Degrees of freedom" and the case study. In D. T. Cook & C. S. Reichardt (Eds.), *Qualitative and quantitative methods in evaluation research* (pp. 49-67). Thousand Oaks, CA: Sage.

Campbell, D. T. (1984). Can we be scientific in applied social science? In R. F. Connor, D. G. Attman, & C. Jackson (Eds.), *Evaluation studies review annual,* (Vol. 9, pp. 26-48). Beverly Hills, CA: Sage.

Campbell, D. T. (1988a). The experimenting society. In *Methodology and epistemology for the social sciences: Selected papers* (E. S. Overman, Ed.; pp. 290-314). Chicago: University of Chicago Press.

Campbell, D. T. (1988b). *Methodology and epistemology for the social sciences: Selected papers* (E. S. Overman, Ed.). Chicago: University of Chicago Press.

Campbell, D. T. (1989). Foreword. In R. K. Yin, *Case study research: Design and methods* (pp. ix-xi). Thousand Oaks, CA: Sage.

Campbell, D. T. (1994). Retrospective and prospective on program impact assessment. *Evaluation Practice, 15*(3), 291-298.

Campbell, D. T., & Fiske, D. W. (1959). Convergent and discriminant validation by the multitrait-multimethod matrix. *Psychological Bulletin, 56*, 81-105.

Campbell, D. T., & Stanley, J. C. (1963). *Experimental and quasi-experimental designs for research.* Chicago: Rand McNally.

Cook, T. D., & Campbell, D. T. (1979). *Quasi-experimentation: Design and analysis issues for field settings.* Chicago: Rand McNally.

Ginsberg, P. E. (1998). Cross-cultural research, ethnography and the multitrait-multimethod matrix. *American Journal of Evaluation, 19*(3), 411-415.

Kuhn, T. S. (1962). *The structure of scientific revolutions.* Chicago: University of Chicago Press.

Mohr, L. B. (1988). *Impact analysis for program evaluation.* Thousand Oaks, CA: Sage.

Patton, M. Q. (1997). *Utilization-focused evaluation: The new century text.* Thousand Oaks, CA: Sage.

Patton, M. Q. (1998). Discovering process use. *Evaluation, 4*(2), 225-233.

Perrin, B. (1994). *"Evaluating" the effectiveness of common strategies to evaluation and the importance of a practical collaborative, future-oriented approach.* Paper presented to the founding conference of the European Evaluation Society, The Hague, The Netherlands.

Perrin, B. (1998). Effective use and misuse of performance measurement. *American Journal of Evaluation, 19*(3), 367-379.

Perrin, B. (1999). Performance measurement: Does the reality match the rhetoric? A rejoinder to Bernstein and Winston. *American Journal of Evaluation, 20*(1), 101-111.

Preskill, H., & Torres, R. T. (1999). *Evaluative inquiry for learning in organizations.* Thousand Oaks, CA: Sage.

Reichardt, C. S. (1996). Obituary for Donald T. Campbell. *Evaluation Practice, 17*(1), 3-5.

Segall, M. H., Campbell, D. T., & Herskovits, M. J. (1966). *The influence of culture on visual perception.* Indianapolis: Bobbs-Merrill.

Shadish, W. R., Cook, T. D., & Leviton, L. C. (1991). *Foundations of program evaluation: Theories of practice.* Thousand Oaks, CA: Sage.

Thomas, R. M., Jr. (1996, May 16). Professor Donald T. Campbell is dead at 79. *New York Times* (Obituaries).

Torres, R. T., Preskill, H., & Piontek, M. E. (1996). *Evaluation strategies for communicating and reporting: Enhancing learning in organizations.* Thousand Oaks, CA: Sage.

Webb, E. J., Campbell, D. T., Schwartz, R. D., & Sechrest, L. (1966). *Unobtrusive measures: Non-reactive research in the social sciences.* Chicago: Rand McNally.

Weiss, C. (1998). Have we learned anything new about the use of evaluation? *American Journal of Evaluation, 19*(1), 21-34.

CHAPTER 11

The Experimenting Society in a Political World

Carol Hirschon Weiss

Donald Campbell was a man of towering gifts. His contributions to the social sciences remain important in field after field. As Overman wrote in the introduction to Campbell's collected papers, his work provides "a comprehensive and consistent map of the major issues and trends in social science during the last forty years" (Campbell, 1988, p. xvii). The centrality of his work in social science is the reason that I am choosing to analyze one issue where his views appear to be unrealistic.

Campbell's work in evaluation has had a lasting effect on the field. One of the spinoffs of his work on evaluation was a paper written in 1971, not published but circulated in manuscript form for many years, "The Experimenting Society." Hundreds of copies were in circulation. He was evidently uneasy about it and asked for critique and suggestions. He did not publish it but wrote a few short pieces about the experimenting society (Campbell, 1973, 1981) and gave an interview about it to *Psychology Today* (Tavris, 1975). The paper did not appear in print until 1988, when it was finally included in a collection of his papers edited by E. Samuel Overman (Campbell, 1988). The original 1971 paper obviously had been revised, because the version in the book includes references from the late 1970s and the 1980s. It even includes reference to a published paper that critiques the original version (Shaver & Staines, 1971). The basic premise, however, is unchanged.

"The Experimenting Society" drew on ideas that Campbell had earlier proffered in his classic 1969 paper, "Reforms as Experiments." In that earlier paper, he wrote:

> The United States and other modern nations should be ready for an experimental approach to social reform, an approach in which we try out new programs designed to cure specific social problems, in which we learn whether or not these programs are effective, and in which we retain, imitate, modify, or discard them on the basis of apparent effectiveness on the multiple imperfect criteria available. (Campbell, 1969, p. 409)

He went on to say that although some people think that the nation is currently making decisions about continuing or discontinuing programs on the basis of evidence (this was 1969!), this was not at all so. One reason, he implied, is that "most ameliorative programs end up with *no* interpretable evaluation" (Campbell, 1969, p. 409). Much of the rest of "Reforms as Experiments" consists of advice to evaluators about research designs that allow valid interpretation of evaluation evidence.

Campbell showed that he was aware of the importance of politics in influencing how feasible his advocacy of experimental reform could be. He discussed the political pressures on officials that almost required them to show that the programs for which they were responsible were successful. He called such officials "trapped administrators" and showed sympathy for their plight. He even gave tongue-in-cheek suggestions for how they could manipulate an evaluation to show positive results. Example: Use as a baseline the very worst year and the very worst social unit; then there is nowhere to go but up. A message of the article was that evaluators had a responsibility to create a demand for what he called "hard-headed" evaluation. They should try to convince administrators and other parties in the political process to advocate the importance of the problem, not the program, and continue to seek workable solutions to the problem until a satisfactory solution is found. Campbell (1969) enjoins evaluators to seek out "*political* inventions that reduce the liability of honest evaluation" (p. 409, italics added).

That brings us to "The Experimenting Society." In this paper, Campbell reiterates his prescription for social reform. When society identifies problems in need of solution, it should implement new programs and meticulously evaluate them. Through evaluation, people will learn which approaches work, which need improvement, and which should be discarded. As in "Reforms as Experiments," Campbell (1988) wants a society that "would vigorously try out possible solutions to recurrent problems and would make hard-headed, multidimensional evaluations of outcomes, and when the evaluation of one reform showed it to have been ineffective or harmful, would move on to try other alternatives" (p. 291). It is a rational prescription for a rational world.

Campbell, along with the rest of us, had been watching the fate of new programs in the War on Poverty and their evaluations. He had seen the constant impediments to sound evaluation and the widespread inattention to the results the evaluation studies produced. As a rational man, he evidently could not understand, or countenance, the frustration of good evaluation design and the neglect of evidence by government officials whose task should be to do the best possible job for the public. Any system that defied logic and rationality was obviously counterproductive, and he yearned for a system that would know how to profit from evidence.

In "The Experimenting Society," Campbell (1988) states:

> As we try to implement high-quality program evaluations, we meet with continual frustration from the existing political system. It seems at times set up just so as to prevent social reality-testing. This leads us to think about alternative political systems better designed for evaluating new programs. (p. 291)

What a portentous leap! It is this shift in Campbell's thinking that fascinates me: from frustration with the current system to a search for an alternative political system. Clearly, he was disenchanted with efforts to educate or reform current political actors. He was now on a utopian quest for a new and better political system.

His paper describes desirable characteristics of an experimenting society. Such a society must be evolutionary, honest, active, scientific, accountable, decentralized where feasible, voluntaristic, egalitarian, and more. This utopian society is obviously not abandoning democratic principles; rather, it seeks to elevate them to an extremely high level. In the process, it adopts some curious inconsistencies (the new system should be responsive to the public and scientific, it should be exploratory and accountable). Campbell does not want to replace the principles of democratic government, but he outlines such a utopian version of democracy that it seems almost more compatible with heaven than with earth.

Oddly, however, the paper gives no further attention to the existing political system. It does not consider the advantages of the system, why it has survived, and why no political system in the world, as Campbell notes, embodies the characteristics of the experimenting society. He does not examine the ways in which policy is actually made in the nation, nor, beyond noting the pressure on "trapped administrators" to show success, does he discuss how agencies actually administer programs. Above all, he ignores key concepts of the policy process, such as interest groups, representation, and negotiation. In his writing, even in his utopia, there are no political parties, mass media, lobbies, or elections either national or local, and hardly any institutions at all. He gives no recognition to the value of accommodations among parties espousing different views and different interests, nor does he have a word for the mechanisms that hold together a sprawling multi-ethnic population with conflicting economic interests.

"The Experimenting Society" has much to say about how to make the utopian society feasible, but the discussion is almost all about methodological issues. Thus, Campbell worries about whether an experimenting posture would ride roughshod over human beings by random assignment. He comes out on the side of voluntary participation in evaluations. He advocates evaluating seemingly successful local programs ("proud programs") and then cross-validating results in multiple sites. He has wise words to say about how to make long-term follow-up a more frequent methodological choice. He recognizes the fallibility of indicator systems and their susceptibility to corruption, and he wants multiple indicators, each necessarily imperfect but with different imperfections. And so on. Some of the ideas are familiar from his earlier writing.

CAMPBELL'S AMBIVALENCE ABOUT HIS UTOPIA

Campbell says several times that once evaluators confront the implications of the experimenting society, they may reject it out of hand. The experimenting society may be dysfunctional. He writes, "As this portrait [of a truly experimenting society] emerges in greater clarity, it will be our duty to continually ask ourselves if we really want to advocate this monster of measurement and experimentation" (1988, p. 314).

Here it is the technical issues, "measurement and experimentation," that are on his mind. He writes that the social sciences have not yet earned the prestige of the natural sciences and are "scientific by intention and effort, but not yet by achievement" (1988, p. 298). The implication is that if the social sciences ever become truly scientific (in the style of physics), then there would be few reasons to oppose the experimenting polity.

Campbell is not advocating technocracy. He is committed to democratic institutions and the necessity of a democratic legislature. An experimenting society, he writes, needs channels for moving evaluation results into the policymaking process. Campbell (1988) styles this requirement "reconcil[ing] our need for facts with democratic decision making" (p. 307). He recognizes that elected representatives of the people make policy, but he has enough of a bent toward the technocratic to state that he considered the idea of delegating control to a scientific elite. He rejects the idea because he does not want to create a government run by social scientists. He does not want to put evaluators in control. He comes up with a compromise:

> New institutions will be needed, such as an auxiliary legislature of quantitative social scientists, each appointed by one real legislator. Such an auxiliary legislature

> could process advisory decisions, with full attention to the scientific evidence. The real legislator could then guide his own decisive vote by the auxiliary legislature's actions and the issues raised. This awkward and expensive procedure seems to me better than delegating this process to an appointed scientific elite. (p. 307)

Awkward? Yes. Calculated to bring the "real legislators" in line with the best scientific evidence? No. Real legislators suffer from no shortage of scientific evidence now. Members of the Congress, for example, have available the resources of the Congressional Research Service, the Congressional Budget Office, and the General Accounting Office (which these days is much given to evaluations). Members are privy to data from the federal executive agencies, including the Census Bureau, the Bureau of Labor Statistics, and the other specialized data agencies of the departments, as well as being inundated by research data from agencies within and outside the executive branch. Members have their own expert staffs and the testimony, solicited and unsolicited, of the best social scientists in the country. None of this extensive apparatus of information and advice has brought about even a pale rendition of the experimenting society. Even when the evidence at hand is as rigorous and authoritative as it now is in the physical sciences, legislators have shown that they can put other considerations first: political advantage, constituent interests, their own values and ideologies, and the constraints of the institutional structures within which policy is adopted and implemented.

THE INTENT OF THIS CHAPTER

In this chapter, I juxtapose Campbell's hyperrational view of the experimenting society with the work of two other giants of American social science writing in the same period: Charles E. Lindblom and James G. March. Lindblom, an economist who became president of the American Political Science Association, devoted much of his attention to the world of public policy making. March specialized in the study of organizations, mainly business organizations but also schools, universities, and government agencies. He wrote about the ways that they went about their work, especially in making decisions. The work of both Lindblom and March has had far-reaching influence on the direction of the social sciences in their efforts to contribute to the practical world.

LINDBLOM'S VIEW: THE NEGOTIATING POLITY

Campbell makes at least one reference to Lindblom. In his paper "Can We Be Scientific in Applied Social Science?" (in Campbell, 1988, originally published 1984), he identifies mistaken beliefs of social scientists as they try to develop research that responds to policy concerns. One mistake is the belief that a single research study is sufficient to provide answers about the effectiveness of a social program. Instead, program decisions should rely on all the prior wisdom and prior science available. In this connection, he cites favorably Lindblom and Cohen's (1979) *Usable Knowledge* book, which makes the same point.

This one idea is only a tiny fraction of Lindblom's formidable oeuvre. Lindblom's career was approximately coterminous with Campbell's (although he, like March, is still alive). Much of what Lindblom wrote over his long and fruitful career had to do with political policy making.

Lindblom has a special interest in the role of analysis in the policy process. By analysis, he means not only policy analysis in the narrow sense but also the array of social scientific tools that can be applied to supplying information for political actors. Where Campbell starts at the research end, Lindblom starts with the policy process.

In a famous book, written at about the same time as "Reforms as Experiments," Lindblom (1968, revised 1980, revised with Woodhouse, 1993) discusses how analysis enters the decision-making system. It is not, he writes, through the prestige and obvious superiority of science and fact. The power of analysis does not sweep all before it by the sheer strength of its conclusions. The primary reason is not, as Campbell would have it, that the applied social sciences lack scientific rigor. Lindblom would agree that they lack rigor, but even if they were as rigorous as physics, they would not prevail, because that is not how political decisions are made in a democracy. In a democratic system of government, analysis does not stand above the fray. When research, evaluation, or analysis is used (and Lindblom believes that each has considerable influence), it is used as a weapon in the fray. All three are used by partisans to strengthen their side in the contest among contending interests. They become part of the partisan struggle.

Lindblom stresses that decisions in a democracy are made through interaction, through discussion and debate among policy actors, that is, through politics. Parties who have standing in the policy-making process are members of the legislature, the chief executive and his staff, appointed officials in government bureaucracies, high civil servants, and the relevant commissions and consultative bodies. They pay attention to, and are strongly influenced by, groups outside government, notably interest groups, the mass media, political parties, business and professional associations, and, in some cases, the judiciary. When citizens mobilize and speak out, they too can get a hearing. On contentious issues, a paramount consideration is not so much to find the "right" answer as to find a policy solution that is at least minimally acceptable to a large majority. Lindblom (1968) has written, "A policy is sometimes the outcome of a political compromise among policy makers, none of whom had in mind quite the problem to which the agreed policy is the solution" (p. 4).

Negotiation is a key process in policy making, and although it can be criticized as "logrolling," "compromising one's principles," or "ignoring evaluation evidence" (all of which it sometimes is), it is essential as a means to accommodate as many interests as possible and maintain the legitimacy of the system. In the negotiations that accompany efforts to win majority support for policy proposals, one of the tools that partisans can use is analysis. They sometimes use research and evaluation to strengthen the commitment of their own supporters to the proposal, to try to convince waverers to join them, and to counter the claims of policy opponents that the proposal is worthless or counterproductive. At its best, Lindblom writes,

> Analysis thereby can serve as an instrument of persuasion—of persuasion in the best sense of the word, meaning the use of information and thought to move people closer to reasoned and voluntary agreement. This is achieved in part by persuading possible opponents that their views are being taken into account. (Lindblom & Woodhouse, 1993, p. 129)

But policymakers do not always pay attention. Lindblom (1968) acknowledges:

> Men turn an indifferent or hostile eye on policy analysis because they are not wholly rational. Because, specifically, it is easier to feel than to think. Because they cling to beliefs that serve the needs of their personalities. Because words or symbols with which they talk about politics come to be more dear to them than the things to which the symbols refer. Because sometimes it pains them to change their minds. Because they have picked up all kinds of beliefs from their families, friends, churches, and other groups—beliefs that give them a comforting orientation to the world about them and which they consequently dare not challenge. Because it may not have occurred to them that policy analysis is of potential great value. (p. 19)

Lindblom is also aware of the limitations in the research evidence. Analysts, evaluators, and social scientists generally all have their own biases and limitations and their own myopia. Their ability to provide an "objective" account is further limited by the biases of the institutions within which they work. These ideas come to the fore in Lindblom's later work, but even in 1968, he wrote:

> Even those people most interested in analysis will know that analysis will always be influenced by the biases of the analysts and by their incompetences, and that hence it is not always to be trusted. And they will know that, since most analysis takes place in organizations, it will always be marred by organizational biases, rigidities, and other incompetence. (p. 19)

Lindblom's work is permeated with the tension between reason and politics in policy making. He has written:

> The conflict between reason, analysis, and science on the one side, and politics and democracy on the other, remains. If a society wants more reason and analysis in policy making, perhaps it must surrender some aspects of democracy. (Lindblom & Woodhouse, 1993, p. 12)

This is precisely the dilemma that Campbell wrestled with in "The Experimenting Society." Campbell struggles to find ways to strengthen the role of reason and logic, even if it means abandoning the whole current political system. Lindblom, too, wishes for reforms in the ways that our system works, but he comes down squarely on the side of politics and democracy.

MARCH'S VIEW: LIMITS ON HUMAN AND ORGANIZATIONAL RATIONALITY

If Campbell barely acknowledges Lindblom, he makes no reference to March, who was writing at the same time. In a series of brilliant books on the ways that organizations function, March makes an even more sweeping critique of the assumptions of rational decision making. He highlights the disconnection between information and decisions in organizational life. One reason he cites is human beings' limited capacity for information processing. People are not very good at thinking and reasoning. Following Herbert Simon's (1945) pioneering analysis, March developed these ideas for decades. In a later summary, he wrote,

> It is well known that individuals are not particularly good interpreters of evidence, and their limitations undermine the intelligence of decision making. They make a variety of mistakes and simplifications. They learn lessons inadequately, recall memories incompletely or incorrectly, estimate futures inaccurately. They learn superstitiously, assigning causal significance to actions that are correlated with effects but not affected by them. They remember history in ways heavily dependent on their current beliefs. They confound their wishes and hopes with their expectations. They ignore useful information out of concern that it may reflect deliberate falsification. (March, 1994, p. 241)

Furthermore, organizations add their own rigidities and conflicting interests. The world of organizations, government bodies, and policy making is fraught with ambiguities and uncertainties. March describes the ways in which organizational action takes place, often through what he terms "garbage can processes" (March & Olsen, 1976, p. 36). In his formulation, organizations face separate streams of problems, solutions, participants, and choice opportunities. Problems flow along and, almost independently, so too do potential solutions.

At intervals, decisions have to be made—budgets adopted, programs initiated, plants built, staff hired, breakdowns fixed. These are choice opportunities. Whatever set of participants are on the scene at the time a choice has to be made define the problem and make a decision. They select one of the solutions streaming along at that moment, attach it to the problem, and thus make the choice. Through such garbage-can processes, decisions are made.[1]

March is not an advocate of decision processes of this kind; he simply states that this is the way the world works. But he goes farther. He states that rational calculation, on the order of Campbell's experiments and evaluations, is only one way, and not a very successful way, to make good decisions. Much of his writing stresses the limits of rationality. He notes the failure of such rationalistic enterprises as long-term planning and computer-assisted decision matrices to improve the quality of organizational decisions. He writes, "Decision makers ignore information they have, ask for more, then ignore the new information" (March, 1994, p. 177).

Such bizarre behavior arises from several sources. One is the frailty of human reasoning, already noted. Another is the common phenomenon of multiple values. Organizations do not seek to reach only one goal or one set of complementary goals. They rarely have a single set of preferences that can be classified in priority order. Often they have a number of goals in mind, some of them mutually inconsistent or even contradictory. It is not only different units within the organization that have inconsistent goals (Cyert & March, 1963); one unit or even one individual can hold a series of conflicting goals (such as security and expansion, excellent service to clients, and avoidance of staff overwork). Because of such inconsistent and conflicting objectives, rationality does not help to resolve the differences among alternative decisions.

Furthermore, goals change. Both organizations and individuals shift their preferences as conditions change; thus, any calculation that leads to a decision today may be outdated by the time results of that decision come home to roost. For evaluators gathering evidence 2, 3, or 5 years after a program started operation, a shift in goals may make their evidence of goal attainment irrelevant to the current situation.

March has a particular complaint about the way that evaluation is performed:

> As nearly as I can determine, there is nothing in a formal theory of evaluation that requires that the criterion function for evaluation be specified in advance. In particular, the evaluation of social experiments need not be in terms of the degree to which they have fulfilled our *a priori* expectations. Rather we can examine what they did in terms of what we now believe to be important. The prior specification of criteria and the prior specification of evaluational procedures that depend on such criteria are common presumptions in contemporary social policy making. They are presumptions that inhibit the serendipitous discovery of new criteria. (March & Olsen, 1976, p. 80)

In fact, only later may we discover that certain outcomes of a social policy, unforeseen at the outset and certainly unintended, were major contributions. As an illustration, I believe that one of the important benefits of the War on Poverty was the opportunities it gave to a generation of people of color to assume leadership positions in social agencies and social action. This was not an objective of the program in any formulation that I ever came across, in writing or discussion, but it was a valuable outcome.

March is impressed by the likelihood of *discovering goals* as a consequence of action. He saw virtue in revising the order of the old command to "ready, fire, aim" and to learn from action what goals to aim for. For him, the quality of the goals is at least as important as whether a program realizes the goals. As Loveless (1998) has recently framed the point, "The success or failure of policy is often measured by how well its stated goals are attained, leaving the wisdom of such goals unexamined" (p. 6). It is true that evaluation rarely queries the wisdom of the goals that lawmakers and administrators define, and March sees this inattention as a shortcoming and a lost opportunity for understanding and redirection.

March also has concerns about the validity of the evidence with which evaluation generally traffics. He is sensitive to the arbitrary nature of much evaluative evidence. He is especially aware of the values that enter into its construction. Some of the values have to do with truth seeking, and some are concerned with self-interest. He writes,

> One standard approach [to dealing with a complicated world] is to deal with summary numerical representations of reality, for example income statements and cost-of-living indexes. . . . There is ambiguity abut the facts and much potential for conflict over them. . . . They have elements of magic about them, pulled mysteriously from a statistician's or a manager's hat. . . . The construction of these magic numbers is partly problem solving. Decision makers and professionals try to find the right answer, often in the face of substantial conceptual and technical difficulties. . . . The construction of magic numbers is also partly political. Decision makers and others try to find an answer that serves their own interests. . . . These simultaneous searches for truth and personal advantage often confound both participants and observers. (March, 1994, pp. 15-17)

March also is concerned about the ambiguity of evidence. Because the world is complicated and fast-moving, people's ability to understand what actually happened in any situation is constrained, however comprehensive the evidence appears to be. Their ability to understand what caused events is even more limited. He writes, "Individuals try to make sense of their experience, even when that experience is ambiguous or misleading. . . . They impose order, attribute meaning, and provide explanations" (March & Olsen, 1976, p. 67). The meanings and explanations may be fantasies.

March's work has been absorbed and elaborated by a generation of scholars. Writing about organizations has been heavily influenced by March's ideas about the process of decision making. So, too, has work in political science, international development, education, and many other fields. I wondered why I could not find a reference to March's work in Campbell's writing. I realized that March's views would be uncongenial. Despite Campbell's approval of qualitative knowing, constructivist positions, and evolutionary epistemology, Campbell did not seem to be comfortable with *untidiness*. To a social scientist of his temperament, acknowledging March's views might look like the acceptance of sloppiness, error, and wrong-headedness.

CONVERGENCES AND DIVERGENCES

Campbell, Lindblom, and March agree on many things. Fundamentally, they are all opposed to the idea of giving analysts (evaluators) the responsibility of deciding which ends or goals are worthwhile for the society to pursue. In policy analysis narrowly defined, this means that they should not be enfranchised to formulate the problems or decide on the goals that a policy should address. In evaluation, it means that evaluators should not be authorized to select the criteria on which a program is judged.

Campbell explicitly recommends that researchers should not attempt to develop policy. Rather, they should simply (or not so simply) tell whether current policy is working.

Campbell's statement does not take into account the fact that most programs aim to achieve multiple goals, stated and unstated. Even when evaluators meet with stakeholders to decide on goal statements, and even when they collect evidence on the attainment of multiple goals, evaluators still have disproportionate influence on the selection of outcome criteria. They have even more influence on the specific definition of outcome criteria, what is and is not counted. If they rely on official statements, they are not always aware that goal statements often mask some of the real purposes behind the program and tend to privilege the aims of official groups and program sponsors over the aims of front-line practitioners and program clients. Lindblom and March are more sensitive to the ambiguity—and to the *changingness*—of goals than is Campbell, and they are more reluctant to cede their definition to researchers.

Furthermore, evaluators tend to treat outcome criteria as fixed and immutable, at least for the length of the study, but decision makers and the public disagree on many things, such as the weight they would assign to the achievement of any one goal, and, over time, their values and perceptions change. As Lindblom has written, "Such judgments, moreover, contain moral components; so the issues cannot be fully settled except by reference to values or interests in conflict in any society. The settlement requires politics more than analysis" (Lindblom & Woodhouse, 1993, p. 22).

In fact, Campbell recognized that evaluation would not and should not replace politics. In a 1984 paper reprinted in his collected papers, he used words reminiscent of Lindblom: "Evaluation reports should enter into political decision-making processes as one component to multiparty argument and negotiation" (Campbell, 1988, p. 331). This acknowledgment represents his closest approach to the theme of this chapter.

I doubt that Lindblom or March would disagree with any of Campbell's advice about improving the scientific validity and relevance of evaluation. In fact, Lindblom has more optimism about evaluation than he does about policy analysis in its more ambitious sense, because evaluation assists the process of learning from what he calls "intelligent trial and error."

Lindblom and March have many points in common. They both explicitly take account of the limitations on human reasoning. They are explicitly aware of value conflicts inside programs and in the policy-making process. They pay attention to organizational and political reasons for neglecting research, analysis, and evaluation. They see more than Campbell's "trapped administrator," who has to defend his program. They are sensitive to the many self-seeking units and coalitions within organizations that may pursue their own aims.

Of the three authors, March probably is the most skeptical about research and evaluation. He is agnostic even about human capacities to learn from trial and error. In discussing learning by trial and error—that is, learning from experience—he is almost as skeptical as he is about learning from systematic inquiry. He notes how difficult it is for people to derive lessons from previous action. The values of people are multiple and ambiguous, and therefore they find it hard to decide what has worked well on the basis of a single strand of outcomes. The connections between previous action and later situations are often almost impossible to fathom. On these grounds, what an organization "learns" through trial and error may be superstition ("I wore my cap backwards when I hit a home run and so the cap is important and I should always wear it backwards"). So much for the vaunted benefits of informal organizational learning. Confronted with the limitations of both calculated rationality and learning from experience, March is not satisfied with either of them.

SOLUTIONS

The differences among the authors are clearest when we analyze the solutions they propose for the neglect of research, evaluation, and analysis in the policy process. Campbell, as we have seen, proposes improvement in the methods used in evaluation. If evaluative evidence were more readily interpretable, and if it could make causal statements with confidence, he apparently believes that policy actors would be more likely to pay attention. He writes about issues of using volunteers in randomized experiments, telling respondents in opinion surveys about the funding and intended uses of the data, the problem of corruption of social indicators when they are used for decision making, and the horrendous burden that multiple measures on multiple programs would create for measurement and information overload. These are important methodological issues, but they are not of an order or a kind that would seem to move toward a society that bases policy on the careful evaluation of past experience. Campbell attended to the methodological roadblocks, but he tended to ignore human frailties, organizational constraints, and political interests.

When Lindblom seeks solutions to failures of "intelligence," he comes up with a different set of prescriptions. In the third edition of his book on the policy-making process, he writes,

> [Analysis is not] explicitly and systematically adapted to being useful in settings characterized by high uncertainty, partisan disagreement, and interactive problem solving. Policy discussions typically slip into behaving as if uncertainty and disagreement could be circumvented by sufficient information or logic. One of the crucial steps in improving policy analysis, therefore, is to adapt it better to the inevitability and the desirability of partisan conflict as the key mechanism in the intelligence of democracy. In other words, analysis should aim to improve the quality of political interaction, not try to substitute for it. (Lindblom & Woodhouse, 1993, p. 127)

This has been a key point for Lindblom all along. Politics, he says, is "desirable." It is the way a democracy deals with differences and reaches decisions. Evidence is used to make a partisan case. Negotiation is necessary, but instead of negotiation based on propaganda and irrational persuasion, or logrolling, bribery, and other means for inducing acquiescence without changing people's minds, evaluators can foster negotiation based on evidence and reason. Lindblom suggests that analysts (and evaluators) might try to align themselves with different groups (and not always government agencies or groups supporting the status quo) and direct their studies to issues of importance to the groups for which they choose to work. The evidence that evaluators provide can then be used in political interactions and debates to try to convince others.

Lindblom's world thus has an important place for knowledge and information. His later books (Lindblom, 1990; Lindblom & Woodhouse, 1993) show disillusionment about the possibility that knowledge (analysis and evaluation) can be unbiased. He sees that all of us, professional researchers and analysts as much as anyone, are socialized into views of the world that are "impaired." We are constrained by our upbringing and the culture around us, and as a consequence we see only a small part of the world's potential. A major consequence is that lay people and professionals alike tend to conform with narrow bands of thought. Inquiry is hobbled by consensus on existing political and economic institutions and the habits of mind that they instill. Professional inquiry, like other forms of decision making, is limited in perspective. To make the system work better, what is needed is "less skewed competition of ideas, less governmental concealment of information, changes in political campaign practices, and less anxious political indoctrination of children in the schools" (Lindblom, 1990, p. 285). Society needs a broader range of ideas from which to choose.

Nevertheless, Lindblom retains faith in the power of inquiry. He wishes it were less constricted, but he still finds applied research a possible route to social improvement. Most relevant to this discussion, he still believes that it has much to contribute to the political interaction that constitutes democracy's form of policy making. Analysis and evaluation can not supplant politics, but they can add to its intelligence. In 1993, he wrote that "by monitoring feedback from experience [e.g., though evaluation of programs and policies], sufficient insight may be gained gradually to shape revised and improved policy trials" (Lindblom & Woodhouse, 1993, p. 30).

March does not believe in the virtues of rational calculation. He does not think that better decisions emerge from research, evaluation, data, computerized decision aids, or any of the armamentarium of applied social science. He writes that "knowledge can be inimical to making decisions. Knowledge seems to increase questions at a faster rate than it increases answers. It provides too many qualifications, recognizes too much complexity" (March, 1994, p. 265). Rather, he advocates "sensible foolishness." He seeks to bring back into scholarly repute some discredited decision-making techniques of ordinary life, such as imitation, coercion, and rationalization. He believes in intuition. He advocates experimentation (as well as rationality), forgetting (as well as memory), and playfulness (as well as consistency). His diagnosis of the faults of much decision making is overemphasis on consistency, history, and rules, and so he advocates breaking out of the cage. People tend to be set on stability of belief and of action. In contrast, he sees virtue in a propensity to innovate, experiment, and search for something better. This course of action is more attractive than any scheme calculated to make better use of information. He wants decision makers to suspend the press toward consistency and to be tolerant—even welcoming—of changes in preferences that come about as a result of action. These ideas have resonance with Lindblom's call to widen the repertoire of ideas in society.

CONCLUSION

I made a good-faith effort to reconcile the differences between Campbell and the other authors. I could not do it. Campbell, who is exquisitely aware of shortcomings in the scientific aspects of evaluation, is apparently tone deaf to the competing forces in the larger society that influence the making of policy. He is inattentive to those elements that insistently press on the awareness of Lindblom and March—evaluator biases, the inevitability of conflicting evaluation conclusions even when the canons of good design are meticulously observed, the ubiquity of politics, the need for negotiation to resolve differences in values and beliefs, failures in human rationality—all the untidy elements that afflict the linkage between evaluation and policy.

Nor does evaluation usually give direction for what to do when the program studied does not work. Suppose the goal of the program is to enhance the development of children in poverty, and agencies provide comprehensive social services toward this end. If evaluation shows little if any result from the program (e.g., St.Pierre, Layzer, Goodson, & Bernstein, 1997), for a long time decision makers can ring changes on the organization and staffing of the services. If, however, no arrangement of comprehensive social services is effective, what can be done next? One possibility, being pursued by the Annie E. Casey Foundation, is to move to a program of providing jobs for adults in the children's families, on the assumption that better family incomes will benefit children. A change of this type is no longer tinkering with internal program design. It is a significant policy change that has wide-ranging ramifications for organizations, staffs, professions, budgets, leadership, and reputations. Decisions of this kind are not made on the basis of data alone. Just because evaluation shows that earlier program approaches did not work, evaluation does not justify embarking on this new course.

Even were the change to be less momentous—say, moving from an emphasis on coordination of social services to inclusion of clients' input in the direction of the services—there are still vital issues of power and control, resources, and direction. Negotiation has to be the order of the day. Negotiation is necessary not just to overcome opposition from the losing side but also to reach accommodations that staff will implement and clients will accept and that stand a reasonable chance of working.[2]

In general, program decisions are value-laden, and evaluation can not resolve conflicts in values. It is not only big policy questions that raise value issues; many smaller issues involve ideological differences as well as issues of power, money, and personal influence.

March and Lindblom make points that deserve serious attention in our expectations for evaluation use. From March, we learn something about the meandering ways in which decisions are often made. Many decisions come about not through careful calculation and ordered consideration but by happenstance. This is true in organizations of all kinds and sizes as well as in councils of the federal government. His analysis also suggests how circumscribed are the contributions that data can confidently make to policy in a changing multivariate world.

Lindblom's emphasis on interaction is another contribution. Policy does not take shape through individual cerebration but through discussion and debate in dozens of offices and legislative chambers. To have an impact, evaluation has to enter the policy debates.

If the decision-making world is much as Lindblom and March describe, how should we think about achieving the benefits of the experimenting society? We have all had experiences that show that evaluation can—and sometimes does—make a difference in policy and program design. We can point to cases where evaluation has changed people's minds and their program decisions. These decisions are not generally policy issues of the highest order, but they can have significant effects on the life chances of men, women, and children. How do we think about making the influence of good evaluation more widespread?

I would suggest four points from this discussion:

1. Avoid evaluation imperialism. Campbell himself did not seek to empower an evaluation elite (especially an elite whose repute was "not yet earned"). On the contrary, he was in favor of liberal reforms that gave more influence to the public, but his paper can be read to give comfort to the technocrats who would like analysts and evaluators to have more power. In many of the applied social science disciplines, there is a coterie of researchers who would like to see their judgment substituted for the political process. Evaluation has its share of such folks, who consciously or subconsciously believe that they know better than politicians and bureaucrats and that the world would be a better place if they were in charge. Their ranks swell when the political process is controlled by a coalition whose priorities violate many of their deeply cherished beliefs. When conservatives are in power, some evaluators are tempted to enlist under the technocratic banner.

The genius of democracy, however, is that the popular will, expressed through its fallible institutions, works to the benefit of the citizenry more effectively than any other political system yet devised. As E. F. Forster (1951) wrote long ago, "two cheers for democracy"—even when it means putting up with the fact that evaluation findings sometimes line the bottom of the bird cage. Evaluators should avoid a technocratic misreading of "The Experimenting Society." They should not assume that an evaluator-run society would necessarily be a better or wiser place or one less riven by conflict.

2. Study actual uses of evaluation. For a time in the 1970s and early 1980s, empirical work on knowledge utilization (including the use of evaluations) flourished. Then researchers realized how complicated the issues were—how difficult it was to define "use," how divergent research results could be, and how dependent utilization was on surrounding conditions (topic, participants, degree of politicization, scope, availability of champions, issue history, and so on). It is time for a new burst of research to explore the conditions under which evaluation has influence. The research will have to be more sophisticated and operate from a stronger and more comprehensive conceptual base, but if we can do such work, it will help us understand the conditions associated with evaluation influence on decisions in a variety of policy domains.

3. Recognize that evaluation cannot be insulated from politics. Evaluation cannot be walled off from politics, whether we are talking about organizational politics or partisan politics. Evaluation itself is a political act (Weiss, 1975). It advantages some groups and disadvantages others, and it is perceived in this light by potential users of its findings. (If case management does not work, as some recent evaluation suggests, what happens to current case managers and their agencies?) Evaluators should know their way around the policy field—know who wants what and why—and they should be aware of the ways in which results can be used, distorted, neglected, or undermined. They should be ready to communicate their results, seek allies, marshal additional evidence, and conduct supplementary studies where possible. They should not think of the use of evaluation to support a position that people already want to take as an abuse of evaluation. If the evaluation shows the superiority of a particular program or policy, evaluators should welcome the existence of precommitted supporters of the cause who will champion the evidence. They should recognize that evaluation can add weight and heft to a position, and if evaluation evidence gains converts to that position, such use of evaluation is legitimate. To the extent that the evidence supports the case of groups underrepresented in the political system, evaluators might choose to disseminate their findings to help those groups make their voices heard in policy chambers.

4. Make evaluations more valid and interpretable. Campbell's vision, if limited, is too luminous to be abandoned. Just as he urged, evaluators should keep improving the validity and reliability of the work they do. Good methodology is the bedrock of evaluators' craft and the reason they were invited into councils of decision. Sound measurement, design, and analysis represent our distinctive calling and are the source of our rewards and our satisfaction (Weiss, 1998a). Evaluators not only need technical skills but also must apply them intelligently and sensitively in sometimes inhospitable contexts. Furthermore, evaluations need to be useful, to respond to the questions that sponsors, practitioners, clients, and critics have about the programs under study. Evaluators also need to adapt good methods and useful responsiveness to another precept that Campbell enunciates in "The Experimenting Society": to respect the dignity and autonomy of the people they study.

When evaluation results turn out to show little effect—or seriously unpleasant side effects—from programs the evaluator strongly supports, the evaluator has to be ready to stand by the results. Expurgated or slanted reports benefit no one, not even the party whom they intend to help. Critics can demolish faulty results too easily, and with them fall the causes they were meant to promote. Still, all evaluation incorporates values. Preferences and values are expressed in defining the key evaluation question, choosing outcome criteria, establishing timing of the inquiry, crafting measures of process, wording items, selecting sites for observation, deciding on modes of analysis, and all the other seemingly technical decisions that go into evaluation (Weiss, 1978). Evaluators should recognize the political cast of these choices and understand that other readings of the situation are possible and legitimate. They cannot assume that they have produced the single and total truth. Finally, evaluators need to become more savvy about communicating effectively in the political sphere. Injunctions about continued face-to-face discussion with decision makers are often irrelevant when opportunities for personal interaction are limited. Evaluators need to learn more about modes of communication that work in political circles (Weiss, 1989, 1991; Weiss & Singer, 1988).

We all long for more reason and analysis in political life. We want governments ruled less by demagoguery and pressure from self-serving interests.[3] If improvements in evaluation alone will not get us to the better world that Campbell envisions, even marginal improvements in methods of evaluation will be valuable. Improvements will help to increase evaluators' confidence in the conclusions they offer, whether the implications are explicit or implicit. Methodological improvements will increase the trust that well-informed audiences give to evaluative evidence. When more attention is paid to evaluations, as it sometimes is, methodologically superior evaluations are more likely to point the way to wise policy.

POSTSCRIPT

A reviewer of an earlier draft of this chapter and the editor of the volume urged me to include my own views and reference to publications I have written on the subject. The use of research and evaluation, it is true, is the topic of my first published paper (Weiss, 1967), which antedates Lindblom's and March's words on the subject, and I've fidgeted and declaimed about social science and policy ever since (Weiss, 1998b, in press). But I think that this chapter has its own coherence. As an uneasy compromise between ignoring my work and ignoring the editor and reviewer, I've included a number of Weiss citations in the list of references.

NOTES

1. It is interesting to note that March came to these views not through experience with government agencies, which might be expected to behave "irrationally," but through extensive experience with business organizations and universities.

2. Another assumption of the experimenting society is that if evaluations show a program to be ineffective, an alternative program is waiting in the wings. In a number of program fields, almost all credible ideas seem to be exhausted. Practice professionals and program designers do not have good ideas for a persuasive alternative.

3. As I write these words in 1998, one of the best ways I can think of to increase the influence of evaluation in public policy making on issues of moment is to work to reduce the influence of special interests on policy. Special interests tend to pursue their single set of aims and ride roughshod over any opposition. They try to bury any evidence that fails to support their cause. Because some of the interest groups are so well funded and well organized, they have a disproportionate influence on elections and issues. Their lobbying activities tend to shut out the views of their opponents. At the national level, this concern would suggest working for campaign finance reform.

REFERENCES

Campbell, D. T. (1969). Reforms as experiments. *American Psychologist, 24*(4), 409-429.

Campbell, D. T. (1973). The social scientist as methodological servant of the experimenting society. *Policy Studies Journal*, 2, 72-75.

Campbell, D. T. (1981). Introduction: Getting ready for the experimenting society. In L. Saxe & M. Fine (Eds.), *Social experiments: Methods for design and evaluation* (pp. 3-18). Beverly Hills, CA: Sage.

Campbell, D. T. (1988). *Methodology and epistemology for social science: Selected papers* (E. S. Overman, Ed.). Chicago: University of Chicago Press.

Cyert, R. M., & March, J. G. (1963). *A behavioral theory of the firm.* Englewood Cliffs, NJ: Prentice Hall.

Forster, E. F. (1951). *Two cheers for democracy.* New York: Harcourt Brace.

Lindblom, C. E. (1968). *The policy-making process.* Englewood Cliffs, NJ: Prentice Hall.

Lindblom, C. E. (1980). *The policy-making process* (2nd ed.). Englewood Cliffs, NJ: Prentice Hall.

Lindblom, C. E. (1990). *Inquiry and change: The troubled attempt to understand and shape society.* New Haven, CT: Yale University Press.

Lindblom, C. E., & Cohen, D. K. (1979). *Usable knowledge: Social science and social problem solving.* New Haven, CT: Yale University Press.

Lindblom, C. E., & Woodhouse, E. J. (1993). *The policy-making process* (3rd ed.). Englewood Cliffs, NJ: Prentice Hall.

Loveless, T. (1998). Uneasy allies: The evolving relationship of school and state. *Educational Evaluation and Policy Analysis, 20*(1), 1-9.

March, J. G. (1994). *A primer on decision making: How decisions happen.* New York: Free Press.

March, J. G., & Olsen, J. P. (1976). *Ambiguity and choice in organizations.* Bergen, Norway: Universitetsforlaget.

Shaver, P., & Staines, G. (1971). Problems facing Campbell's "experimenting society." *Urban Affairs Quarterly, 7*(2), 173-186.

Simon, H. (1945). *Administrative behavior: A study of decision-making processes in administrative organization.* New York: Free Press.

St.Pierre, R. G., Layzer, J. I., Goodson, B. D., & Bernstein, L. S. (1997). *National impact evaluation of the Comprehensive Child Development Program: Final report.* Cambridge, MA: Abt Associates.

Tavris, C. (1975, September). The experimenting society: To find programs that work, government must measure its failures. *Psychology Today,* pp. 47-56.

Weiss, C. H. (1967). Utilization of evaluation: Toward comparative study. In U.S. House of Representatives, *The use of social research in federal domestic programs* (Part III, pp. 426-432). Washington, DC: Government Printing Office.

Weiss, C. H. (1975). Evaluation research in the political context. In E. Struening & M. Guttentag (Eds.), *Handbook of evaluation research* (Vol. 1, pp. 13-26). Beverly Hills, CA: Sage.

Weiss, C. H. (1978). Improving the linkage between social research and public policy. In L. E. Lynn, Jr. (Ed.), *Knowledge and policy: The uncertain connection* (pp. 23-81). Washington, DC: National Academy of Sciences.

Weiss, C. H. (1989). Congressional committees as users of analysis. *Journal of Policy Analysis and Management, 8*(3), 411-431.

Weiss, C. H. (Ed.). (1991). *Organizations for policy analysis: Helping government think.* Newbury Park, CA: Sage.

Weiss, C. H. (1998a). *Evaluation: Methods of studying programs and policies* (2nd ed.). Upper Saddle River, NJ: Prentice Hall.

Weiss, C. H. (1998b). Have we learned anything new about the use of evaluation? *American Journal of Evaluation, 19*(1), 21-33.

Weiss, C. H. (in press). The interface between evaluation and public policy. *Evaluation: The International Journal of Theory, Research and Practice.*

Weiss, C. H., & Singer, E. (1988). *Reporting of social science in the national media.* New York: Russell Sage Foundation.

Index

About the Editor

Leonard Bickman is Professor of Psychology, Psychiatry, and Public Policy at Vanderbilt University, where he directs the Center for Mental Health Policy. He is coeditor of the *Applied Social Research Methods* series and is the editor of a new journal, *Mental Health Services Research.* He has published more than 15 books and monographs and more than 140 articles and chapters. He has received several awards recognizing the contributions of his research, the most recent being the 1998 American Psychological Association's Public Interest Award for Distinguished Contribution to Research in Public Policy. He is a past president of the American Evaluation Association and the Society for the Psychological Study of Social Issues. He was a Senior Policy Advisor at the U.S. Substance Abuse and Mental Health Services Administration, where his contribution was recognized with the 1997 Secretary's Award for Distinguished Service.

About the Contributors

Robert F. Boruch is University Trustee Chair Professor in the Graduate School of Education and the Statistics Department of the Wharton School at the University of Pennsylvania. A Fellow of the American Statistical Association, he has received awards for his work on research and evaluation methods and research policy from the American Educational Research Association, the American Evaluation Association, and the Policy Studies Association. He is the author of nearly 150 scholarly papers and author or editor of 15 books on topics ranging from evaluation of AIDS prevention programs and social experiments to ensuring confidentiality of data in social research. His most recent book, *Controlled Experiments: A Practical Guide*, was published by Sage in 1997. His most recent papers concern process and impact evaluation in social services, issued in the *Scandinavian Journal of Social Work Research*; coupling model-based surveys and controlled experiments for educational productivity research in the *Annual Review of Advances in Educational Productivity*; and tax expenditures evaluations in *New Directions for Program Evaluation.*

Thomas D. Cook is Professor of Sociology, Psychology, Education, and Public Policy at Northwestern University. He has a B.A. from Oxford University and a Ph.D. from Stanford University. He was an Academic Visitor at the London School of Economics in 1973-1974, a Visiting Scholar at the Russell Sage Foundation in 1987-1988, and a Fellow at the Center for Advanced Study for Behavioral Sciences in 1997-1998. He has received the Myrdal Prize for Science of the American Evaluation Association, the Donald T. Campbell Prize for Innovative Methodology of the Policy Sciences Organization, and the Distinguished Research Scholar Prize of Division 5 of the American Psychological Association. He is a Trustee of the Russell Sage Foundation and serves on the Advisory Panel of the Joint Center for Poverty Research. He has also served on many federal committees and scientific advisory boards dealing with child and adolescent development, preschools and school education, and the evaluation of social programs of many kinds.

Elvira Elek-Fisk works as a senior research analyst in the Office of Institutional Research and Assessment at Suffolk County Community College, where her efforts focus on academic program review/evaluation and outcomes assessment. She most recently initiated a study of the diversity climate. She received her doctorate in social psychology from the State University of New York at Stony Brook in 1998. As a graduate student, she conducted correlational, experimental, and meta-analytic research on the relationship between ethnic identification and outgroup acceptance.

Ellen Foley is a Research Associate at the Consortium for Policy Research in Education (CPRE) at the University of Pennsylvania. She recently received her Ph.D. from the University of Pennsylvania. While completing her M.S. Ed. and Ph.D. coursework, she worked in a variety of teaching and research positions, focusing primarily on urban education reform. Her primary research interests include integrated services for children and families and the role of noninstructional supports for students in increasing educational equity and student achievement. At CPRE, she is further pursuing these interests as a research team member on the Evaluation of Children Achieving, the School District of Philadelphia's systemic reform effort.

Mark W. Lipsey is Professor of Public Policy at Vanderbilt University's Peabody College, where he serves as Co-Director of the Center for Evaluation Research and Methodology at the Vanderbilt Institute for Public Policy Studies. He received a Ph.D. in psychology from The Johns Hopkins University in 1972 following a B.S. in applied psychology from The Georgia Institute of Technology in 1968. His research and teaching interests are in the areas of public policy, program evaluation research, social intervention, field research methodology, and research synthesis. He is a coauthor (with P. Rossi and H. Freeman) of *Evaluation: A Systematic Approach* (6th ed.); is former editor-in-chief of *New Directions for Program Evaluation*, a journal of the American Evaluation Association; and has served on the editorial boards of *Evaluation Review*, *Evaluation Studies Review Annual*, *Evaluation and Program Planning*, and *American Journal of Community Psychology*.

Melvin M. Mark is Professor of Psychology at Pennsylvania State University and editor of *American Journal of Evaluation*. His publications include the coedited volumes *Social Science and Social Policy*, *Multiple Methods in Pro-*

gram Evaluation, and *Realist Evaluation: An Emerging Theory in Support of Practice*. He is coauthor of a forthcoming book presenting a new theory of evaluation (with Gary Henry and George Julnes).

Norman Miller holds the Mendel B. Silberberg Chair in the Department of Psychology at the University of Southern California. He previously held positions at the University of Minnesota; the University of California, Riverside; and Yale University. He received his B.A. degree from Antioch College and his M.S. and Ph.D. degrees from Northwestern University. During his years of graduate study, he worked with Don Campbell, primarily on the "projection project," an ambitious correlational study of the social perception of personality in which Don hoped to achieve a theoretical integration of principles from behavioristic psychology, gestalt psychology, and clinical psychology. Over the course of his career, Professor Miller has received Guggenheim, Fulbright, National Institutes of Mental Health, and Haynes Research Fellowships, and he is a Fellow in Divisions 8 and 9 of the American Psychological Association, as well as the Western Psychological Association. In 1999, he received the Outstanding Contribution to Cooperative Learning Award from the American Educational Research Association. His current research primarily centers on intergroup relations and on aggression. He has published about 150 articles and chapters.

William C. Pedersen is a Graduate Fellow at the University of Southern California. His research interests include aggression, intergroup relations, stereotype processing, evolutionary psychology, suboptimal priming, and methodological issues. He is a Haynes Foundation Fellow and a member of Phi Kappa Phi. His most recent work is concerned with both experimental and theoretical investigations of displaced aggression and ruminative thought.

Burt Perrin is a consultant in evaluation, policy, and organizational development. His working career has been spent largely in Canada, but he recently moved to France, where he continues to consult internationally as well as to write and to present. His preoccupation is with identifying how evaluation actually can be used to make a difference in public policies and programs. He studied with Don Campbell at Northwestern University in the 1960s and maintained contact with him throughout the remainder of Don's life. He owes the direction of his career, indeed his intellectual development and manner of thinking, to Don's influence and direction, inspiration, encouragement, ideas, and writings.

Vicki E. Pollock is Associate Professor of Psychiatry and the Behavioral Sciences at the University of Southern California. After receiving her A.B. at Washington University in St. Louis, Missouri, she completed her graduate studies at the University of Southern California, earning degrees in psychology (M.A., 1982; Ph.D., 1984). She finished her internship in clinical psychology at the Neuropsychiatric Institute at the University of California, Los Angeles, in 1985 and obtained licensure in 1987. Her research has centered on the utility of psychophysiological measures in the study of abnormal behavior. In particular, much of her published work concerns the electroencephalogram (EEG) in substance use disorders and depression. Her work with meta-analysis has addressed a variety of specific substantive topics. She also assesses psychological factors that might serve as risk factors for the development of alcoholism. Her contributions in these areas were recognized in a Research Scientist Development Award from the National Institute on Alcohol Abuse and Alcoholism, and by the Sigma Xi Excellence in Research Award.

Michael J. Puma, Principal Research Associate at the Urban Institute, has 25 years of professional experience in policy analysis and program evaluation, having directed major national studies in education, child development, nutrition, income security, and employment and training. Prior to joining the Urban Institute, he was a Managing Vice President of Abt Associates Inc., where he directed one of the company's largest research practices and also launched the company's international education practice with initial projects in Egypt and South Africa. Most recently, he completed one of the largest national longitudinal studies of American schoolchildren for the Department of Education and has written extensively on the topic of federal education policy for disadvantaged children. His work formed the basis for the 1994 congressional reform of the federal Title I program, and he is currently involved with senior policymakers as they consider options for the 1999 reauthorization.

Lanette A. Raymond is a graduate student in the Social/Health Program in the Department of Psychology at SUNY—Stony Brook. She is working toward her doctoral degree with Paul M. Wortman. Her interests focus on quantitative research methods and program evaluation in the areas of health and education. She is currently working with Wortman to evaluate the psychosocial impact of research-synthesis information on decision making for prostate cancer patients. Her recent work, in conjunction with the Office of Institutional Research at Suffolk County Community College, has addressed academic program evaluation and the psychometric properties of evaluation instruments.

Robert Rosenthal received his A.B. (1953) and Ph.D. (1956) in psychology from UCLA and is a Diplomate in Clinical Psychology. He has taught at the University of North Dakota (where he was director of the Ph.D. program in clinical psychology), Harvard University (where he was Chair of the Department of Psychology, 1992-1995), and the University of California, Riverside.

His research has centered for more than 40 years on the role of self-fulfilling prophecies in everyday life and in laboratory situations and on the effects of teachers' and researchers' expectations on their students and research subjects. Longtime interests also include the role of nonverbal communication in the mediation of interpersonal expectancy effects and in the relationship between members of small social and work groups. He also has strong interests in sources of artifact in behavioral research and in various quantitative procedures. In the realm of data analysis, his special interests are in experimental design and analysis, contract analysis, and meta-analysis. Among numerous awards for his work is the Donald Campbell Award of the Society for Personality and Social Psychology (1988). He has lectured widely in the United States, Canada, and other countries around the world. He is the author or coauthor of more than 350 articles in the journals of his field and is the author, coauthor, editor, or coeditor of 29 books.

Robert G. St.Pierre, Ph.D., is Vice President and principal associate in the Education and Family Support Area of Abt Associates Inc., a research and consulting firm in Cambridge, MA. Since 1975, he has been principal investigator for educational research, evaluation, and policy analysis projects spanning diverse areas such as family literacy, family support, child development, compensatory education, curricular interventions, school health education, and child nutrition. He currently directs national evaluations of the Even Start Family Literacy Program and the Comprehensive Child Development Program, is serving a 4-year term as a member of the Advisory Committee on Head Start Research and Evaluation, and reviews articles for several evaluation journals. He has recently published in the *American Journal of Evaluation*, the *Future of Children*, and the *SRCD Social Policy Monograph* series.

Carol Hirschon Weiss is Professor of Education at Harvard University in the Graduate School of Education. Her main research interests are the evaluation of public programs and the uses of research and evaluation in policy making. She has published 11 books and about 100 articles and book chapters. The most recent book is *Evaluation: Methods of Studying Programs and Policies* (1998). She has been a Fellow at the Center for Advanced Study in the Behavioral Sciences, a Guest Scholar at the Brookings Institution, Senior Fellow at the Office of Educational Research and Improvement in the Department of Education, a Visiting Scholar at the U.S. General Accounting Office, a Congressional Fellow with the Senate Committee on Education and Labor, and a Fellow of the American Association for the Advancement of Science. She has been a member of eight panels of the National Academy of Sciences, served on a dozen government advisory bodies, consulted with research organizations and operating agencies, served on editorial boards for more than a dozen periodicals, and spoken widely both in the United States and abroad. She currently directs a postdoctoral fellowship program in evaluation at the Harvard Children's Initiative.

Paul M. Wortman, Ph.D., is a professor in the social/health program in the Department of Psychology at SUNY–Stony Brook. He obtained master's and doctoral degrees in psychology from Carnegie-Mellon University and completed postdoctoral training in evaluation research with Donald Campbell at Northwestern University. He has conducted research on various evaluation methods in the areas of education and health. His work has focused primarily on meta-analysis or research synthesis of health technologies. He is a past president of the American Evaluation Association. His most recent work examines the utilization of health syntheses in human decision making. That research has demonstrated the superiority of synthesis to consensus panels of expert physicians. He is currently extending this work to evaluate the psychosocial impact of such syntheses for prostate cancer patients.

Robert K. Yin has been doing applied social research and evaluations for nearly 30 years, initially at the New York City–Rand Institute, then with the RAND Corporation, and now with COSMOS Corporation, where he has been president for 20 years. Much of his work has focused on community-based initiatives in substance abuse and violence prevention, education, community development, and related law enforcement and social services. He also has studied how research results are used, including research utilization, dissemination of findings, and technology transfer. Among the methodologies relevant to these investigations, he has defined alternatives that complement and augment quasi-experimental and experimental methods. The most prominent is his work on the case study method, widely shared through a textbook that has had three versions and numerous reprintings since 1984. He also has advised ongoing projects as a Visiting Scholar to the U.S. General Accounting Office (Program Evaluation and Methodology Division); served on several social science panels sponsored by the National Academy of Sciences, the National Science Foundation, and the National Institute of Mental Health; and published numerous books and journal articles. He received a Ph.D. from the Massachusetts Institute of Technology in brain and cognitive sciences (where he designed and conducted *true* experiments) and a B.A. (magna cum laude) from Harvard College.